T0418695

Paper-based Analytical Devices for Chemical Analysis and Diagnostics

Paper-based Analytical Devices for Chemical Analysis and Diagnostics

Edited by

William R. de Araujo
Department of Analytical Chemistry, Institute of Chemistry, State University of Campinas—UNICAMP, Campinas, Brazil

Thiago R.L.C. Paixão
Departament of Fundamental Chemistry, Institute of Chemistry, University of Sao Paulo, Sao Paulo, Brazil

Elsevier
Radarweg 29, PO Box 211, 1000 AE Amsterdam, Netherlands
The Boulevard, Langford Lane, Kidlington, Oxford OX5 1GB, United Kingdom
50 Hampshire Street, 5th Floor, Cambridge, MA 02139, United States

Copyright © 2022 Elsevier Inc. All rights reserved.

No part of this publication may be reproduced or transmitted in any form or by any means, electronic or mechanical, including photocopying, recording, or any information storage and retrieval system, without permission in writing from the publisher. Details on how to seek permission, further information about the Publisher's permissions policies and our arrangements with organizations such as the Copyright Clearance Center and the Copyright Licensing Agency, can be found at our website: www.elsevier.com/permissions.

This book and the individual contributions contained in it are protected under copyright by the Publisher (other than as may be noted herein).

Notices
Knowledge and best practice in this field are constantly changing. As new research and experience broaden our understanding, changes in research methods, professional practices, or medical treatment may become necessary.

Practitioners and researchers must always rely on their own experience and knowledge in evaluating and using any information, methods, compounds, or experiments described herein. In using such information or methods they should be mindful of their own safety and the safety of others, including parties for whom they have a professional responsibility.

To the fullest extent of the law, neither the Publisher nor the authors, contributors, or editors, assume any liability for any injury and/or damage to persons or property as a matter of products liability, negligence or otherwise, or from any use or operation of any methods, products, instructions, or ideas contained in the material herein.

British Library Cataloguing-in-Publication Data
A catalogue record for this book is available from the British Library

Library of Congress Cataloging-in-Publication Data
A catalog record for this book is available from the Library of Congress

ISBN: 978-0-12-820534-1

For Information on all Elsevier publications
visit our website at https://www.elsevier.com/books-and-journals

Publisher: Susan Dennis
Acquisition Editor: Kathryn Eryilmaz
Editorial Project Manager: Lena Sparks
Production Project Manager: Bharatwaj Varatharajan
Cover Designer: Greg Harris

Typeset by MPS Limited, Chennai, India

Dedication

Thiago R.L.C. Paixão dedicates this book to his wife and son, who have inspired, encouraged, and helped him in everything he has done. Additionally, he would like to thank all the funding agencies that supported his research group (FAPESP, CNPq, INCTBio, and CAPES) and his students and collaborators who have contributed with inspiration and knowlegde. Additionally, he also would like to thank the Institute of Chemistry of the University of São Paulo.

William R de Araujo dedicates this book to his parents, Cleina and Antonio, and to the memory of his grandparents, Assis and Maria, who gave all their support and love. Additionally, he is grateful to his students and collaborators who have contributed with inspiring works and knowledge. He also would like to thank the Institute of Chemistry of the State University of Campinas and the Brazilian funding agencies (FAPESP, CNPq, and CAPES) that supported his research group.

Contents

List of contributors

Hugo Águas i3N/CENIMAT, Department of Materials Science, Faculty of Science and Technology, Universidade NOVA de Lisboa and CEMOP/UNINOVA, Campus de Caparica, Caparica, Portugal

Waleed Alahmad Chemical Approaches for Food Applications Research Group, Faculty of Science, Chulalongkorn University, Bangkok, Thailand; Department of Chemistry, Faculty of Science, Chulalongkorn University, Bangkok, Thailand

Wilson A. Ameku Departamento de Química Fundamental, Instituto de Química, Universidade de São Paulo, São Paulo, Brazil

Marylyn Setsuko Arai São Carlos Institute of Physics, University of São Paulo, São Carlos, Brazil

Iana V.S. Arantes Departamento de Química Fundamental, Instituto de Química, Universidade de São Paulo, São Paulo, Brazil

Vanessa N. Ataide Departamento de Química Fundamental, Instituto de Química, Universidade de São Paulo, São Paulo, Brazil

Emanuel Carrilho São Carlos Institute of Chemistry, University of São Paulo, São Carlos, Brazil; National Institute of Science and Technology in Bioanalytics, INCTBio, Campinas, Brazil

Cyro L.S. Chagas Instituto de Química, Universidade de Brasília, Brasília, Brazil

Chao-Min Cheng Institute of Biomedical Engineering, National Tsing Hua University, Hsinchu, Taiwan

Wendell K.T. Coltro Instituto de Química, Universidade Federal de Goiás, Goiânia, Brazil; Instituto Nacional de Ciência e Tecnologia de Bioanalítica (INCTBio), Campinas, Brazil

Bruno Costa-Silva Champalimaud Research, Champalimaud Centre for the Unknown, Lisbon, Portugal

William R. de Araujo Portable Chemical Sensors Lab, Department of Analytical Chemistry, Institute of Chemistry, State University of Campinas—UNICAMP, Campinas, Brazil

Vanessa N. de Ataide Departamento de Química Fundamental, Instituto de Química, Universidade de São Paulo, São Paulo, Brazil

Andrea Simone Stucchi de Camargo São Carlos Institute of Physics, University of São Paulo, São Carlos, Brazil

Elvira Fortunato i3N/CENIMAT, Department of Materials Science, Faculty of Science and Technology, Universidade NOVA de Lisboa and CEMOP/UNINOVA, Campus de Caparica, Caparica, Portugal

Carlos D. Garcia Department of Chemistry, Clemson University, Clemson, SC, United States

Paulo T. Garcia Institute of Exact Sciences, Faculty of Chemistry, Federal University of South and Southeast of Pará (UNIFESSPA), Marabá, Brazil

Juliana L.M. Gongoni Departamento de Química Fundamental, Instituto de Química, Universidade de São Paulo, São Paulo, Brazil

Erin M. Gross Creighton University Department of Chemistry and Biochemistry, Omaha, NE, United States

Bárbara G.S. Guinati Instituto de Química, Universidade Federal de Goiás, Goiânia, Brazil

Samaya Kallepalli Creighton University Department of Chemistry and Biochemistry, Omaha, NE, United States

Takashi Kaneta Department of Chemistry, Graduate School of Natural Science and Technology, Okayama University, Okayama, Japan

Ana Carolina Marques i3N/CENIMAT, Department of Materials Science, Faculty of Science and Technology, Universidade NOVA de Lisboa and CEMOP/UNINOVA, Campus de Caparica, Caparica, Portugal; Champalimaud Research, Champalimaud Centre for the Unknown, Lisbon, Portugal; BioMark@UC, Department of Chemical Engineering, Faculty of Science and Technology, Coimbra University, Coimbra, Portugal

Rodrigo Martins i3N/CENIMAT, Department of Materials Science, Faculty of Science and Technology, Universidade NOVA de Lisboa and CEMOP/UNINOVA, Campus de Caparica, Caparica, Portugal

Letícia F. Mendes Departamento de Química Fundamental, Instituto de Química, Universidade de São Paulo, São Paulo, Brazil

Nikaele S. Moreira Instituto de Química, Universidade Federal de Goiás, Goiânia, Brazil

Thiago R.L.C. Paixão Department of Fundamental Chemistry, Institute of Chemistry, University of Sao Paulo, Sao Paulo, Brazil; National Institute of Bioanalitical Science and Technology, Campinas, Brazil

Maria Goreti Sales BioMark@UC, Department of Chemical Engineering, Faculty of Science and Technology, Coimbra University, Coimbra, Portugal; BioMark@ISEP, School of Engineering, Polytechnic Institute Porto, Porto, Portugal; CEB–Centre of Biological Engineering, University of Minho, Braga, Portugal

Ching-Fen Shen Department of Pediatrics, National Cheng Kung University Hospital, College of Medicine, National Cheng Kung University, Tainan, Taiwan

Habdias A. Silva-Neto Instituto de Química, Universidade Federal de Goiás, Goiânia, Brazil

Lucas R. Sousa Instituto de Química, Universidade Federal de Goiás, Goiânia, Brazil

Yao-Hung Tsai Institute of Biomedical Engineering, National Tsing Hua University, Hsinchu, Taiwan

Pakorn Varanusupakul Chemical Approaches for Food Applications Research Group, Faculty of Science, Chulalongkorn University, Bangkok, Thailand; Department of Chemistry, Faculty of Science, Chulalongkorn University, Bangkok, Thailand

Ting Yang Institute of Biomedical Engineering, National Tsing Hua University, Hsinchu, Taiwan

Preface

Paper-based Analytical Devices for Chemical Analysis and Diagnostics have been a hot topic in the literature since 2007, enabling the fabrication of low-cost devices together with a well-known substrate with exciting possibilities for point-of-care and point-of-need applications without expensive instrumentation. Hence, this book provides a review of the scientific and technological progress of paper-based devices combining different detection techniques on the paper platform and future trends in the field of portable paper-based sensors for chemical analysis and diagnostics. The detection techniques focus on colorimetric, electrochemical, surface-enhanced Raman spectroscopy, chemiluminescence, fluorescence, electrochemiluminescence, and immunoassays. Collectively, this book focuses on the analytical methods of each device, highlights some tips and challenges in this research field, shows a plethora of clever approaches and strategies to develop accessible, accurate, and convenient portable analytical devices, and provides a practical framework for any researcher to use them while learning how to use new ones. Additionally, the chemistry of paper aiming at its unique properties and modification strategies for (bio)analytical chemistry applications is presented.

We want to thank all the contributing authors for their enthusiasm and participation in preparing this book. We also would like to express our gratitude to the editorial and production staff of Elsevier, in particular, Lena Sparks, for her assistance in bringing this book to print and publication.

Chapter 1

Introduction remarks for paper-based analytical devices and timeline

William R. de Araujo[1] and Thiago R.L.C. Paixão[2,3]
[1]Portable Chemical Sensors Lab, Department of Analytical Chemistry, Institute of Chemistry, State University of Campinas—UNICAMP, Campinas, Brazil, [2]Department of Fundamental Chemistry, Institute of Chemistry, University of Sao Paulo, Sao Paulo, Brazil, [3]National Institute of Bioanalitical Science and Technology, Campinas, Brazil

1.1 Introduction

A single piece of paper has now been calling attention from academic research groups and industry in the last 13 years more than in the previous centuries to develop analytical devices for point-of-need applications. From the past to 2007, different random contributions from the academic literature had attempted to highlight the potentiality of paper as a substrate for analytical purposes, as chromatographic and electrophoretic separations, and in rapid spot tests. However, only in 2007, this point-of-view started to change to put paper as "new"-old material for the development of analytical devices, as shown by the number of articles and citations reported in Fig. 1.1. In addition, it is essential to highlight how this development spread globally, Fig. 1.2, showing the engagement of different research groups around the world in the development of paper devices to be applied to decrease costs of real-time analytical tests for various on-site applications.

Fig. 1.3 highlights some recent applications of paper-based analytical devices (PADs) in the main fields of analytical chemistry. Note that the significant contributions were applied in healthcare/clinical and environmental areas due to the extraordinary appeal for new point-of-care testing (POCT) devices and the continuous need for portable sensors to check the quality of our surroundings (air, water, and soil), respectively. It is worth mentioning that the areas of food quality and forensics have attracted significant attention and, in the near future, will benefit from the paper-based sensor technology.

Paper-Based Analytical Devices for Chemical Analysis and Diagnostics.
DOI: https://doi.org/10.1016/B978-0-12-820534-1.00004-9
© 2022 Elsevier Inc. All rights reserved.

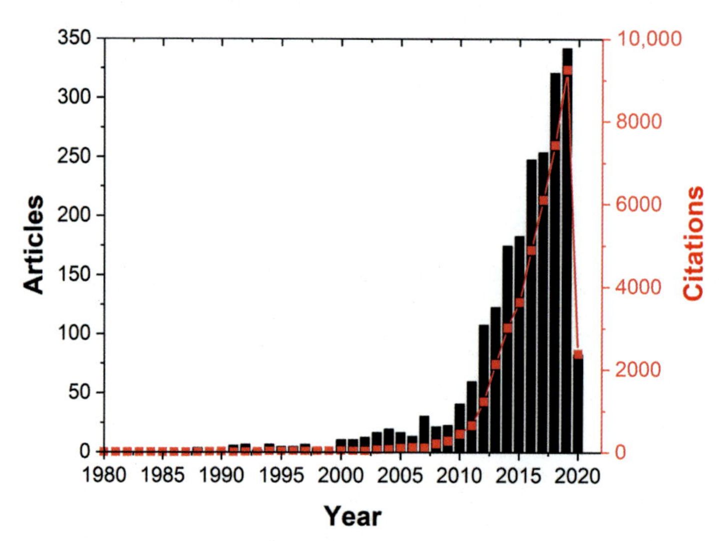

FIGURE 1.1 Number of articles published and citation over the year extracted from the Web of Science using the following keywords: "paper patterned," "microfluidic paper-based," "paper-based analytical," "electrochemical paper-based," "colorimetric paper-based," "paper-based analytical devices," "sensing paper," "paper substrate," or "paper as a platform."

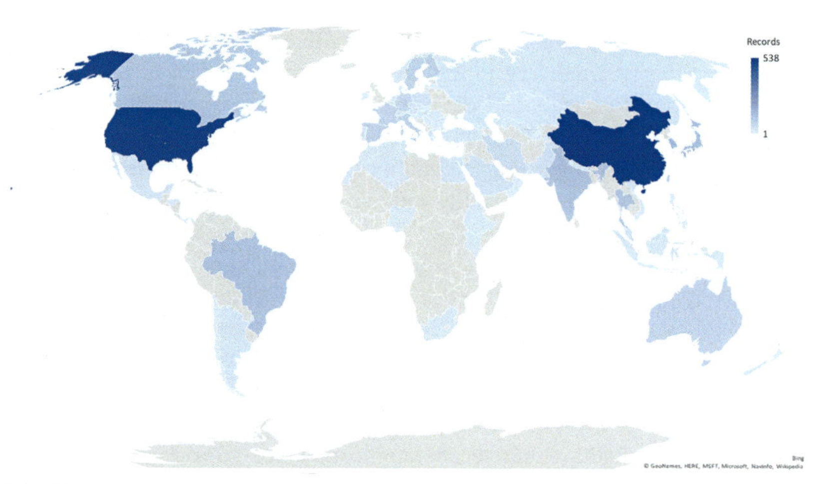

FIGURE 1.2 World map shows the number of total article records during the same period of the data reported in Fig. 1.1 and using the same Fig. 1.1 keywords. Data were extracted from the Web of Science.

The use of paper material for chemical analysis refers to several eras. Maybe the first known account was associated with Pliny studies in 23 BCE, where a test "paper" method for detecting the presence of ferrous sulfate adulteration in verdigris was described. Papyrus strips dipped in an extract from gallnuts

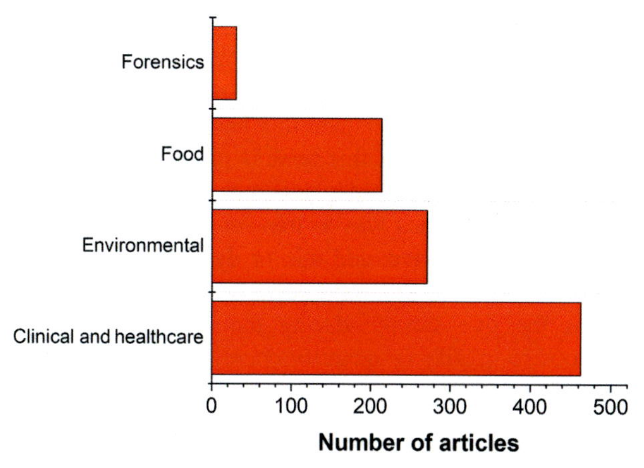

FIGURE 1.3 Number of articles published using PADs in different areas of analytical chemistry in the same period of Fig. 1.1. Data extracted from the Web of Science using the following keywords: "paper-based devices and clinical" or "paper-based devices and healthcare," "paper-based devices and environmental," "paper-based devices and food," and "paper-based devices and forensics."

blackens immediately in the presence of the iron adulterant [1,2]. Nowadays, many kinds of paper materials are available, varying in composition, morphology, fiber alignment, porosity, etc., and these differences provide special performances. These aspects can improve the performance of paper-based sensors, like their microfluidics properties or capability to immobilize (bio)compounds. Recently, filter and chromatographic papers are the significant substrates employed in the development of microfluidic devices due to their superior wicking ability [3]; however, different physical and/or chemical characteristics may be required biosensors, other categories of paper have been explored depending on the target application. Nitrocellulose membranes have been used due to their chemical functional groups that enable the quickly covalent immobilization of biomolecules [3].

Other nonlaboratory-based paper substrates have been used for the fabrication of inexpensive PADs, such as the use of glossy paper [4], office paper [5], towel paper [6], and paperboard [7]. These types of paper, in general, provide lower wettability and higher physical stability. However, they are less pure presenting fillers in your composition, whose compounds could interfere in the bioassays or, in some cases, enhance the measured properties [7]. In addition, the chemistry of paper surface is very vast, and several approaches to change functional groups to enhance chemical interactions, microfluidics, and other properties of paper sensing have been reported. A deep discussion about the main strategies and types of paper will be presented in Chapter 2, Chemistry of Paper—Properties, Modification Strategies, and Uses in Bioanalytical Chemistry.

Since the invention of the paper, and its fabrication process, by Chinese Cai Lun in 105 CE [8], its main advantage was to storage acknowledgment from generation to generation without using old substrates like writing in stone and animal skins. To understand this new search for the development of analytical devices using paper, this book is proposed. We will start a brief historical aspect discussion of this development, aiming at analytical devices.

Since its invention, the idea was to transfer history from one civilization to another, keep memories alive, add ink, or dyes, to paper. The use of analytical devices was based on this same idea. The paper matrix stores chemical reagents, as the dye storage during the writing process, to perform chemical reactions to qualify or quantify chemical species. This feature was probably not thought of by its inventor. This capability to storage chemical compounds is based on the paper production, which came from a dilute aqueous suspension of cellulose fibers that, through different processes like sieve drain, press, and dry process, yield a sheet resulted by randomly interwoven fibers orientation [9], a reservoir for chemical compounds. In addition, this 3D structure obtained will be essential to result in the spontaneous capillary transport of solution inside that, which is essential for the microfluidics development [10].

Historically, the litmus paper (mixture of dyes extracted from lichens absorbed onto filter cellulose paper) is probably the first analytical test found for acidity and basicity of different solutions as a fast point-of-need experiment because the chemical compound (dyes) stored in the cellulose paper changes the color in the presence of proton without the necessity of chemical equipment for measurement. Its invention was attributed to Irish chemist Robert Boyle in the 17th century, and it is still in use nowadays [11]. At this point, the first essential characteristic of paper could be noted, the absorbency, the capability to store chemical reagents inside the paper matrix, and the disposability of the device because the paper strip could be used for a single time.

Intentionally or not, the introduction of paper substrate solved an essential trade-off in the sensor area and is well-explored by paper-based researchers nowadays. One of the requirements for developing the sensor is that the proposed device needs to be robust. However, robustness requires weak interactions only, which implies low sensitivity and low chemical specificity. The disposability characteristics of the paper-based sensors, and considering the reproducibility of the fabrication process, allowed us to probe a wide range of chemical interaction and reaction, including strong sensor-analyte binding because the device will not be used more than once. No requirement will need to perform a chemical or physical treatment to use the device for a second time. Another critical aspect of the litmus paper is related to the visual reporter. The proposed analytical device converts chemical interaction responses to an optical output, which is essential for low-cost spread of the paper-based sensors and appeals for the development of diagnostic for the developing world [12].

After this first found paper-based test, in 1850, a urine test strip was proposed by Maumené, which uses a brown-black color formation for a colorimetric reaction between sugar in urine and the proposed reagent [9,13,14]. In another manuscript [15], the author used the previous urine test strip reporting that devices have essential advantages for medical applications due to their simplicity and on-site applicability compared to solution-based assays.

In 1937 Yagoda [1] described several achievements of that time regarding the development of spot tests on paper platform and highlights that "The use of paper impregnated with suitable reagents in establishing the presence of chemical constituents is probably one of the earliest developments in the art of Analytical Chemistry" using the pioneering Fritz Feigl's discovery about "spot analysis" (spot test) [36]. In addition, your pioneering studies to confine the spot test with the aid of a water-repellent barrier (paraffin) embedded in the fibers of the paper allowed performing colorimetric tests on porous filter paper without loss in sensitivity owing to the spreading of the spot over a large surface. This point was necessary because the uniformity in the area and tint of the spot provides the correlation between the concentration of the species with the intensity of color produced by the drop of solution. This was the first step to progress in semiquantitative analyses instead of qualitative results generally performed.

An interesting historical point regarding the pattern of paper substrate reported by Yagoda [1] for microanalysis was the description of many substances such as waxes, resins, and cellulose esters to standard water-repellent zones on filter paper. Highlighting mainly the use of paraffin wax due to its general inertness to chemical reagents and to the ease with which it can be embedded in diverse patterns on the paper, as the use of a hot metal tube for melt the paraffin and transfer to a sheet of paper in the ring-shaped zones (Fig. 1.4A) [1]. More than 70 years later, Carrilho et al. [16][16] described the technological advances that allowed the use of a commercial screen-printer controlled by computer to print a thin layer of wax in any desired design to pattern the paper reproducibly, which moves the paper-based sensors toward accurate quantitative results (Fig. 1.4B). Several strategies based on the Yagoda approach were published in the 2000s, like using a handheld stamping process to transfer paraffin from paraffined paper to another using a preheated metal stamp to fabricate microfluidic PADs (μPADs) (Fig. 1.4C) [17].

In light of the analytical applications, the capability to perform physical and chemical separations of an analyte or group of analytes from a complex matrix is essential for accurate determinations. Regarding this aspect, the paper substrate composition and structure allow the spontaneous transport of aqueous solution by capillarity and differential flow rate of the components due to physicochemical interactions with the cellulosic phase. The first indication on the use of paper in separation science was given by Von Klobusitzky and König [18], who employed this platform to isolates yellow pigments from snake venom. In the 1940s and 1950s, paper began to make

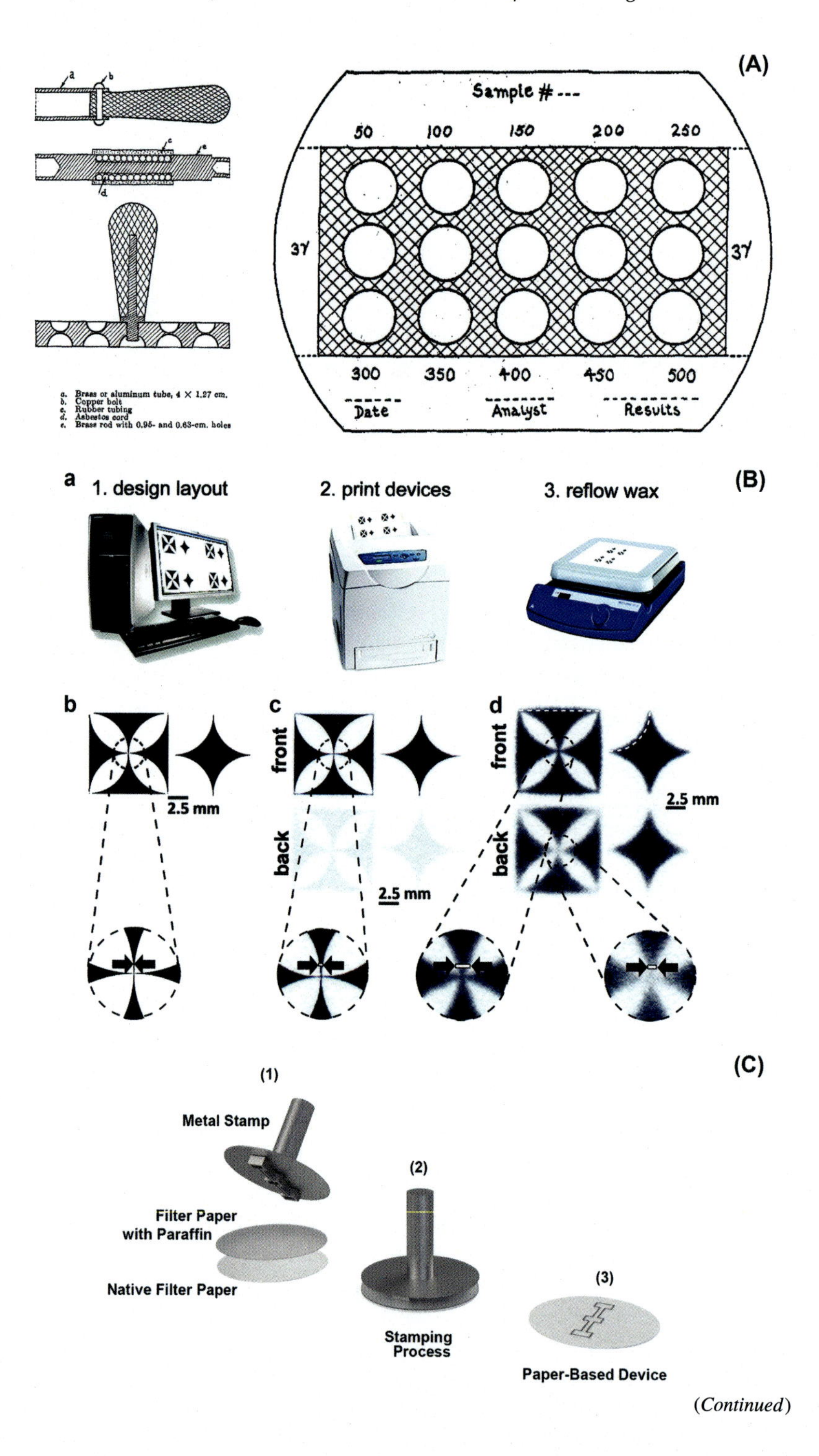
(A)
a
b
c
d
e
a. Brass or aluminum tube, 4 × 1.27 cm.
b. Copper bolt
c. Rubber tubing
d. Asbestos cord
e. Brass rod with 0.95- and 0.63-cm. holes
Sample # ---
50
100
150
200
250
37γ
37γ
300
350
400
450
500
Date
Analyst
Results
(B)
a
1. design layout
2. print devices
3. reflow wax
b
2.5 mm
c
front
back
2.5 mm
d
front
back
2.5 mm
(C)
(1)
Metal Stamp
Filter Paper
with Paraffin
Native Filter Paper
(2)
Stamping
Process
(3)
Paper-Based Device

(*Continued*)

several contributions to separation sciences, being explored as a substrate for chromatographic and electrophoretic separations, mainly for plant extract analyses and biochemical applications that aimed to separate amino acids, peptides, and proteins [19,20]. A further discussion and milestones of paper-based separation devices can be found in Chapter 3, Paper-Based Separation Devices.

In 1957 when it is reported the first paper-based biosensor in the literature [21] which used color reported from a chemical reaction to detect glucose to circumvent the issue reported by Benedict's test. Another important fact in developing paper-based sensors and microfluidics in 1982 appeared in the literature using another colorimetric visual reporter and the capillarity properties of paper originally introduced by Müller and Clegg [22]. The lateral flow originated by the capillarity properties was used to fabricate an assay to screen hybridomas based on antigen–antibody interaction on paper to produce a visible color change to the naked eye [23]. These articles opened the basis for the pregnancy test kits widely used nowadays and started implementing microfluidics research. From that to 2007, no revolutionary development seemed to have occurred in terms of paper-based chemical sensing approaches, only some essential industries started to work in the development of paper-based test strips such as Merck [24] and others.

The next breakthrough in PADs development was reported in 2007. It was responsible for the enhancement of articles published, as shown in Fig. 1.1, when the Whitesides and collaborators reported a very clever way to introduce microfluidic patterns on a paper structure using a photoresist to control the fluid flow direction [25], and the research field related with the μPADs was open publishing more than 1000 articles until now. In addition,

FIGURE 1.4 (A) Paraffin patterned spot tests on paper. (B) Patterning hydrophobic barriers in paper by wax printing. (a) Schematic representation of the basic steps [1–3] required for wax printing. (b) Digital image of a test design. The central area of the design was magnified to show the more minor features. (c) Images of the test design printed on Whatman no. 1 chromatography paper using the solid ink printer. The front and back faces of the paper were imaged using a desktop scanner. (d) Images of the test design after heating the paper. The dashed white lines indicate the original edge of the ink. The white bars in the insets highlight the width of the pattern at the position indicated by the arrows. (C) Scheme of the fabrication process of μPADs based on stamping. In (1), a paraffinized filter paper is placed over the native filter paper surface; in (2), the metal stamp is heated at 150 °C and brought into contact with the layered paper pieces; step (3) represents a typical μPAD fabricated by the proposed method. *(A) Reprinted adapted with permission from H. Yagoda, Applications of confined spot tests in analytical chemistry: preliminary paper, Ind. Eng. Chem. Anal. Ed. 9 (2) (1937) 79–82. Available from: https://pubs.acs.org/doi/abs/10.1021/ac50106a012. Copyright 2020 American Chemical Society. (B) Reprinted with permission from E. Carrilho, A.W. Martinez, G.M. Whitesides, Understanding wax printing: a simple micropatterning process for paper-based microfluidics, Anal. Chem. 81 (16) (2009) 7091–7095. Available from: https://pubs.acs.org/doi/10.1021/ac901071p. Copyright 2020 American Chemical Society. http://xlink.rsc.org/?DOI = C4RA07112C*

in 2009 the same group reported in our opinion the real article responsible for the popularization of the PADs in analytical chemistry due to low cost and commercially available way to create the microfluidic channel in the paper structure using wax printing technology without laborious steps or any additional laboratory-based requirements [16].

Afterward, numerous μPADs using colorimetric detection started to be fabricated. The number of articles published in this field began to increase as reported again in Fig. 1.1, creating now the scientific environment for paper gain popularity to be the "new" material in analytical chemistry to produce and create novel commercially available devices for clinical applications, in the first moment due to its advantageable characteristics as of low cost, availability, equipment-free pumping of a solution, simplicity, and suitability in disposable devices as compared over the other materials.

Based on the advantage characteristics reported here, PADs appear in a good moment in analytical chemistry. One of the research lines in this analytical chemistry field is the miniaturization of the devices to have small devices for point-of-need applications to help medical experts. Fig. 1.5

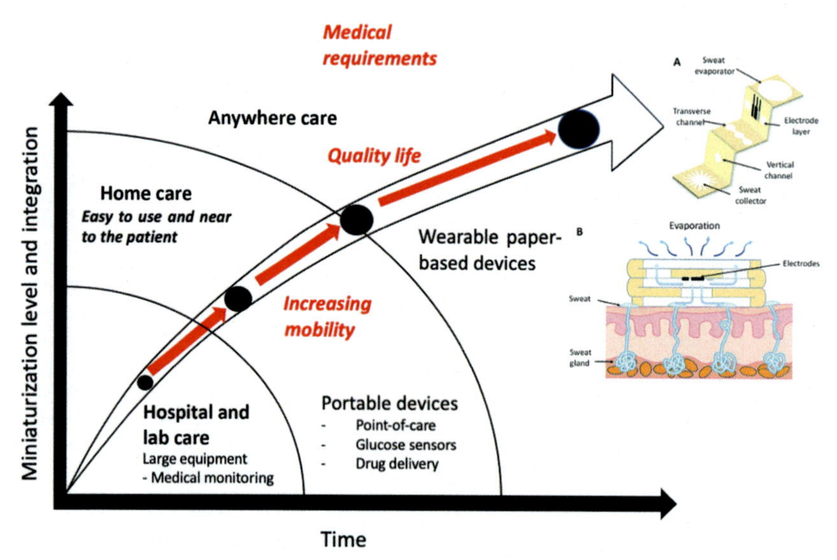

FIGURE 1.5 Evolution in analytical chemistry to miniaturize the instruments for a complete point-of-need application anywhere. Fig. 1.1A and B showed an example of a paper wearable device for sweat measurements [29]. *Reprinted with permission by Creative Commons Attribution-Non Commercial 3.0 Unported Licence and published by The Royal Society of Chemistry from Q. Cao, B. Liang, T. Tu, J. Wei, L. Fang, X. Ye, Three-dimensional paper-based microfluidic electrochemical integrated devices (3D-PMED) for wearable electrochemical glucose detection, RSC Adv. 9 (10) (2019) 5674–5681. Available from: http://xlink.rsc.org/?DOI = C8RA09157A. Copyright 2020 The Royal Society of Chemistry.*

reports the evolution of portable devices to have real and small, point-of-need equipment. We cross the first boundary with the first glucometer proposed in the literature [26], moving the diagnostics for our home. It is probably the first successful commercial case of electrochemical POCT sensors in the industry point-of-need application. The wearable equipment came to stay [27]. Now we are moving to the next boundary in Fig. 1.5, where paper [28] appears as one alternative for this evaluation and for characteristics discussed in this book.

Note that the first reports of using paper for chemical tests were primarily used as a substrate for carrying out presumptive tests. However, there is currently a great effort to integrate multiple analytical steps in the same device. The paper has different functions with sampling steps, sample treatment, constituent separation, and even multiple detection zones for multiplexed determinations. Thus there is a trend toward the development of lab-on-a-paper devices to operate in-field analysis truly.

The smartphone technology has a significant contribution to this research field because these devices allow coupling several detectors to be used on PADs directly at the point-of-need, enhancing the user-friendly of the portable chemical tests and allowing the communication by wireless to transfer in real-time the results to the user or a manager center. An important point was the use of the camera of the smartphones allied to several apps to extract color patterns from images of chemical tests. This approach allows conducting quantitative analyses and minimizing possible mistakes due the dubious naked-eye color interpretation. Hence, as already described, Chapter 2, Chemistry of Paper—Properties, Modification Strategies, and Uses in Bioanalytical Chemistry, will start to discuss the paper chemistry to understand its particular characteristic, which turns this old material as the "new" platform for analytical devices and the unique attributes for separations sciences (Chapter 3: Paper-Based Separation Devices). After that, different detectors were coupled to the PADs after the colorimetric (Chapter 4: Colorimetric Paper-Based Analytical Devices) one to enhance the sensitivity, selectivity or expand the range of sensing analytes. Other standard detection methods such as electrochemical, surface-enhanced Raman scattering, chemiluminescence, fluorescence, electrochemiluminescence, and using immunoassays will be discussed from Chapters 5–10, reporting different ways to produce these devices.

A brief recent timeline of the evolution of PADs for quantitative analysis after 2007 is displayed in Fig. 1.6, where it is highlighted the different detection methods occurrence coupled to portable paper sensors. However, it is essential to emphasize that its qualitative use is older than this proposed timeline reported in Fig. 1.6, as previously mentioned in this chapter.

Finally, the final chapter of this book summarizes the main achievements and drawbacks found in this fantastic research field, highlighting the tremendous challenges for the continuous progress of this area.

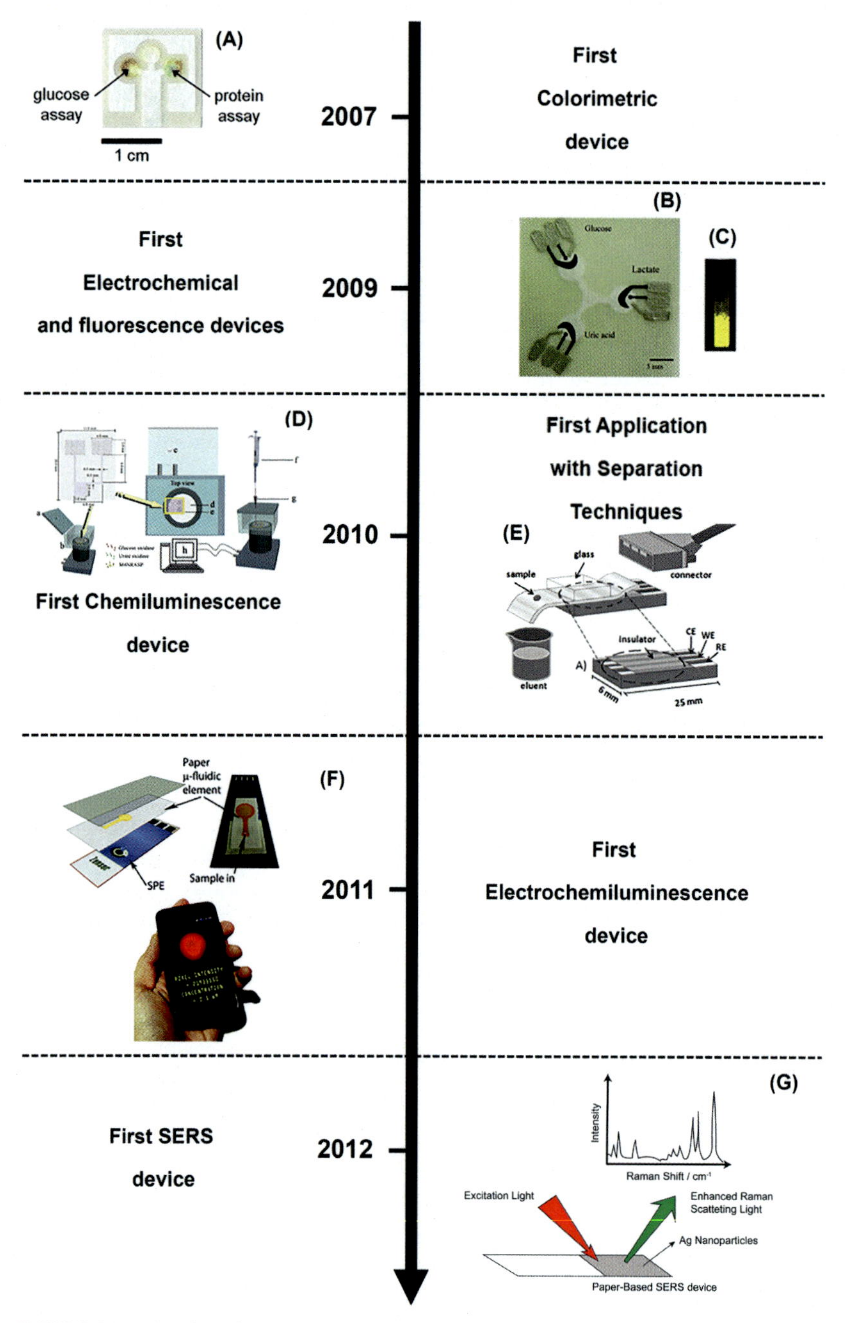

FIGURE 1.6 Timeline of PADs after 2007 using different detectors, or separation techniques, coupled to the PADs: (A) colorimetric [25], (B) electrochemical [30], (C) fluorescent [31], (D)

(*Continued*)

chemiluminescence [32], (E) paper-based separation device [33], (F) electrogenerated chemiluminescence detection [34], (G) Surface-enhanced Raman spectroscopy paper-based device. *(A) Reprinted with permission from A.W. Martinez, S.T. Phillips, M.J. Butte, G.M. Whitesides, Patterned paper as a platform for inexpensive, low-volume, portable bioassays, Angew. Chem. Int. Ed.46 (8) (2007) 1318–1320. Available from: http://doi.wiley.com/10.1002/anie.200603817. Copyright John Wiley and Sons. (B) Reprinted with permission from W. Dungchai, O. Chailapakul, C.S. Henry, Electrochemical detection for paper-based microfluidics. Anal. Chem. 81 (14) (2009) 5821–5826. Available from: https://pubs.acs.org/doi/10.1021/ac9007573. Copyright 2020 American Chemical Society. (C) Reprinted with permission from M.M. Ali, S.D. Aguirre, Y. Xu, C.D.M. Filipe, R. Pelton, Y. Li, Detection of DNA using bioactive paper strips. Chem. Commun. (43) (2009) 6640. Available from: http://xlink.rsc.org/?DOI = b911559e. Copyright 2020 The Royal Society of Chemistry. (D) Reprinted with permission from J. Yu, L. Ge, J. Huang, S. Wang, S. Ge, Microfluidic paper-based chemiluminescence biosensor for simultaneous determination of glucose and uric acid. Lab. Chip 11 (7) (2011) 1286. Available from: http://xlink.rsc.org/?DOI = c0lc00524j. Copyright 2020 The Royal Society of Chemistry. (E) Reprinted with permission from R.F. Carvalhal, M. Simão Kfouri, M.H. de Oliveira Piazetta, A. L. Gobbi, L.T. Kubota, Electrochemical detection in a paper-based separation device. Anal. Chem. 82 (3) (2010) 1162–1165. Available from: https://pubs.acs.org/doi/10.1021/ac902647r. Copyright 2020 American Chemical Society. (F) Reprinted with permission from J.L. Delaney, C.F. Hogan, J. Tian, W. Shen, Electrogenerated chemiluminescence detection in paper-based microfluidic sensors, Anal. Chem. 83 (4) (2011) 1300–1306. Available from: https://pubs.acs.org/doi/10.1021/ac102392t. Copyright 2020 American Chemical Society. http://xlink.rsc.org/?DOI = C2AN36116G*

References

[1] H. Yagoda, Applications of confined spot tests in analytical chemistry: preliminary paper, Ind. Eng. Chem. Anal. Ed. 9 (2) (1937) 79–82. Available from: https://pubs.acs.org/doi/abs/10.1021/ac50106a012.

[2] O. Sonntag, Seralyzer, Laboratory Techniques in Biochemistry and Molecular Biology, vol. 25, Elsevier, 1993, pp. 435–549. Available from: http://www.sciencedirect.com/science/article/pii/S007575350870398X.

[3] A. Singh, D. Lantigua, A. Meka, S. Taing, M. Pandher, G. Camci-Unal, Paper-based sensors: emerging themes and applications, Sensors 18 (9) (2018) 2838. Available from: http://www.mdpi.com/1424-8220/18/9/2838.

[4] A. Arena, N. Donato, G. Saitta, A. Bonavita, G. Rizzo, G. Neri, Flexible ethanol sensors on glossy paper substrates operating at room temperature, Sens. Actuators B Chem. 145 (1) (2010) 488–494. Available from: https://linkinghub.elsevier.com/retrieve/pii/S0925400509010065.

[5] W.R. de Araujo, T.R.L.C. Paixão, Fabrication of disposable electrochemical devices using silver ink and office paper, Analyst 139 (11) (2014) 2742–2747. Available from: http://xlink.rsc.org/?DOI = C4AN00097H.

[6] S. Cinti, V. Mazzaracchio, I. Cacciotti, D. Moscone, F. Arduini, Carbon black-modified electrodes screen-printed onto paper towel, waxed paper and Parafilm M®, Sensors 17 (10) (2017) 2267. Available from: http://www.mdpi.com/1424-8220/17/10/2267.

[7] W.R. de Araujo, C.M.R. Frasson, W.A. Ameku, J.R. Silva, L. Angnes, T.R.L.C. Paixão, Single-step reagentless laser scribing fabrication of electrochemical paper-based analytical devices, Angew. Chem. 129 (47) (2017) 15309–15313. Available from: http://doi.wiley.com/10.1002/ange.201708527.

[8] T.H. Barrett, The Woman Who Discovered Printing, Great Britain, Yale University Press, 2008, p. 34.

[9] J. Credou, T. Berthelot, Cellulose: from biocompatible to bioactive material, J. Mater. Chem. B 2 (30) (2014) 4767–4788. Available from: http://xlink.rsc.org/?DOI = C4TB00431K.

[10] E. Elizalde, R. Urteaga, C.L.A. Berli, Rational design of capillary-driven flows for paper-based microfluidics, Lab. Chip 15 (10) (2015) 2173–2180. Available from: http://xlink.rsc.org/?DOI = C4LC01487A.

[11] R. Carlisle, Scientific American Inventions and Discoveries, John Wiley & Sons, Inc, 2004, p. 502.

[12] A.W. Martinez, S.T. Phillips, G.M. Whitesides, E. Carrilho, Diagnostics for the developing world: microfluidic paper-based analytical devices, Anal. Chem. 82 (1) (2010) 3–10. Available from: https://pubs.acs.org/doi/10.1021/ac9013989.

[13] J. Maumené, No title, Philos. Mag. Ser. 3 (1980) 482.

[14] V.N. Ataide, L.F. Mendes, L.I.L.M. Gama, W.R. de Araujo, T.R.L.C. Paixão, Electrochemical paper-based analytical devices: ten years of development, Anal. Methods 12 (8) (2020) 1030–1054. Available from: http://xlink.rsc.org/?DOI = C9AY02350J.

[15] G. Oliver, On bedside urinary tests, Lancet 121 (3100) (1883) 139–140. Available from: https://linkinghub.elsevier.com/retrieve/pii/S0140673602372258.

[16] E. Carrilho, A.W. Martinez, G.M. Whitesides, Understanding wax printing: a simple micropatterning process for paper-based microfluidics, Anal. Chem. 81 (16) (2009) 7091–7095. Available from: https://pubs.acs.org/doi/10.1021/ac901071p.

[17] P. de Tarso Garcia, T.M. Garcia Cardoso, C.D. Garcia, E. Carrilho, W.K. Tomazelli Coltro, A handheld stamping process to fabricate microfluidic paper-based analytical devices with chemically modified surface for clinical assays, RSC Adv. 4 (71) (2014) 37637–37644. Available from: http://xlink.rsc.org/?DOI = C4RA07112C.

[18] D. Klobusitzky, P. König, Biochemische Studien über die Gifte der Schlangengattung Bothrops, Naunyn Schmiedebergs Arch Exp Pathol Pharmakol 192 (2–5) (1939) 271–275. Available from: http://link.springer.com/10.1007/BF01924816.

[19] R. Consden, A.H. Gordon, A.J.P. Martin, Qualitative analysis of proteins: a partition chromatographic method using paper, Biochem. J. 38 (3) (1944) 224–232. Available from: https://portlandpress.com/biochemj/article/38/3/224/41178/Qualitative-analysis-of-proteins-a-partition.

[20] D.F. Evered, Ionophoresis of acidic and basic amino acids on filter paper using low voltages, Biochim. Biophys. Acta 36 (1) (1959) 14–19. Available from: https://linkinghub.elsevier.com/retrieve/pii/0006300259900630.

[21] A.H. Free, E.C. Adams, M.L. Kercher, H.M. Free, M.H. Cook, Simple specific test for urine glucose, Clin. Chem. 3 (3) (1957) 163–168. Available from: http://www.ncbi.nlm.nih.gov/pubmed/13437464.

[22] R.H. Müller, D.L. Clegg, Automatic paper chromatography, Anal. Chem. 21 (9) (1949) 1123–1125. Available from: https://pubs.acs.org/doi/abs/10.1021/ac60033a032.

[23] R. Hawkes, E. Niday, J. Gordon, A dot-immunobinding assay for monoclonal and other antibodies, Anal. Biochem. 119 (1) (1982) 142–147. Available from: https://linkinghub.elsevier.com/retrieve/pii/0003269782906777.

[24] Merk, Visual tests for semi-quantitative analyses. Analytics and sample preparation. (2020) <https://www.merckmillipore.com/BR/pt/products/analytics-sample-prep/test-kits-and-photometric-methods/visual-tests-for-semi-quantitative-analyses/mt2b.qB.69oAAAE_OAl3.Lxj,nav> (cited 06.04.20).

[25] A.W. Martinez, S.T. Phillips, M.J. Butte, G.M. Whitesides, Patterned paper as a platform for inexpensive, low-volume, portable bioassays, Angew. Chem. Int. Ed. 46 (8) (2007) 1318–1320. Available from: http://doi.wiley.com/10.1002/anie.200603817.

[26] S.F. Clarke, J.R. Foster, A history of blood glucose meters and their role in self-monitoring of diabetes mellitus, Br. J. Biomed. Sci. 69 (2) (2012) 83–93. Available from: http://www.ncbi.nlm.nih.gov/pubmed/22872934.

[27] P.C. Ferreira, V.N. Ataíde, C.L.S. Chagas, L. Angnes, W.K.T. Coltro, T.R.L.C. Paixo, et al., Wearable electrochemical sensors for forensic and clinical applications, TrAC. Trends Anal. Chem. 119 (2019) 115622. Available from: https://linkinghub.elsevier.com/retrieve/pii/S0165993619301177.

[28] M. Pandey, K. Shahare, M. Srivastava, S. Bhattacharya, Paper-based devices for wearable diagnostic applications, in: Paper Microfluidics, Springer, Singapore, 2019, pp. 193–208. Available from: https://doi.org/10.1007/978-981-15-0489-1_12.

[29] Q. Cao, B. Liang, T. Tu, J. Wei, L. Fang, X. Ye, Three-dimensional paper-based microfluidic electrochemical integrated devices (3D-PMED) for wearable electrochemical glucose detection, RSC Adv. 9 (10) (2019) 5674–5681. Available from: http://xlink.rsc.org/?DOI = C8RA09157A.

[30] W. Dungchai, O. Chailapakul, C.S. Henry, Electrochemical detection for paper-based microfluidics, Anal. Chem. 81 (14) (2009) 5821–5826. Available from: https://pubs.acs.org/doi/10.1021/ac9007573.

[31] M.M. Ali, S.D. Aguirre, Y. Xu, C.D.M. Filipe, R. Pelton, Y. Li, Detection of DNA using bioactive paper strips, Chem. Commun. 43 (2009) 6640. Available from: http://xlink.rsc.org/?DOI=b911559e.

[32] J. Yu, L. Ge, J. Huang, S. Wang, S. Ge, Microfluidic paper-based chemiluminescence biosensor for simultaneous determination of glucose and uric acid, Lab. Chip 11 (7) (2011) 1286. Available from: http://xlink.rsc.org/?DOI = c0lc00524j.

[33] R.F. Carvalhal, M. Simão Kfouri, M.H. de Oliveira Piazetta, A.L. Gobbi, L.T. Kubota, Electrochemical detection in a paper-based separation device, Anal. Chem. 82 (3) (2010) 1162–1165. Available from: https://pubs.acs.org/doi/10.1021/ac902647r.

[34] J.L. Delaney, C.F. Hogan, J. Tian, W. Shen, Electrogenerated chemiluminescence detection in paper-based microfluidic sensors, Anal. Chem. 83 (4) (2011) 1300–1306. Available from: https://pubs.acs.org/doi/10.1021/ac102392t.

[35] W.W. Yu, I.M. White, Inkjet-printed paper-based SERS dipsticks and swabs for trace chemical detection, Analyst 138 (4) (2013) 1020–1025. Available from: http://xlink.rsc.org/?DOI = C2AN36116G.

[36] Friedrich Feigl, Rosa Stern, Uber (lie Verwendung yon Tiipfelreaktionen in der qualitativen Analyse, Zeitschrift Für Analytische Chemie 69 (1) (1921) 1–43. doi:alyse. Fresenius, Zeitschrift f. anal. Chemie 60, 1–43 (1921). [16].

Chapter 2

Chemistry of paper—properties, modification strategies, and uses in bioanalytical chemistry

Thiago R.L.C. Paixão[1,2] and Carlos D. Garcia[3]
[1]*Department of Fundamental Chemistry, Institute of Chemistry, University of Sao Paulo, Sao Paulo, Brazil,* [2]*National Institute of Bioanalitical Science and Technology, Campinas, Brazil,* [3]*Department of Chemistry, Clemson University, Clemson, SC, United States*

2.1 Introduction

Due to its unique chemical characteristics, paper has changed the way knowledge was transferred from generation to generation. Before its widespread use, transferring information required the use of heavy-weight materials like rocks, bones, stones, wood, and clay. Through the development of new technologies and creating new cognitive knowledge, it was possible to create more refined (and lightweight) materials to use as a support to transfer information, including skin, leather, baked clay (ceramic), metal, bark trees, and other vegetable fibers [1]. However, despite their lower resistance, the materials developed from vegetable fibers became consecrated as a support for writing due to their greater lightness and practicality. These two last characteristics were important too for the resurgence of paper-based chemical sensors.

The paper (as we know it today) was developed around 105 AD (2nd century) in China, and historians attribute the invention to the Chinese Ts'aiLun [2,3]. The manufacturing process was done by cooking vegetable nonwoody fibers, mainly cotton, leading to a paste that was then sieved and dried. This paper was of relatively good quality, and the monopoly of its production occurred around the year 751 AD when the Arab army attacked the Chinese-dominated city of Samarkanda. The prisoners were taken to Baghdad and forced to produce paper. But it was only in the 11th century that the papermaking technique was taken by the Arabs to Spain and, consequently, to the West [2,4].

Paper-Based Analytical Devices for Chemical Analysis and Diagnostics.
DOI: https://doi.org/10.1016/B978-0-12-820534-1.00008-6
© 2022 Elsevier Inc. All rights reserved.

Paper consumption increased significantly within the Industrial Revolution and the emergence of printed media (books, newspapers, and magazines), promoting significant advances in the area. Thus, essential technologies emerged from the 8th century, which allowed manufacturing paper from wood, considerably increasing the production capacity due to the greater availability of these raw materials, compared to the traditional fibers used. At the beginning of the 19th century, with the discovery of chlorine, bleaching processes increased the number of materials used in the manufacture of paper, mitigating the constant scarcity of other raw materials. The equipment was modernized and reached a high degree of automation and productivity [2,4].

Hence, it is essential to understand the fabrication process of paper, which will be the sensor-based material for the sensor manufacture in this book. Additionally, it is important to highlight that in this chapter, nonreactive paper-based devices fabrication methods that use soaking, writing, printing, sputtering, etc. will not be discussed. However, readers are strongly encouraged to learn about those approaches in recent reviews by Noviana et al. [5] and by Tang [6].

2.2 Paper fabrication

The primary raw material for the paper's production is cellulose, a long-chain polymer very abundant in wood, which can be also found in hardwoods and fruits, such as cotton. It is well-known that the paper purity can be correlated with the origin of the raw cellulose. This polymer's characteristic is to contain an -OH group attached to a carbon atom next to a carbonyl group, an aldehyde, or a ketone group. These groups are essential for paper modification, as we discuss in more details later in this chapter. In addition to cellulose, other compounds can be found in wood, such as hemicelluloses, lignin, and wood extracts. Cellulose represents 40%–50% of the chemical composition of wood, hemicelluloses represent 15%–25%, lignin 10%–30%, and extracts 0.5%–5% [7]. Hemicelluloses are branched polysaccharides formed by at least two units of sugars, such as hexoses and pentoses and oxyhexoses, and uronic acids. They have a low degree of polymerization, are white solid materials, and have a crystalline nature, characteristics that give hemicelluloses hygroscopic properties [8]. Also, it has a fibrous structure, which is associated with cellulose fibers. The main difference between wood and non-wood fibers is not associated with cellulose content. However, it does appear in the contents of hemicellulose and lignin. Non-wood fibers such as wheat, corn, rice contain up to 40% of the significant hemicellulose found in grasses, whereas wood fibers are composed of 25%–35% of hemicellulose [8].

Lignin is a macromolecule with a phenolic characteristic, whose main structure comes from the condensation of compounds that have -OH groups of phenol and alcohol. The main precursors of lignin are trans-synaphyl alcohol, *trans*-coniferyl alcohol, and *trans-p*-cumaric alcohol [9]—the condensation of these compounds, forming ether groups and eliminating water, similar to the cellulose formation. The presence of lignin in wood is of no value to the pulp and paper industry. The main objective is removing that compound in the pulping process, thereby freeing the fibers and removing impurities that cause discoloration and possible future disintegration of the paper.

The composition of these extracts is highly variable, typically including compounds with high molecular weight that are soluble in either organic solvents or water. These compounds provide color, odor, flavor, and resistance to some species' degradation to vary greatly in each species of wood. Typically, the extracts found in wood are terpenes [10], and their derivatives, triglycerides (oils and fats and their derivatives) and phenolic compounds, amino acids, soluble sugars, and alkaloids, can also be found [11].

The pulp and paper industry implements physical and chemical processes intending to obtain the final solid composed of cellulose and hemicelluloses, solubilizing lignin, which can be included in the following operations: (1) material preparation (2) extraction of cellulose or pulping, (3) bleaching of cellulose, (4) refining, and (5) drying and finished paper.

2.2.1 Material preparation

The paper production starts with a washing step of the wood logs to remove sand or dirt accumulated during the logs' handling in the forest or the industry yard. After washing, the process is followed by the removal of the wood bark, also known as debarking. The bark is removed from the logs by mechanical friction, in which the trunk is subjected to high-pressure water jets. It can also be done by placing the logs in cylindrical drums containing water and, by shaking and collision, the bark is separated from the wood. [12–15] After debarking, the logs are sent to chipping, a process that cuts the logs into small pieces of uniform thickness, improving the reagents' consumption and consequent pulp yield cellulosic.

Cotton is another source of cellulose, and it is mainly used in the fabrication of high-purity papers like filter paper. This source was the first one used for the paper-based devices in 2007 [16], and the number of the paper industry using non-wood plants is increasing mainly in China and India [17]. Cellulose from non-wood sources have long and thin fibers (better hygroscopicity), good tenacity, elasticity, and smooth surface. However, these fibers typically feature poor mechanical strength. Additionally, it has a high content of cellulose (95%–97%), and the residual lignin and hemicellulose contents are lower than the wood pulp [17]. Before entering into paper and pulp mills, the cotton preparation starts with open and break up the treated cotton linter with a cotton slitter, feed the cotton linter into carding machine, and remove the cottonseed hull stalk and other impurities.

A very interesting concept, recycling used clothing as a source for paper fabrication, was recently discussed by Sanchis-Sebastiá et al. [18].

2.2.2 Pulping process and extraction of the cellulose

For the formation of the paper sheets, cellulose fibers are used from wood and non-wood sources. From the wood and non-wood sources, cellulose fibers need to be separated from hemicellulose, lignin, and extracts, using mechanical or chemical steps to separate them from the other constituents and form a cellulosic mass the pulp.

The chemical pulping process is the most used in the cellulose and paper industry, promoting delignification from cellulose fibers. Mainly this process occurs using alkaline methods. In the alkaline pulping process, the wood chips are digested in a chemical solution, under pressure and temperature conditions. They can be applied to various wood species, divided into oxygen alkali, sulfate, and caustic soda methods. The change in each process depends on the cooking agent using in each one. In this process, delignification occurs without cellulose degradation or removal, which makes the pulp obtained highly resistant. The chemical solution used in the digestion process consists of aqueous solutions of either sodium hydroxide/sodium carbonate ($NaOH/Na_2CO_3$ for caustic soda pulping) or sodium sulfide ($NaOH/Na_2S$, for kraft pulping). Unfortunately, the sulfide process is not fully selective and a substantial fraction of the hemicelluloses and cellulose that make up the cellulose fibers also undergo degradation [19], decreasing the cellulose content on final paper.

The total hydroxide ions are called effective alkali. Hydrosulfide ions react selectively with lignin, promoting the breakdown of its chain, and preserving cellulose molecules. Hydroxide ions, on the other hand, exhibit temperature-dependent selectivity, promoting hydrolysis of lignin, hemicelluloses, and cellulose, reacting with the latter occurring on a smaller scale. Thus, the temperature and concentration of the white liquor needs to be adjusted in order to promote the achievement of an ideal cooking or pulping, removing a greater amount of lignin without degrading the cellulose fibers of the pulp. [20] After the reaction, the hydrolysis products lignin and hemicelluloses, as well as extractables, remain in solution, forming what is called "black liquor." The cellulose pulp obtained is widely used in the production of very resistant papers, such as cement bags, corrugated sheets, cardboard used in boxes, among others. [21] Antraquinone could be added to the process in order to accelerate the cooking process and to protect the carbohydrates [17]. Additionally, oxygen also can be added to improve the lignin remotion [17], based on phenolic reaction of the lignin groups with oxygen in alkaline conditions.

In the acidic chemical pulping process, also known as the sulfite process, the wood chips are digested in a mixture of sulfurous acid and bisulfite ions at a pressure and temperature-controlled conditions. The pulp obtained in this process is widely used to produce white papers, which are used for the production of books and toilet papers. [21]

The pulping process is critical given the porosity properties of the paper. It is responsible for the paper assays' lateral-flow properties since this step removes the lignin from the fiber walls, leaving the porous structure inside the paper matrix. Additionally, it could change the fiber's morphology [22]. One important aspect to discuss is the concept of pore-volume fraction. The Whatman No. 1 paper (commonly used by many groups) [16] has a 0.69 value, calculated according to Eq. 2.1, [23]

$$\text{Pore-volume fraction} = 1 - (d_{\text{paper}}/d_{\text{fiber}}) \tag{2.1}$$

where d_{fiber} is the density of the solid component of the wood fiber (1540 kg m^{-3}) and d_{paper} is the density of the paper, obtained by ratio between the basis weight (87 g m^{-2} for Whatman No. 1 paper) and thickness (180 mm for Whatman No. 1 paper). Equally important to this value is the fiber orientation, which is essential for lateral flow assays. These parameters have to be taken into account where flow paper-based devices have to be manufactured. As an example, office paper is cut parallel to the papermaking direction in the longest dimension. Hence, the fluid transport along the strip is in the order of the cut process. Other parameters of commercial papers, and their impact on the analytical performance of devices, can be found elsewhere. [24]

2.2.3 Bleaching

After the pulping process, the cellulosic pulp obtained has a brown color due to dissolved lignin being still impregnated in the cellulose. Thus, it can be subjected to a bleaching process to remove residual lignin, which consists of bleaching agents, such as sodium hydroxide, chlorine, and their compounds (hypochlorite and chlorine dioxide), and ozone. This bleaching process can be considered as a continuation of delignification; that is, it completes the removal of the lignin initiated in the pulping process [17], and this step is responsible for the harmful substances generated in the paper manufacture due to the formation of toxic organic chloride compounds. Based on the low content of lignin, the non-wood pulp's bleachability is better than of wood pulp, achieving brightness close to 70%, which is essential mainly for colorimetric paper-based devices.

2.2.4 Refining, drying, and paper manufacturing

The next step is mechanical treatment, in which the fibers of the cellulose pulp are broken into fibrils, thus increasing the surface area. As a result, when the paper is formed, a greater contact surface and a greater entanglement between the fibrils are obtained. Thus, refining allows changing the pulp fibers' structure, resulting in modified properties in the finished paper, such as softness, opacity, receptivity to ink, and resistance to moisture and tear.

Since all cellulose pulp is obtained in water presence, it is mandatory to remove that from the pulp to obtain the paper finally. Hence, the pulp must pass through drying stages by heated air or medium pressure steam, followed by conditioning in bales to eventually be sent to the production of the finished paper, which is this process is done in a continuous regime, using conveyor belts and rollers that produce the paper in the form of large reels. These reels are subsequently cut to the desired sizes and shapes. Additionally, some additives could be added to provide further resistance to the paper and change the texture and brightness, among others. One example is carbonate [23] that is added to office paper and could change the pH of the solutions added to paper-based devices, indicating the necessity of characterization of the paper used for the manufacture of the analytical purpose.

2.3 Strategies using the dissolution of cellulose to manufacture sensing inks

Cellulose could be used to develop sensing inks containing mixtures of CNTs and cellulose [25,26], carboxymethyl cellulose [27], or cellulose xanthate [28]. In these cases, the process typically starts by dissolving cellulose in a mixture of lithium chloride and *N,N*-dimethylacetamide (DMAc), leading to a solution than can be deposited in a number of substrates by spin coating [25]. Other groups have also reported the possibility to dissolve cellulose in 7 wt.% NaOH/12 wt.% urea, [29,30] developing materials to print sensors on paper substrates.

2.4 Properties of existing unmodified papers

The paper properties of paper are the essential factors affecting paper-based devices' performance accordingly with their composition and fillers [31,32]. From the engineering standpoint, cellulose is organized by a number of hierarchical structures supported by a network of hydrogen bonds and hydrophobic interactions [33], resulting in a material that features a coefficient of thermal expansion as low as $0.1\ \text{ppm}\ K^{-1}$, a Young's modulus of (114 GPa, single fibril), high degree of crystallinity, and high specific surface area ($37\ m^2\ g^{-1}$). [34] Additionally, the pore size, specific surface area, and hydrophobicity/hydrophilicity of paper affect the capillary forces [35], adsorption capability [36], and paper flow rate [24,37,38].

Since cellulose is the primary paper component and almost all the impurities, like lignin and hemicellulose, are removed as discussed previously, hydroxyl (–OH) and carboxyl groups (–COOH) will be present on paper surfaces, and they will affect and help to immobilize chemical reagents on its surface based on chemical reactions [39]. Based on that, the choice of the paper source for preparing paper-based devices will be discussed here.

As discussed previously, filter paper and chromatography paper sources were the dominant sources in the literature since their fabrication does not contain structure-reinforce additives like the paper sources [23]. These additives could interfere in the analytical measures and chemical modification for the wrong or right side, as will be discussed.

Some paper could have a surface coating with optical brightness compounds to increase the whiteness. However, such treatment will result in a very high background for fluorescence-detection assays [23].

Printing or office paper was first introduced for the electrochemical paper-based analytical devices (ePADs) by Araujo and Paixão in 2014 [40]. Usually, this type of paper could contain up to 30 wt.% of mineral fillers [23]. Kaolin is one of these fillers used, where calcium carbonate is one of the significant components, and it could work also as a pH buffer interfering in some analytical applications [23,41]. Despite these reports, kaolin was reported as a beneficial material when cardboard boxes were used to fabricate ePADs. It helps forming carbon nanoparticles when a laser is used to convert the cellulose into carbon tracks [42]. This observation was attributed to the content of Al and Si, present in the kaolin material. Cardboard, corrugated boxes, and grocery bags were paper examples where the bleaching process was not used and some content of lignin still on the paper matrix which results in the brown color [23].

Additionally, the inclusion of fillers is able to decrease the printing/office papers' strength, issue that can be partically addressed by cationic starch, as it promotes interfiber adhesion. Nevertheless, the applicability of this approach toward wet on paper-based devices is relatively limited [23].

2.5 Paper modification approaches for chemical sensing

Due to the potential benefits of addressing the chemical properties of cellulose without significantly changing its physical properties, several chemical modification approaches have been described and reviewed [43–45]. Despite the target functionality, all approaches aim to either take advantage of the existing –OH groups (three per anhydroglucose unit), to activate them to other functionalities, or to graft new groups. It is important to mention that the native alcohol groups are rather inert and neutral below pH = 12, often resembling the behavior of the glucose units and thus limiting the type of chemistry to be applied. Considering these aspects, this section aims to provide a brief summary of those procedures that seem to be more relevant for developing chemical sensors.

As the simplest case, one should consider that cellulose provides a versatile platform to implement various chemical derivatization strategies, although the dense hydrogen bond network and the bulk nature of paper strongly decrease their reactivity compared to soluble or highly dispersed polysaccharides [46] To address this limitation, several groups have integrated an activation step, prior to the surface functionalization [47]. In the most common case, cellulose samples are immersed in an aqueous solution of 8–10 wt.% NaOH for 16–18 h, resulting in a material that provides 18 mmol of active –OH/g cellulose [48,49].

One step further, but without changing the chemical functionality of cellulose, several groups have performed physical modifications using only the retention capabilities of cellulose simply by adding additional materials, which become entangled within the cellulose fibers [50]. This method, often referred to as casting modification, has been applied to deposit carbon nanotubes to develop a chemiresistive sensor for ammonia [51] or an amperometric sensor for prostate specific antigen [52]. Similarly, Evans et al. applied silica nanoparticles to improve the immobilization of enzymes in paper-based devices [53], Ruecha deposited a graphene–polyaniline nanocomposite for the simultaneous detection of heavy metals [54], and Citerio's group deposited ion-selective optode nanoparticles using an ink-jet printer [55]. Taking advantage of the wettability of cellulose, several composites have also been developed by soaking paper in conductive polymers [56] or by applying layer-by-layer constructs of polyelectrolytes (such as polyethyleneimine [57,58]) electrostatically assembled. In a similar way, Andrescu's group [59] described a paper-based colorimetric sensor with immobilized tyrosinase for the naked-eye detection of phenolic compounds. The method used alternative layers of chitosan, tyrosinase, and sodium alginate, deposited onto a filter paper disk that constitutes the sensing platform. While these methods provide a versatile, fast, and straightforward avenue to modify paper, the applicability of sensors produced may be limited by their stability.

Perhaps the most straightforward approach to chemically modify paper substrates is to use the structure of the cellulose as nucleation centers to promote the condensation of other materials on the surface of the fibers, again, taking advantage of the native –OH groups. Probably one of the simplest examples of this approach has been the silanization of paper to reduce its hydrophilicity, a strategy that can be applied to pattern cellulose. As described by Koga (see Fig. 2.1), paper samples can be simply immersed in 2.5% (v/v) 3-aminopropyltrimethoxysilane (APTMS) dissolved in ethanol. After removing the solvent, a hydrolyzation step was implemented to form reactive silanol groups that were then attached to the –OH groups of the cellulose. [60]

Other groups have reported similar procedures based on the immersion of the paper in 0.1% (v/v) octadecyltrichlorosilane (OTS) in *n*-hexane at room temperature for 5 min and etching with UV-light (35 mW cm^{-1} at 254 nm) for 90 min [61] or the immersion of paper into a 2.0% trimethoxyoctadecylsilane (TMOS)-heptane solution for 10 s [62] and etching using a paper mask impregnated with NaOH solution (containing 30% glycerol) [63] Additional considerations to these reactions include the possibility to improve the efficiency by depositing a layer of Al_2O_3 (by atomic layer deposition) to increase the density of hydroxyl groups [64], apply a mixture

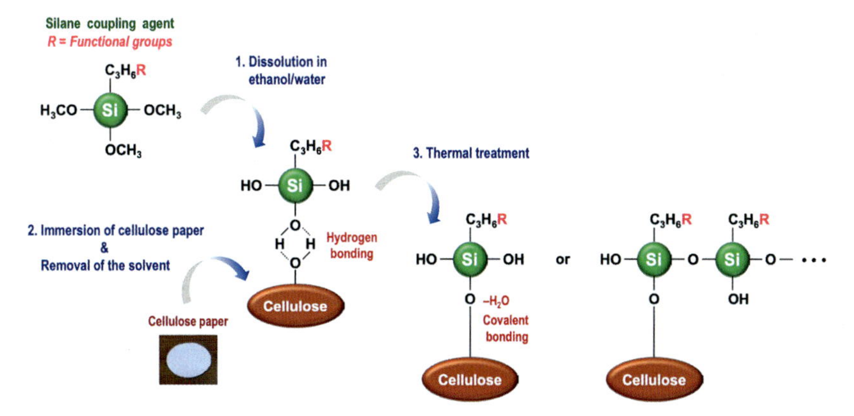

FIGURE 2.1 Putative scheme of the silane coupling process on the cellulosic substrate. *"Reproduced from Koga, H., Kitaoka, T., and Isogai, A. In situ modification of cellulose paper with amino groups for catalytic applications.* J. Mater. Chem. *21 (25) (2011) 9356–9361 with permission from The Royal Society of Chemistry."*

of two silanes (OTS and methyltrichlorosilane, MTS) to increase the contact angles above 125° [65], or controlling the amount of water added to complete the hydrolysis and condensation steps [66]. Additional examples of silanizing agents and secondary reactions (based on the terminal groups) are summarized in a recent review by Shakeri [67]. This is also the strategy applied to accomplish the direct growth of ZnO rods to preconcentrate myoglobin, a cardiac biomarker [68] and $NiCo_2S_4$ nanostructures on free-standing cellulose samples. The latter case is exciting because it involved sequentially soaking paper strips in solutions containing $SnCl_2$/HCl, $PdCl_2$/HCl (forming Pd particles) and $NiCl_2$/citric acid + NH_4OH (to then cover the fibers with Ni nanoparticles). These substrates were finally immersed in a mixture $NiCl_2$, $CoCl_2$, and thiourea to form electrochemically active $NiCo_2S_4$ nanostructures [69].

Regardless of the density of –OH groups, one of the most common transformations applied to paper is the chemical modification to make it more hydrophobic. As noted by Kumar et al. [70], this strategy (known as internal sizing) is routinely applied to improve printability. Two main reagents are used for this purpose: alkyl ketene dimer (AKD) and alkenyl succinic anhydride (ASA). AKD emulsions contain a waxy solid material (14–20 carbons with a melting point around 50°C) and a stabilizer (cationic polyelectrolyte [71,72]) applied to cellulose using various strategies. For example, Lee et al. soaked filter paper in 0.1% AKD for 2 min, then rinsed it with DW to eliminate the unbound AKD, and finally treated the paper using a drum drier (at 120°C) [73]. One of the advantages provided by AKD is that it can afford patterning and re-dissolution under relatively mild conditions such as plasma treatment [74] or various alcohols [75,76]. One drawback of AKD is that unbound fractions of the emulsion can remain in the paper, significantly affecting the properties of the material and thus requiring a careful selection of the experimental conditions (concentration, time, removal, etc.) for its application. A second avenue to turn the paper hydrophobic is by the use of ASA. ASA combines the hydrophobicity of the alkyl chain (14–20 carbons) with the reactivity of the succinic anhydride, a moiety that shares some similarities with *N*-hydroxysuccinimide (NHS), used to couple primary amines with carboxylic groups. As shown in Fig. 2.2, ASA undergoes a ring-opening reaction in the presence of water, forming two carboxylic groups that can lead to the subsequent formation of an ester bond with the –OH in the cellulose [77].

As mentioned before, the oxidation step is also used to enable coupling primary amines to cellulose using NHS and 1-ethyl-3-(3-dimethylaminopropyl)carbodiimide (EDC). This approach has been applied to immobilize various proteins [79] to cellulose and develop a glucose sensor for cell culture monitoring [80], to immobilize cellulose to cystine [81], or covalently attach ligands or target analytes for highly specific binding [82]. An additional alternative to immobilize proteins (also using EDC/NHS) was implemented by Zhu et al. [83], following a procedure initially described by Anna Carlmark and Eva Malmström [84]. In these experiments, paper samples were converted into atom transfer radical polymerization initiators using the reaction with 2-bromoisobutyryl bromide, triethylamine, and 2-dimethyl aminopryridine (DMAP, catalyst), affording the subsequent derivatization with acrylates. Although this option presents a tremendous versatility, the reaction is also quite time consuming (up to 24 h), making it comparable to the functionalization with benzophenone-based substituents recently reported [85], which provides free primary amines that can be further modified with an

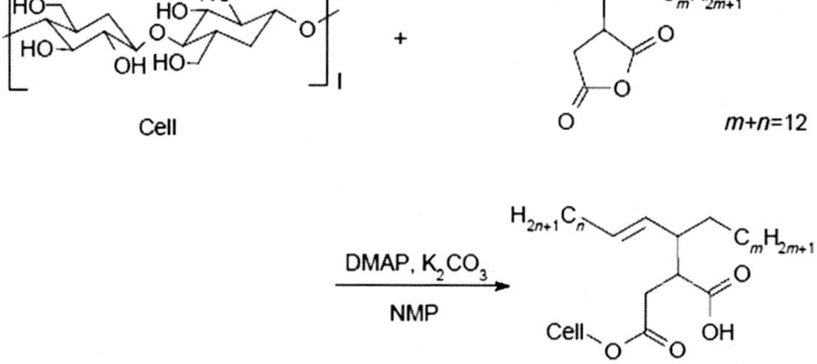

FIGURE 2.2 Reaction of cellulose (nanofibers) with ASA. *Reproduced from Sato, A., Kabusaki, D., Okumura, H., Nakatani, T., Nakatsubo, F., Yano, H. Surface modification of cellulose nanofibers with alkenyl succinic anhydride for high-density polyethylene reinforcement.* Compos. Part. A Appl. Sci. Manuf. *83 (2016) 72–79 with permission from Elsevier [78].*

NHS-activated benzophenone ester to introduce a light-sensitive functional group to the cellulose chains.

Cellulose fibers can also be modified to introduce carboxymethyl groups, making the surface negatively charged [45] and leading to improved hydrophilicity [86] and absorption capacity [87]. This material is also widely available from various commercial courses and enables its modification via standard –COOH routes. Carboxylic groups can also be introduced in cellulose via oxidation of the alcohols, process that has multiple implications for sensing applications [88,89]. Perhaps one of the most commonly used routes is the oxidation by NaClO and mediated by 2,2,6,6-tetramethylpiperidine-1-oxyl (TEMPO) [57,72]. As described by Eyley [44], the reaction is typically performed at pH 10 and in the presence of sodium bromide (to increase the rate of oxidation via in situ formation of sodium hypobromite). Alternatively, cellulose can be also oxidized using periodate, reaction that introduces two aldehydes/carboxylic groups by breaking the glucopyranose ring at the C2 and C3 of the anhydro glucose units [90,91]. Livia Sgobbi's group used this approach to develop an electrochemical sensor for lactate [92]. Here, the paper sample was soaked in a solution of sodium m-periodate (0.5 mol L^{-1}) for 30 min in the absence of light and then sequentially rinsed with ultrapure water and dried at room temperature. Other groups have also applied similar approaches to stabilize fluorescent copper nanoclusters and quantify mercury ions [93], to develop a sensor for glucose [30], or to measure silver ions after grafting thiourea by a Schiff base reaction [94]. It is essential to mention that reaction seems to follow two different rates, including a first rapid process followed by a slower one with the corresponding increase of the carbonyl content a careful balance between the [IO_4^-] and the reaction time is required to preserve the structure of the paper. Other groups have also combined this approach with other oxidants such as chlorite [95], which facilitated the conversion of C2,C3 dialdehydes to dicarboxylates. As a cleaner way to modify paper samples, Carlsson and Stroem used a cold plasma in the presence of either hydrogen or oxygen, to induce reductions or oxidation reactions, respectively [96]. They noted that the hydrogen plasma could reduce the density of hydroxyl groups, induce the formation of low molecular weight materials, and reduce the wettability of cellulose. As expected, the oxygen plasma had the opposite effects.

Another clever way to derivatize paper was recently presented by Rull-Barrull et al. [46]. This report described the possibility of functionalizing cellulose with benzaldehyde groups by esterification of hydroxyl functions with 4-formylbenzoyl chloride (Fig. 2.3). Authors presented the hypothesis that the esterification mainly took place on primary alcohols, which are more reactive than the remaining secondary alcohols of glucose units, and reached a degree of substitution of about 4%, meaning that one hydroxyl group is functionalized every 25 glucose units, leading to an optical rhodamine-based sensor for hydrogen sulfate in water.

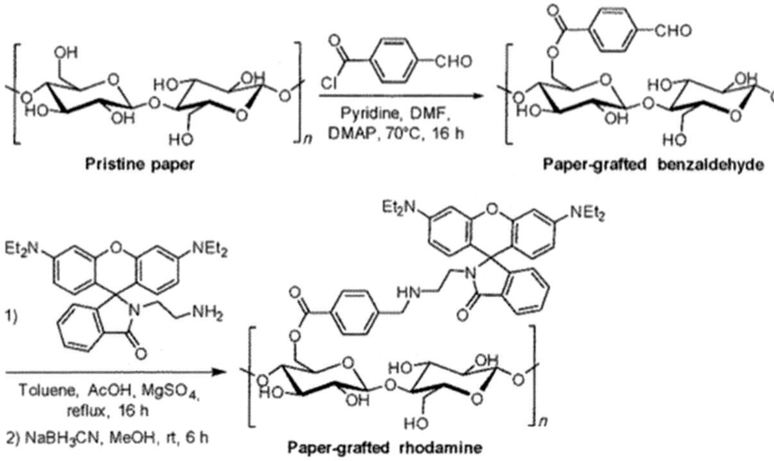

FIGURE 2.3 Preparation of the paper-based analytical device. *Reproduced from Rull-Barrull, J., d'Halluin, M., Le Grognec, E., and Felpin, F.-X. Chemically-modified cellulose paper as smart sensor device for colorimetric and optical detection of hydrogen sulfate in water.* Chem. Commun. *52 (12) (2016) 2525–2528 with permission from The Royal Society of Chemistry.*

A strategy to produce a photoresponsive cellulose paper covalently grafted with dithiodiglycolic acid was described by Bretel et al. [97]. The process, presented as photothiol-X chemistry, is based on the esterification of the pristine paper with dithiodiglycolic acid, using *p*-toluenesulfonic acid (PTSA) as a catalyst in refluxing toluene for at least 16 h. The main advantage of this approach is that the modified paper would feature a disulfide bond that can undergo homolytic cleavage into the corresponding thiyl radical upon light irradiation at 365 nm in the presence of 2,2-dimethoxy-2-phenylacetophenone.

2.6 Thermal treatment of cellulose

As one of the most abundant polymers on the planet, cellulose represents a suitable material to develop analytical technologies. Besides the examples previously described, and as noted by Huber's group [98], the pyrolysis of cellulose is a complicated process involving multiphase reactions, complex chemical pathways, highly unstable intermediates, and heat and mass transfer effects (see summary in Fig. 2.4).

Considering that carbon has the highest sublimation point of all elements (3825°C), this intricate reaction map has allowed the development of various thermal approaches to produce carbon-based materials for analytical purposes. As a reference point, it is important to mention that this section will be focused on processes that are significantly different from chemical vapor deposition (CVD), where a precursor (such as methane and acetylene) is

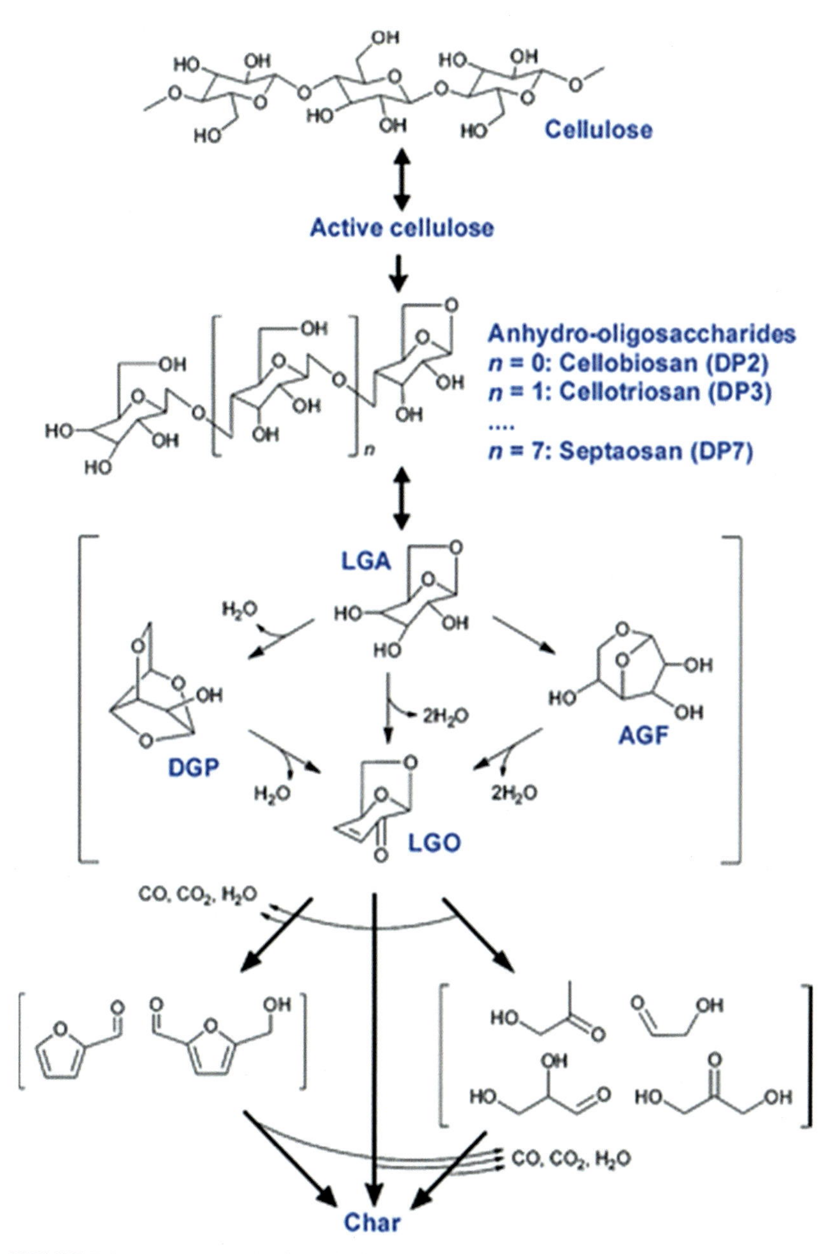

FIGURE 2.4 Proposed mechanism of cellulose pyrolysis (DP indicates "degree of polymerization"). *From Lin, Y.-C., Cho, J., Tompsett, G.A., Westmoreland, P.R., and Huber, G.W. Kinetics and mechanism of cellulose pyrolysis.* J. Phys. Chem. C, *113 (46) (2009) 20097–20107.*

thermally decomposed, releasing carbon atoms that then recombine over a surface. When cellulose is used as the starting material, the overall pyrolysis mechanism involves the gradual heating of the material to first release low-MW groups (methane, carboxyl, carbonyls, and hydrogen), cross-link the polymer chains, and finally form graphitic domains. [99] Under certain conditions, this approach yields what most articles refer to as glassy carbon, [100] a carbon material structured as layers of tangled ribbons, [101,102] similar to those found in graphite but without the corresponding extensive, oriented sheets of sp^2 carbon. Probably the main advantage of this approach over CVD is the possibility to select a much more diverse group of starting materials and reach a source-to-product efficiency as high as 70%. [103,104]. It is also critical to indicate that because the C–C bonds in the polymer backbone do not break under the inert atmosphere and temperatures usually selected for this process (<1200°C), only small graphitic planes are typically formed [105] and that different substrates typically yield carbon materials with very similar properties. Several recent reports, developed using theoretical and experimental methods [106–109], have identified specific compounds formed during the process and the corresponding reactions involved.

Pyrolysis of cellulose is typically performed in a tube furnace by heating up the sample up to 1000°C (typically within 30 min) under an atmosphere of 5% H_2–95% Ar. As the temperature increases from room temperature, a slight weight loss (<115°C) is typically observed, a process that is attributed to the release of chemical compounds (such as water) weakly bound to the paper. Release of methane, carboxyl, carbonyls, and hydrogen (favoring the formation of volatile compounds) occurs at temperatures in the range of 250°C–500°C and yields significant losses in weight, shrinking, and coalescence of polymeric chains. Within this phase, the most considerable decomposition step is observed, where the formation of levoglucosan and other volatile compounds occurs [110]. At temperatures in the 500°C–700°C range, the weight of the product is stabilized and the material undergoes crosslinking of polymeric chains, formation of graphitic domains, and a reduction in pore size. It is noted that a key component in this mechanism is the dehydration of the glucose units, leading to the formation of (more) reactive species including furfural, levoglucosenone (LGO), 1,4:3,6-dianhydro-α-D-glucopyranose (DGP), levoglucosan (LGA), and 1,6-anhydro-β-D-glucofuranose (AGF) [98,111]. Some of the advantages of the material resulting from the pyrolysis of cellulose include electrical conductivity and a 3D structure that resembles the structure of the original material, thus enabling the development of electrochemical sensors upon the modification enzymes [103], nanoparticles [112], or films [113]. It is also important to state that although a number of variables can affect the contact angle of carbon materials made by pyrolysis [114], the as-produced substrates feature a contact angle of approximately 80° (hydrophobic) [115,116], allowing their use as adsorbents via nonspecific hydrophobic interactions. One example of the applicability of this material include the removal of proteins from biological samples prior to their analysis [117].

While the above-described process enables the fabrication of rather large amounts of material, it does not enable patterning. Addressing this limitation, Paixao's group pioneered laser engraving to promote the carbonization of cellulose-based materials [42]. In this case (shown in Fig. 2.5), a sample of cardboard (1 mm thick) was exposed to the IR radiation of the CO_2 laser (under a stream of N_2), transferring enough heat to the selected area to induce carbonization. Authors noted that besides the formation of carbon, a

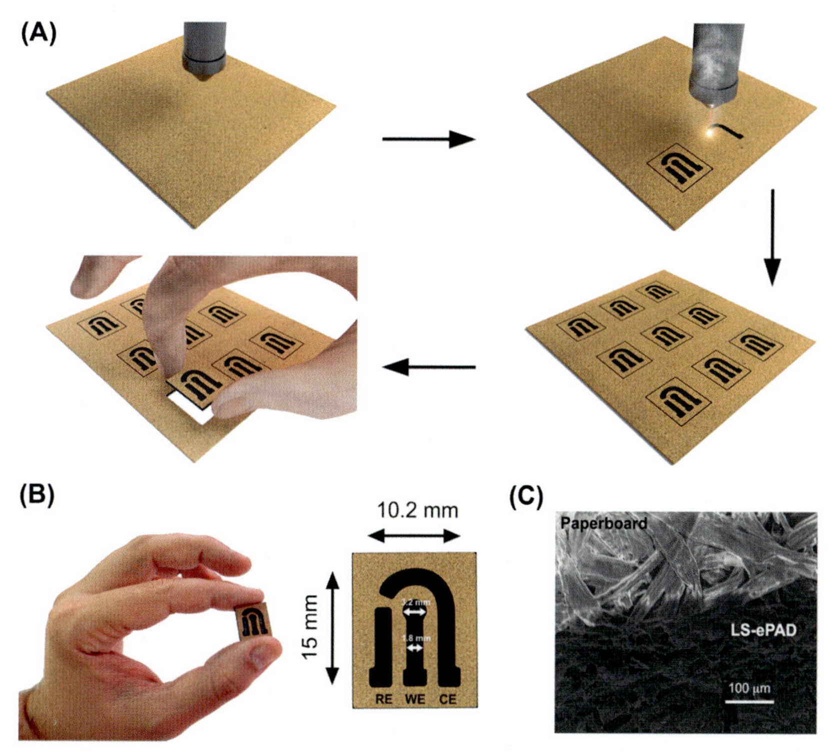

FIGURE 2.5 (A) Representation of the fabrication process. (B) Dimensions of the LS-ePADs and the working, counter, and reference electrodes (WE, CE, and RE, respectively). Also shown is the patterning step (area delimitation using Cascola glue) and silver ink painting processes on the top of the RE scribed carbon film. (C) The dimensions of LS-ePADs in manual handling. (D) Scanning electron microscopy (SEM) image of the interface between the paperboard and the nongraphitizing carbon film formed by CO_2 laser radiation. *Reproduced from de Araujo, W.R., Frasson, C.M.R., Ameku, W.A., Silva, J.R., Angnes, L., and Paixão, T.R.L.C., Single-step reagentless laser scribing fabrication of electrochemical paper-based analytical devices.* Angew. Chem. Int. (Ed.), *56 (47) (2017), 15113–15117 with permission from Wiley-VCH Verlag GmbH & Co. KGaA, Weinheim.*

fraction of the material was always removed by ablation, the issue that can be (partially) addressed by pre-soaking the material in 0.1 mol L^{-1} sodium borate for 10 min and drying it overnight. The borate acts as a flame retardant, decreasing the amount of material ablated during the engraving process and promoting the formation of char (instead of tar).

These electrodes have already been used to develop a number of electrochemical sensors [118], including those integrating a gas-diffusion microextraction (GDME) unit [119], gold nanoparticles [120], and graphite from pencils [121]. In addition, a similar approach has been used to develop a conductivity-based sensor for ammonia in gas phase [122]. The latter example is important because it highlights the interdependence of the variables involved in the engraving process (incident power, borate concentration, and lateral speed) regarding the resistance of the resulting material and the sensor's performance. It is also important to mention that this sensor incorporated a natural deep eutectic solvent (NADES) prepared from a mixture of lactic acid (L) and glucose (G) with a 5:1 ratio with 15% of H_2O (v/v) (H). This NADES was selected due to its physical properties (vapor pressure, viscosity, and polarity) and its ability to interact with the target analyte (NH_3). It is worth mentioning that there is an additional, hybrid strategy known as catalytic graphitization [123,124] that allows the preferential conversion of paper into structured carbon electrodes by amorphous carbon in direct contact with a molten metallic catalyst. This method, which only leads to small amounts of graphite formation, was recently used by Giordano's group [125] to form nanostructures using a catalytic ink composed of $Fe(NO_3)_3$.

One exciting aspect of the reaction pathway involved in the carbonization of paper is that regardless of the process (hydrothermal [126], pyrolysis, laser, etc.), the first reaction that occurs is the dehydration of the glucose units. This process, which has tremendous industrial implications [127–129], is typically performed at relatively low temperatures ($<200°C$) and can be described by various mechanisms, including 1,2-dehydration, Pinacol rearrangement, Tandem Alkaline Pinacol Rearrangement/Retro Aldol Fragmentation, cyclic Grob fragmentation, and generalized alcohol condensation [130]. Regardless of the route, this initial formation of C5 sugars and furfurals can lead to various sp^2-hybridized structures with very specific optical responses [131,132]. These molecules have allowed the development of multiple sensors with specific recognition capabilities, as described by Mohan et al. [133,134]. Also taking advantage of the thermal reactivity of cellulose, Clark et al. recently described the possibility to use laser engraving (using low power intensity) to promote the formation of a combination of small organic molecules (including 5-(hydroxymethyl)furfural, see Fig. 2.6A) with fluorescence emission in the 300–540 nm range [41]. As expected, the fluorescence intensity of these compounds was sensitive to redox processes, enabling their application to determine the concentration of sodium hypochlorite in the range between 0 and 0.1 mM (Fig. 2.6B).

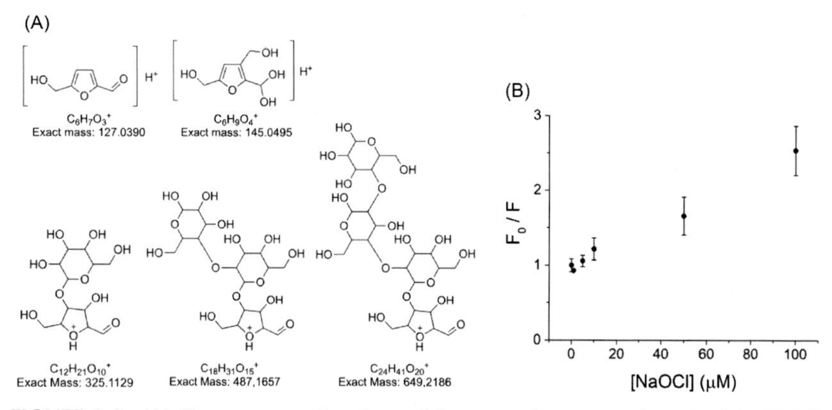

FIGURE 2.6 (A) Fluorescent species observed in engraved paper, as determined by HPLC-HRMS-PDA-FD. (B) Dependence of the fluorescence quenching with respect to the concentration of sodium hypochlorite. *From Clark, K.M., Skrajewski, L., Benavidez, T.E., Mendes, L.F., Bastos, E.L., Dörr, F.A., et al., Fluorescent patterning of paper through laser engraving.* Soft Matter *2020, 16 (33), 7659–7666.*

As discussed in [41], the formation of these compounds is in line with previous reports describing the thermal formation of chromophores from paper [135,136] and highlight the differences with respect to the orange-red luminescence from dots formed by aggregates of furfural derivatives after a 10 min synthesis in the presence of H_3PO_4 [137]. This observation is critically important because distinguishes the short reaction times involved in the laser treatment with respect to the longer processes that allow the growth of these nuclei into carbon dots [138,139]. Although analytical applications of carbon dots abound in the literature and will not be discussed in this chapter, the following references provide specific examples of the application of carbon dots fabricated form cellulose [140–142].

2.7 Conclusion

This chapter's objective is to call the readers' attention to the importance of paper chemistry to rationally develop paper-based chemical devices discussed in this book. To accomplish this, the chapter presented some chemical insights about the fabrication and modification of the paper's surface including grafting, coatings, and more aggressive treatments based on covalent modification of the –OH groups.

Acknowledgments

Financial support for this project has been provided in part by Clemson University and the Brazilian agencies FAPESP (Grant Numbers: 2018/08782-1 and 2019/16491-0), CAPES, CNPq (grant number: 302839/2020-8), and INCTBio (573672/2008-3).

References

[1] V. Hughes, Were the first artists mostly women? Natl. Geographic (2013).

[2] J. Pickering, The history of paper money in china, J. Am. Orient. Soc. 1 (2) (1844) 136–142.

[3] Britannica, T. E. o. t. E., Cai Lun. 2021.

[4] R.I. Burns, The paper revolution in Europe: crusader valencia's paper industry: a technological and behavioral breakthrough, Pac. Hist. Rev. 50 (1) (1981) 1–30.

[5] E. Noviana, D.B. Carrão, R. Pratiwi, C.S. Henry, Emerging applications of paper-based analytical devices for drug analysis: a review, Anal. Chim. acta 1116 (2020) 70–90.

[6] R.H. Tang, L.N. Liu, S.F. Zhang, X.C. He, X.J. Li, F. Xu, et al., A review on advances in methods for modification of paper supports for use in point-of-care testing, Microchim. Acta 186 (8) (2019) 521.

[7] R.M. Rowell, Handbook of wood chemistry and wood composites, CRC Press, 2012.

[8] O. Benaimeche, N.T. Seghir, Ł. Sadowski, M. Mellas, The utilization of vegetable fibers in cementitious materials, in: S. Hashmi, I.A. Choudhury (Eds.), Encyclopedia of Renewable and Sustainable Materials, Elsevier, Oxford, 2020, pp. 649–662.

[9] N.D. Patil, N.R. Tanguy, N. Yan, 3 Lignin interunit linkages and model compounds, in: O. Faruk, M. Sain (Eds.), Lignin in Polymer Composites, William Andrew Publishing, 2016, pp. 27–47.

[10] J.M. Martin-Martinez, Chapter 13 Rubber base adhesives, in: D.A. Dillard, A.V. Pocius, M. Chaudhury (Eds.), Adhesion Science and Engineering, Elsevier Science B.V, Amsterdam, 2002, pp. 573–675.

[11] A. Gutiérrez, C. José, Á.T. Martínez, Chemical analysis and biological removal of wood lipids forming pitch deposits in paper pulp manufacturing, Environmental Microbiology, Springer, 2004, pp. 189–202.

[12] C.J. Biermann, Essentials of Pulping and Papermaking, Academic press, 1993.

[13] L.A. Bell, Plant Fibers for Papermaking, Liliaceae Press, 1981.

[14] K. Ferguson, New Trends & Developments in Papermaking, Miller Freeman Books, 1994.

[15] J. Munsell, 1876–1990 Chronology and Process of Papermaking, Albert Saifer Publisher, 1992.

[16] A.W. Martinez, S.T Phillips, M.J. Butte, G.M. Whitesides, Supporting information for patterned paper as a platform for inexpensive, low volume, portable bioassays (2007).

[17] Z. Liu, H. Wang, L. Hui, Pulping and papermaking of non-wood fibers, Pulp Pap. Process. 1 (2018) 4–31.

[18] M. Sanchis-Sebastiá, V. Novy, L. Stigsson, M. Galbe, O. Wallberg, RSC Adv. 11 (4) (2021) 12321–12329.

[19] R. Alen, R. Andersson, G. Annergreen, C.-G. Berg, C. Chirat, J. van Dam, et al., *Chemical pulping part 1, fibre chemistry and technology*. Paper Engineers' Association/ Paperi ja Puu Oy; Helsinki: 2011; Vol. 6.

[20] M.L.O. D'Almeida, M.E.T. Koga, D.C. Ferreira, R.J.B. Pigozzo, R. Toucini, H.M. dos Reis, et al., Cellulose, SENAI, São Paulo, 2013.

[21] R.N. Shreve, J.A. Brink, Chemical process industries, Fourth (ed.), Indústrias de processos químicos, 1, Guanabara Koogan, Rio de Janeiro, 1997, p. 717.

[22] C. Duan, Y. Long, J. Li, X. Ma, Y. Ni, Changes of cellulose accessibility to cellulase due to fiber hornification and its impact on enzymatic viscosity control of dissolving pulp, Cellulose 22 (4) (2015) 2729–2736.

[23] R. Pelton, Bioactive paper provides a low-cost platform for diagnostics, Trends Anal. Chem. 28 (8) (2009) 925–942.

[24] E. Evans, E.F.M. Gabriel, W.K.T. Coltro, C.D. Garcia, Rational selection of substrates to improve color intensity and uniformity on microfluidic paper-based analytical devices, Analyst 139 (9) (2014) 2127–2132.

[25] S.K. Mahadeva, S. Yun, J. Kim, Flexible humidity and temperature sensor based on cellulose–polypyrrole nanocomposite, Sens. Actuators A: Phys. 165 (2) (2011) 194–199.

[26] H. Qi, E. Mäder, J. Liu, Unique water sensors based on carbon nanotube–cellulose composites, Sens. Actuators B Chem. 185 (2013) 225–230.

[27] F. Loghin, A. Rivadeneyra, M. Becherer, P. Lugli, M. Bobinger, A facile and efficient protocol for preparing residual-free single-walled carbon nanotube films for stable sensing applications, Nanomaterials 9 (3) (2019) 471.

[28] B. Wei, P. Guan, L. Zhang, G. Chen, Solubilization of carbon nanotubes by cellulose xanthate toward the fabrication of enhanced amperometric detectors, Carbon 48 (5) (2010) 1380–1387.

[29] X. Luo, H. Zhang, Z. Cao, N. Cai, Y. Xue, F. Yu, A simple route to develop transparent doxorubicin-loaded nanodiamonds/cellulose nanocomposite membranes as potential wound dressings, Carbohydr. Polym. 143 (2016) 231–238.

[30] X. Luo, J. Xia, X. Jiang, M. Yang, S. Liu, Cellulose-based strips designed based on a sensitive enzyme colorimetric assay for the low concentration of glucose detection, Anal. Chem. 91 (24) (2019) 15461–15468.

[31] H.T. Sahin, M.B. Arslan, A study on physical and chemical properties of cellulose paper immersed in various solvent mixtures, Int. J. Mol. Sci. 9 (1) (2008) 78–88.

[32] L. Senni, C. Casieri, A. Bovino, M.C. Gaetani, F. De Luca, A portable NMR sensor for moisture monitoring of wooden works of art, particularly of paintings on wood, Wood Sci. Technol. 43 (1) (2009) 167–180.

[33] D.C. Malaspina, J. Faraudo, Molecular insight into the wetting behavior and amphiphilic character of cellulose nanocrystals, Adv. Colloid Interface Sci. 267 (2019) 15–25.

[34] S. Ummartyotin, H. Manuspiya, A critical review on cellulose: from fundamental to an approach on sensor technology, Renew. Sustain. Energy Rev. 41 (2015) 402–412.

[35] M.F. Mora, C.D. Garcia, F. Schaumburg, P.A. Kler, C.L.A. Berli, M. Hashimoto, et al., Patterning and modeling three-dimensional microfluidic devices fabricated on a single sheet of paper, Anal. Chem. 91 (13) (2019) 8298–8303.

[36] L. McCann, T.E. Benavidez, S. Holtsclaw, C.D. Garcia, Addressing the distribution of proteins spotted on μPADs, Analyst 142 (20) (2017) 3899–3905.

[37] R. Tang, H. Yang, Y. Gong, Z. Liu, X. Li, T. Wen, et al., Improved analytical sensitivity of lateral flow assay using sponge for HBV nucleic acid detection, Sci. Rep. 7 (1) (2017) 1360.

[38] S.-G. Jeong, J. Kim, S.H. Jin, K.-S. Park, C.-S. Lee, Flow control in paper-based microfluidic device for automatic multistep assays: a focused minireview, Korean J. Chem. Eng. 33 (10) (2016) 2761–2770.

[39] D.R. Ballerini, X. Li, W. Shen, Patterned paper and alternative materials as substrates for low-cost microfluidic diagnostics, Microfluid. Nanofluid. 13 (5) (2012) 769–787.

[40] W.R. de Araujo, T.R.L.C. Paixão, Fabrication of disposable electrochemical devices using silver ink and office paper, Analyst 139 (11) (2014) 2742–2747.

[41] K.M. Clark, L. Skrajewski, T.E. Benavidez, L.F. Mendes, E.L. Bastos, F.A. Dörr, et al., Fluorescent patterning of paper through laser engraving, Soft Matter 16 (33) (2020) 7659–7666.

[42] W.R. de Araujo, C.M.R. Frasson, W.A. Ameku, J.R. Silva, L. Angnes, T.R.L.C. Paixão, Single-step reagentless laser scribing fabrication of electrochemical paper-based analytical devices, Angew. Chem. Int. (Ed.) 56 (47) (2017) 15113–15117.

[43] K. Missoum, M.N. Belgacem, J. Bras, Nanofibrillated cellulose surface modification: a review, Materials 6 (5) (2013).

[44] S. Eyley, W. Thielemans, Surface modification of cellulose nanocrystals, Nanoscale 6 (14) (2014) 7764–7779.

[45] D. Miyashiro, R. Hamano, K. Umemura, A review of applications using mixed materials of cellulose, nanocellulose and carbon nanotubes, Nanomaterials 10 (2) (2020) 186.

[46] J. Rull-Barrull, M. d'Halluin, E. Le Grognec, F.-X. Felpin, Chemically-modified cellulose paper as smart sensor device for colorimetric and optical detection of hydrogen sulfate in water, Chem. Commun. 52 (12) (2016) 2525–2528.

[47] G. Bali, X. Meng, J.I. Deneff, Q. Sun, A.J. Ragauskas, The effect of alkaline pretreatment methods on cellulose structure and accessibility, ChemSusChem 8 (2) (2015) 275–279.

[48] D. Roy, J.T. Guthrie, S. Perrier, Graft polymerization: grafting poly(styrene) from cellulose via reversible addition – fragmentation chain transfer (RAFT) polymerization, Macromolecules 38 (25) (2005) 10363–10372.

[49] A.S. Goldmann, T. Tischer, L. Barner, M. Bruns, C. Barner-Kowollik, Mild and modular surface modification of cellulose via hetero diels – alder (HDA) cycloaddition, Biomacromolecules 12 (4) (2011) 1137–1145.

[50] S.K. Mahadeva, K. Walus, B. Stoeber, Paper as a platform for sensing applications and other devices: a review, ACS Appl. Mater. Interfaces 7 (16) (2015) 8345–8362.

[51] L.R. Shobin, S. Manivannan, Carbon nanotubes on paper: flexible and disposable chemiresistors, Sens. Actuators B Chem. 220 (2015) 1178–1185.

[52] S. Ji, M. Lee, D. Kim, Detection of early stage prostate cancer by using a simple carbon nanotube@paper biosensor, Biosens. Bioelectron. 102 (2018) 345–350.

[53] E. Evans, E.F. Moreira Gabriel, T.E. Benavidez, W.K. Tomazelli Coltro, C.D. Garcia, Modification of microfluidic paper-based devices with silica nanoparticles, Analyst 139 (21) (2014) 5560–5567.

[54] N. Ruecha, N. Rodthongkum, D.M. Cate, J. Volckens, O. Chailapakul, C.S. Henry, Sensitive electrochemical sensor using a graphene–polyaniline nanocomposite for simultaneous detection of Zn(II), Cd(II), and Pb(II), Anal. Chim. Acta 874 (2015) 40–48.

[55] Y. Soda, H. Shibata, K. Yamada, K. Suzuki, D. Citterio, Selective detection of K + by ion-selective optode nanoparticles on cellulosic filter paper substrates, ACS Appl. Nano Mater. 1 (4) (2018) 1792–1800.

[56] J.K. Pandey, H. Takagi, A.N. Nakagaito, D.R. Saini, S.-H. Ahn, An overview on the cellulose based conducting composites, Compos. Part. B Eng. 43 (7) (2012) 2822–2826.

[57] F.D. Guerra, M.L. Campbell, M.F. Attia, D.C. Whitehead, F. Alexis, Capture of aldehyde VOCs using a series of amine-functionalized cellulose nanocrystals, ChemistrySelect 3 (20) (2018) 5495–5501.

[58] I. Bravo, F. Figueroa, M.I. Swasy, M.F. Attia, M. Ateia, D. Encalada, et al., Cellulose particles capture aldehyde VOC pollutants, RSC Adv. 10 (13) (2020) 7967–7975.

[59] R.S.J. Alkasir, M. Ornatska, S. Andreescu, Colorimetric paper bioassay for the detection of phenolic compounds, Anal. Chem. 84 (22) (2012) 9729–9737.

[60] H. Koga, T. Kitaoka, A. Isogai, In situ modification of cellulose paper with amino groups for catalytic applications, J. Mater. Chem. 21 (25) (2011) 9356–9361.

[61] Q. He, C. Ma, X. Hu, H. Chen, Method for fabrication of paper-based microfluidic devices by alkylsilane self-assembling and UV/O3-patterning, Anal. Chem. 85 (3) (2013) 1327–1331.

[62] L. Cai, Y. Wang, Y. Wu, C. Xu, M. Zhong, H. Lai, et al., Fabrication of a microfluidic paper-based analytical device by silanization of filter cellulose using a paper mask for glucose assay, Analyst 139 (18) (2014) 4593–4598.
[63] L. Cai, C. Xu, S. Lin, J. Luo, M. Wu, F. Yang, A simple paper-based sensor fabricated by selective wet etching of silanized filter paper using a paper mask, Biomicrofluidics 8 (5) (2014) 056504.
[64] L. Kong, Q. Wang, S. Xiong, Y. Wang, Turning low-cost filter papers to highly efficient membranes for oil/water separation by atomic-layer-deposition-enabled hydrophobization, Ind. Eng. Chem. Res. 53 (42) (2014) 16516–16522.
[65] L. Zhang, H. Kwok, X. Li, H.-Z. Yu, Superhydrophobic substrates from off-the-shelf laboratory filter paper: simplified preparation, patterning, and assay application, ACS Appl. Mater. Interfaces 9 (45) (2017) 39728–39735.
[66] L. Zhang, A.G. Zhou, B.R. Sun, K.S. Chen, H.-Z. Yu, Functional and versatile superhydrophobic coatings via stoichiometric silanization, Nat. Commun. 12 (1) (2021) 982.
[67] A. Shakeri, N.A. Jarad, A. Leung, L. Soleymani, T.F. Didar, Biofunctionalization of glass- and paper-based microfluidic devices: A review, Adv. Mater. Interfaces 6 (19) (2019) 1900940.
[68] S. Tiwari, M. Vinchurkar, V.R. Rao, G. Garnier, Zinc oxide nanorods functionalized paper for protein preconcentration in biodiagnostics, Sci. Rep. 7 (1) (2017) 43905.
[69] K.J. Babu, T. Raj kumar, D.J. Yoo, S.-M. Phang, G. Gnana kumar, Electrodeposited nickel cobalt sulfide flowerlike architectures on disposable cellulose filter paper for enzyme-free glucose sensor applications, ACS Sustain. Chem. Eng. 6 (12) (2018) 16982–16989.
[70] S. Kumar, V.S. Chauhan, S.K. Chakrabarti, Separation and analysis techniques for bound and unbound alkyl ketene dimer (AKD) in paper: a review, Arab. J. Chem. 9 (2016) S1636–S1642.
[71] K. Ashish, N.K. Bhardwaj, S.P. Singh, Cationic starch and polyacrylamides for alkenyl succinic anhydride (ASA) emulsification for sizing of cellulosic fibers, Cellulose 26 (18) (2019) 9901–9915.
[72] J. Qin, L. Chen, C. Zhao, Q. Lin, S. Chen, Cellulose nanofiber/cationic conjugated polymer hybrid aerogel sensor for nitroaromatic vapors detection, J. Mater. Sci. 52 (14) (2017) 8455–8464.
[73] M. Lee, K. Oh, H.-K. Choi, S.G. Lee, H.J. Youn, H.L. Lee, et al., Subnanomolar sensitivity of filter paper-based SERS sensor for pesticide detection by hydrophobicity change of paper surface, ACS Sens. 3 (1) (2018) 151–159.
[74] X. Li, J. Tian, T. Nguyen, W. Shen, Paper-based microfluidic devices by plasma treatment, Anal. Chem. 80 (23) (2008) 9131–9134.
[75] G.I. Salentijn, N.N. Hamidon, E. Verpoorte, Solvent-dependent on/off valving using selectively permeable barriers in paper microfluidics, Lab Chip 16 (6) (2016) 1013–1021.
[76] N.N. Hamidon, Y. Hong, G.I.J. Salentijn, E. Verpoorte, Water-based alkyl ketene dimer ink for user-friendly patterning in paper microfluidics, Anal. Chim. Acta 1000 (2018) 180–190.
[77] A. Kumar, N.K. Bhardwaj, S.P. Singh, Sizing performance of alkenyl succinic anhydride (ASA) emulsion stabilized by polyvinylamine macromolecules, Colloids Surf. A Physicochem. Eng. Asp. 539 (2018) 132–139.
[78] A. Sato, D. Kabusaki, H. Okumura, T. Nakatani, F. Nakatsubo, H. Yano, Surface modification of cellulose nanofibers with alkenyl succinic anhydride for high-density polyethylene reinforcement, Compos. Part. A Appl. Sci. Manuf. 83 (2016) 72–79.

[79] O. Haske-Cornelius, S. Weinberger, F. Quartinello, C. Tallian, F. Brunner, A. Pellis, et al., Environmentally friendly covalent coupling of proteins onto oxidized cellulosic materials, N. J. Chem. 43 (36) (2019) 14536–14545.

[80] Y. Tang, K. Petropoulos, F. Kurth, H. Gao, D. Migliorelli, O. Guenat, et al., Screen-printed glucose sensors modified with cellulose nanocrystals (CNCs) for cell culture monitoring, Biosensors 10 (9) (2020) 125.

[81] Q. Bi, S. Dong, Y. Sun, X. Lu, L. Zhao, An electrochemical sensor based on cellulose nanocrystal for the enantioselective discrimination of chiral amino acids, Anal. Biochem. 508 (2016) 50–57.

[82] C.-H. Yang, C.-A. Chen, C.-F. Chen, Surface-modified cellulose paper and its application in infectious disease diagnosis, Sens. Actuators B: Chem. 265 (2018) 506–513.

[83] Y. Zhu, X. Xu, N.D. Brault, A.J. Keefe, X. Han, Y. Deng, et al., Cellulose paper sensors modified with zwitterionic poly(carboxybetaine) for sensing and detection in complex media, Anal. Chem. 86 (6) (2014) 2871–2875.

[84] A. Carlmark, E.E. Malmström, ATRP grafting from cellulose fibers to create block-copolymer grafts, Biomacromolecules 4 (6) (2003) 1740–1745.

[85] P.B. Groszewicz, P. Mendes, B. Kumari, J. Lins, M. Biesalski, T. Gutmann, et al., N-Hydroxysuccinimide-activated esters as a functionalization agent for amino cellulose: synthesis and solid-state NMR characterization, Cellulose 27 (3) (2020) 1239–1254.

[86] J. Li, J. Zhang, H. Sun, Y. Yang, Y. Ye, J. Cui, et al., An optical fiber sensor based on carboxymethyl cellulose/carbon nanotubes composite film for simultaneous measurement of relative humidity and temperature, Opt. Commun. 467 (2020) 125740.

[87] J. Wang, M. Dang, C. Duan, W. Zhao, K. Wang, Carboxymethylated cellulose fibers as low-cost and renewable adsorbent materials, Ind. Eng. Chem. Res. 56 (51) (2017) 14940–14948.

[88] T.G. Cordeiro, M.S. Ferreira Santos, I.G.R. Gutz, C.D. Garcia, Photochemical oxidation of alcohols: simple derivatization strategy for their analysis by capillary electrophoresis, Food Chem. 292 (2019) 114–120.

[89] M.S. Ferreira Santos, T. Gomes Cordeiro, Z. Cieslarova, I.G.R. Gutz, C.D. Garcia, Photochemical and photocatalytic degradation of 1-propanol using UV/H2O2: identification of malonate as byproduct, Electrophoresis 40 (18–19) (2019) 2256–2262.

[90] N. Guigo, K. Mazeau, J.-L. Putaux, L. Heux, Surface modification of cellulose microfibrils by periodate oxidation and subsequent reductive amination with benzylamine: a topochemical study, Cellulose 21 (6) (2014) 4119–4133.

[91] T. Nypelö, B. Berke, S. Spirk, J.A. Sirviö, Review: periodate oxidation of wood polysaccharides—Modulation of hierarchies, Carbohydr. Polym. 252 (2021) 117105.

[92] N.O. Gomes, E. Carrilho, S.A.S. Machado, L.F. Sgobbi, Bacterial cellulose-based electrochemical sensing platform: a smart material for miniaturized biosensors, Electrochim. Acta 349 (2020) 136341.

[93] X. Feng, J. Zhang, J. Wang, A. Han, G. Fang, J. Liu, et al., The stabilization of fluorescent copper nanoclusters by dialdehyde cellulose and their use in mercury ion sensing, Anal. Methods 12 (24) (2020) 3130–3136.

[94] L. Wang, C. Zhang, H. He, H. Zhu, W. Guo, S. Zhou, et al., Cellulose-based colorimetric sensor with N, S sites for Ag+ detection, Int. J. Biol. Macromolecules 163 (2020) 593–602.

[95] G. Patterson, Y.-L. Hsieh, Tunable dialdehyde/dicarboxylate nanocelluloses by stoichiometrically optimized sequential periodate–chlorite oxidation for tough and wet shape recoverable aerogels, Nanoscale Adv. 2 (12) (2020) 5623–5634.

[96] C.G. Carlsson, G. Stroem, Reduction and oxidation of cellulose surfaces by means of cold plasma, Langmuir 7 (11) (1991) 2492–2497.

[97] G. Bretel, J. Rull-Barrull, M.C. Nongbe, J.-P. Terrier, E. Le Grognec, F.-X. Felpin, Hydrophobic covalent patterns on cellulose paper through photothiol-X ligations, ACS Omega 3 (8) (2018) 9155–9159.

[98] Y.-C. Lin, J. Cho, G.A. Tompsett, P.R. Westmoreland, G.W. Huber, Kinetics and mechanism of cellulose pyrolysis, J. Phys. Chem. C. 113 (46) (2009) 20097–20107.

[99] M.M. Tang, R. Bacon, Carbonization of cellulose fibers—I. low temperature pyrolysis, Carbon 2 (3) (1964) 211–220.

[100] E. Fitzer, K.-H. Kochling, H.P. Boehm, H. Marsh, Recommended terminology for the description of carbon as a solid (IUPAC Recommendations 1995), Pure Appl. Chem. 67 (3) (2009) 473–506.

[101] T. Noda, M. Inagaki, The Structure of glassy carbon, Bull. Chem. Soc. Jpn. 37 (10) (1964) 1534–1538.

[102] G.M. Jenkins, K. Kawamura, Structure of glassy carbon, Nature 231 (5299) (1971) 175–176.

[103] J.G. Giuliani, T.E. Benavidez, G.M. Duran, E. Vinogradova, A. Rios, C.D. Garcia, Development and characterization of carbon based electrodes from pyrolyzed paper for biosensing applications, J. Electroanal. Chem. 765 (2016) 8–15.

[104] T.E. Benavidez, R. Martinez-Duarte, C.D. Garcia, Analytical methodologies using carbon substrates developed by pyrolysis, Anal. Methods 8 (21) (2016) 4163–4176.

[105] R.L. McCreery, Advanced carbon electrode materials for molecular electrochemistry, Chem. Rev. 108 (7) (2008) 2646–2687.

[106] G. Dai, K. Wang, G. Wang, S. Wang, Initial pyrolysis mechanism of cellulose revealed by in-situ DRIFT analysis and theoretical calculation, Combust. Flame 208 (2019) 273–280.

[107] Z. Gao, N. Li, M. Chen, W. Yi, Comparative study on the pyrolysis of cellulose and its model compounds, Fuel Process. Technol. 193 (2019) 131–140.

[108] X. Yang, Z. Fu, D. Han, Y. Zhao, R. Li, Y. Wu, Unveiling the pyrolysis mechanisms of cellulose: experimental and theoretical studies, Renew. Energy 147 (2020) 1120–1130.

[109] Q. Wang, H. Song, S. Pan, N. Dong, X. Wang, S. Sun, Initial pyrolysis mechanism and product formation of cellulose: an experimental and density functional theory(DFT) study, Sci. Rep. 10 (1) (2020) 3626.

[110] A.G.W. Bradbury, Y. Sakai, F. Shafizadeh, A kinetic model for pyrolysis of cellulose, J. Appl. Polym. Sci. 23 (11) (1979) 3271–3280.

[111] J.M.R. Gallo, M.A. Trapp, The chemical conversion of biomass-derived saccharides: an overview, J. Braz. Chem. Soc. 28 (9) (2017) 1586–1607.

[112] G.M. Duran, T.E. Benavidez, J.G. Giuliani, A. Rios, C.D. Garcia, Synthesis of CuNP-modified carbon electrodes obtained by pyrolysis of paper, Sens. Actuators B: Chem. 227 (2016) 626–633.

[113] L.A.J. Silva, W.P. da Silva, J.G. Giuliani, S.C. Canobre, C.D. Garcia, R.A.A. Munoz, et al., Use of pyrolyzed paper as disposable substrates for voltammetric determination of trace metals, Talanta 165 (2017) 33–38.

[114] B. Hsia, M.S. Kim, M. Vincent, C. Carraro, R. Maboudian, Photoresist-derived porous carbon for on-chip micro-supercapacitors, Carbon 57 (2013) 395–400.

[115] S.A. Alharthi, T.E. Benavidez, C.D. Garcia, Ultrathin optically transparent carbon electrodes produced from layers of adsorbed proteins, Langmuir 29 (10) (2013) 3320–3327.

[116] B.A. Lagasse, L. McCann, T. Kidwell, M.S. Blais, C.D. Garcia, Decomposition of chemical warfare agent simulants utilizing pyrolyzed cotton balls as wicks, ACS Omega 5 (32) (2020) 20051–20061.

[117] P.A. Reed, R.M. Cardoso, R.A.A. Muñoz, C.D. Garcia, Pyrolyzed cotton balls for protein removal: analysis of pharmaceuticals in serum by capillary electrophoresis, Anal. Chim. Acta 1110 (2020) 90–97.

[118] L.F. Mendes, A. de Siervo, W. Reis de Araujo, T.R. Longo Cesar Paixão, Reagentless fabrication of a porous graphene-like electrochemical device from phenolic paper using laser-scribing, Carbon 159 (2020) 110–118.

[119] A. Bezerra Martins, A. Lobato, N. Tasić, F.J. Perez-Sanz, P. Vidinha, T.R.L.C. Paixão, L. Moreira Gonçalves, Laser-pyrolyzed electrochemical paper-based analytical sensor for sulphite analysis, Electrochem. Commun. 107 (2019) 106541.

[120] W.A. Ameku, W.R. de Araujo, C.J. Rangel, R.A. Ando, T.R.L.C. Paixão, Gold nanoparticle paper-based dual-detection device for forensics applications, ACS Appl. Nano Mater. 2 (9) (2019) 5460–5468.

[121] V.N. Ataide, W.A. Ameku, R.P. Bacil, L. Angnes, W.R. de Araujo, T.R.L.C. Paixão, Enhanced performance of pencil-drawn paper-based electrodes by laser-scribing treatment, RSC Adv. 11 (3) (2021) 1644–1653.

[122] M. Reynolds, L.M. Duarte, W.K.T. Coltro, M.F. Silva, F.J.V. Gomez, C.D. Garcia, Laser-engraved ammonia sensor integrating a natural deep eutectic solvent, Microchem. J. 157 (2020) 105067.

[123] A. Ōya, H. Marsh, Phenomena of catalytic graphitization, J. Mater. Sci. 17 (2) (1982) 309–322.

[124] M. Sevilla, A.B. Fuertes, Fabrication of porous carbon monoliths with a graphitic framework, Carbon 56 (2013) 155–166.

[125] S. Glatzel, Z. Schnepp, C. Giordano, From paper to structured carbon electrodes by inkjet printing, Angew. Chem. Int. (Ed.) 52 (8) (2013) 2355–2358.

[126] N. Shi, Q. Liu, R. Ju, X. He, Y. Zhang, S. Tang, et al., Condensation of α-carbonyl aldehydes leads to the formation of solid humins during the hydrothermal degradation of carbohydrates, ACS Omega 4 (4) (2019) 7330–7343.

[127] B.M. Matsagar, S.A. Hossain, T. Islam, H.R. Alamri, Z.A. Alothman, Y. Yamauchi, et al., Direct production of furfural in one-pot fashion from raw biomass using brønsted acidic ionic liquids, Sci. Rep. 7 (1) (2017) 13508.

[128] Y. Luo, Z. Li, X. Li, X. Liu, J. Fan, J.H. Clark, et al., The production of furfural directly from hemicellulose in lignocellulosic biomass: a review, Catal. Today 319 (2019) 14–24.

[129] D.A. Gonzalez-Casamachin, J. Rivera De la Rosa, C.J. Lucio–Ortiz, L. Sandoval-Rangel, C.D. García, Partial oxidation of 5-hydroxymethylfurfural to 2,5-furandicarboxylic acid using O_2 and a photocatalyst of a composite of ZnO/PPy under visible-light: electrochemical characterization and kinetic analysis, Chem. Eng. J. 393 (2020) 124699.

[130] M.W. Easton, J.J. Nash, H.I. Kenttämaa, Dehydration pathways for glucose and cellobiose during fast pyrolysis, J. Phys. Chem. A 122 (41) (2018) 8071–8085.

[131] S.-T. Yang, L. Cao, P.G. Luo, F. Lu, X. Wang, H. Wang, et al., Carbon dots for optical imaging in Vivo, J. Am. Chem. Soc. 131 (32) (2009) 11308–11309.

[132] Y. Fang, S. Guo, D. Li, C. Zhu, W. Ren, S. Dong, et al., Easy synthesis and imaging applications of cross-linked reen Fluorescent hollow carbon nanoparticles, ACS Nano 6 (1) (2012) 400–409.

[133] B. Mohan, H.K. Sharma, Synthesis of calix[6]arene and transduction of its furfural derivative as sensor for Hg(II) ions, Inorg. Chim. Acta 486 (2019) 63–68.

[134] B. Mohan, S. Kumar, H.K. Sharma, Synthesis and characterizations of flexible furfural based molecular receptor for selective recognition of Dy(III) ions, Polyhedron 183 (2020) 114537.

[135] A.M. Emsley, G.C. Stevens, Kinetics and mechanisms of the low-temperature degradation of cellulose, Cellulose 1 (1) (1994) 26–56.

[136] H. Tylli, I. Forsskåhl, C. Olkkonen, The effect of heat and IR radiation on the fluorescence of cellulose, Cellulose 7 (2) (2000) 133–146.

[137] V. Gude, A. Das, T. Chatterjee, P.K. Mandal, Molecular origin of photoluminescence of carbon dots: aggregation-induced orange-red emission, Phys. Chem. Chem. Phys. 18 (40) (2016) 28274–28280.

[138] M.L. Liu, B.B. Chen, C.M. Li, C.Z. Huang, Carbon dots: synthesis, formation mechanism, fluorescence origin and sensing applications, Green Chem. 21 (3) (2019) 449–471.

[139] N. Papaioannou, M.-M. Titirici, A. Sapelkin, Investigating the effect of reaction time on carbon dot formation, structure, and optical properties, ACS Omega 4 (26) (2019) 21658–21665.

[140] D.R. da Silva Souza, L.D. Caminhas, J.P. de Mesquita, F.V. Pereira, Luminescent carbon dots obtained from cellulose, Mater. Chem. Phys. 203 (2018) 148–155.

[141] H. Su, Z. Bi, Y. Ni, L. Yan, One-pot degradation of cellulose into carbon dots and organic acids in its homogeneous aqueous solution, Green. Energy Environ. 4 (4) (2019) 391–399.

[142] B. Sui, Y. Zhang, L. Huang, Y. Chen, D. Li, Y. Li, et al., Fluorescent nanofibrillar hydrogels of carbon dots and cellulose nanocrystals and their biocompatibility, ACS Sustain. Chem. Eng. 8 (50) (2020) 18492–18499.

Chapter 3

Paper-based separation devices

Cyro L.S. Chagas[1], Nikaele S. Moreira[2], Bárbara G.S. Guinati[2] and Wendell K.T. Coltro[2,3]

[1]*Instituto de Química, Universidade de Brasília, Brasília, Brazil,* [2]*Instituto de Química, Universidade Federal de Goiás, Goiânia, Brazil,* [3]*Instituto Nacional de Ciência e Tecnologia de Bioanalítica (INCTBio), Campinas, Brazil*

3.1 Introduction

The use of paper for analytical applications dates back to 1700, when the first tests for the determination of pH using a piece of paper impregnated with chromophore reagents were reported [1]. Since then, many researchers have dedicated their studies using paper in the most diverse applications, mainly in spot tests for metal analysis, paper electrophoresis for protein analysis, and paper chromatography for immunoassays [2]. However, the use of paper was restricted due to the emergence of new, more efficient, automated, and better performing analytical techniques, such as high-performance liquid chromatographs and commercial capillary electrophoresis equipment.

After many years of being left out, paper returned to be in evidence when, in 2007, Martinez et al. proposed the use of paper for the development of microfluidic paper-based analytical devices (μPADs), which was a milestone for analytical chemistry [3]. The manufacturing and use of μPADs have several advantages, such as simplicity, low cost, global abundance and affordability, and disposability, when compared to conventional platforms structured in glass, quartz, silicon, and polymers. Another important point to be highlighted is the paper's ability to transport solution through capillarity, without the need for an external pumping system, quite common in other microfluidic devices.

Since the publication of this pioneering work, proposed by Martinez et al. [3], several studies involving the use of μPADs in multiple areas have been reported in the literature, mainly exploring the integration and use of electrochemical, colorimetric, chemiluminescent, and fluorescent detectors [4–7]. More recently, examples of chromatographic and electrophoretic separations on paper-based platforms have successfully demonstrated the great potential of this substrate for this purpose [8]. In this chapter, analytical

Paper-Based Analytical Devices for Chemical Analysis and Diagnostics.
DOI: https://doi.org/10.1016/B978-0-12-820534-1.00006-2
© 2022 Elsevier Inc. All rights reserved.

separations in μPADs are discussed through recent publications, which are ordered in electrophoretic and chromatographic separations in paper-based devices.

3.2 Paper-based devices for electrophoretic separations

Although paper substrate has been extensively explored for electrophoretic separations in the 1950–60s [9–11], it was no longer an interesting substrate for analytical applications in subsequent years due to the low separation efficiency, high sample and reagent consumption, and long analysis time. However, in the decade of 2010, with the advance of miniaturization of analytical devices, paper has returned as an attractive and promising platform for microfluidic applications [8]. Thus, this topic will demonstrate outstanding studies reported in the last two decades involving analytical separations in paper-based platforms. To provide a better and clear visualization, the selected reports will be divided according to the type of electromigration technique used to separate the analytes in the microdevices.

3.2.1 Electrophoresis

Capillary electrophoresis is an analytical separation technique where the analytes are separated according to their electrophoretic mobilities when subjected to an electric field [12]. Since the pioneering study showing electrophoretic separations on chip-based devices [13], different materials have been explored as platforms for the development of electrophoresis microchips [14]. Among the electrophoretic modes, capillary zone electrophoresis is by far the most popular method employed for performing separations on chip-based devices. However, other modes including micellar electrokinetic chromatography, gel electrophoresis, and electrochromatography has also been explored for microscale applications. The latter separation mode makes use of a stationary phase and the separation is achieved based on their electrophoretic migration and differential partitioning in the stationary phase [8,15]. In this way, electrophoresis on paper involves the interaction between analytes and paper fibers [16]. Consequently, the cellulose fibers may act as pseudo stationary phase thus promoting electrochromatographic separations.

Electrophoretic separations were first demonstrated on μPADs by Ge et al. in 2014 (Fig. 3.1A). In this pioneering study, the amino acids serine, aspartic acid, and lysine were successfully separated and detected using an electrogenerated chemiluminescence detector, which was fully integrated into the paper microdevice [17]. From this work, several authors began to study the electrophoretic separations in μPADs, exploring the versatility of paper to design analytical devices in different layouts. Luo et al. reported the development of an origami-shaped paper device (Fig. 3.1B) for rapid

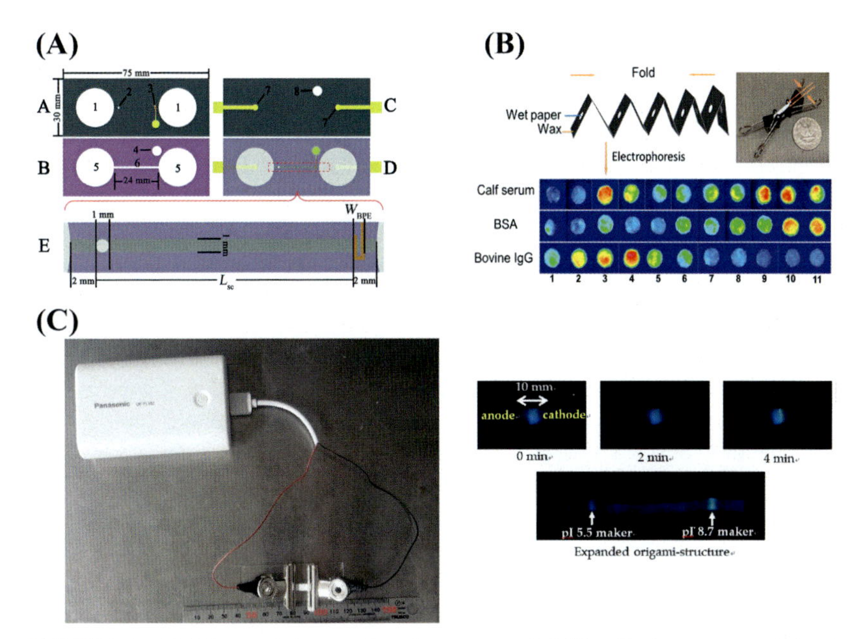

FIGURE 3.1 (A) Scheme of the manufacturing process of μPAD with electrogenerated chemiluminescence detector. (B) Representation of the origami-paper-based electrophoretic device for the separation of bovine serum proteins. (C) Origami-paper-based electrophoretic device coupled to a power bank through the USB port, for the separation of two isoelectric point markers. *(Figure 3.1A) Reproduced with permission from L. Ge, S. Wang, S. Ge, J. Yu, M. Yan, N. Li, et al., Electrophoretic separation in a microfluidic paper-based analytical device with an on-column wireless electrogenerated chemiluminescence detector, Chem. Commun. 50 (2014) 5699–5702. (Figure 3.1B) Reproduced with permission from L. Luo, X. Li, R.M. Crooks, Low-voltage origami-paper-based electrophoretic device for rapid protein separation, Anal. Chem. 86 (2014) 12390–12397. (Figure 3.1C) Reproduced with permission from Y. Matsuda, K. Sakai, H. Yamaguchi, T. Niimi, Electrophoretic separation on an origami paper-based analytical device using a portable power bank, Sensors (Switz.). 19 (2019).*

electrophoretic separations of proteins present in bovine serum combined with a fluorescence detection system. The origami device has advantages such as a short separation time of different proteins (~5 min) and the use of low voltages (~10 V), without requiring a high voltage source, conventionally used for electrophoretic separations. In addition, the low cost of the manufacturing process and portability make the origami paper device very attractive for the analysis of proteins in clinical samples [18]. Origami devices were also studied by Matsuda et al., who used a folded paper strip to separate two isoelectric point markers (pI 5.5 and 8.7), using a portable power bank as a voltage source, connected to the device through a USB connection (Fig. 3.1C). The proposed device presented an excellent separation time of the isoelectric point markers (~4 min) and a great separation resolution, because when the device was unfolded, the analytes were separated at a distance of approximately 35 mm from each other, thus demonstrating the viability of the proposed method [19].

Typically, electrophoresis microchips are usually manufactured in glass or polymeric materials exploiting a common cross-shaped geometry. In this sense, some authors have investigated the possibility of using the same geometry in paper electrophoresis. In 2016, independently, Xu et al. [20] and Chagas et al. [16] demonstrated the fabrication of paper-based analytical devices containing channels designed in a cross format, which were produced from a thermo laminating process (Fig. 3.2A and B). Xu et al. studied the electrokinetic injection and separation of two dyes, carmine and yellow sunset, in filter paper channels defined with width and length of 3 and 55 mm, respectively. Although separation has been achieved on this low-cost platform, the long analysis time (10 min), the low applied potential (300 V) and the detection through means of a camera are points to be improved [20]. Similarly, Chagas et al. performed a study of electrokinetic injection and electrophoretic separation of two biomolecules of clinical interest, bovine serum albumin and creatinine, in channels made on chromatographic paper [16].

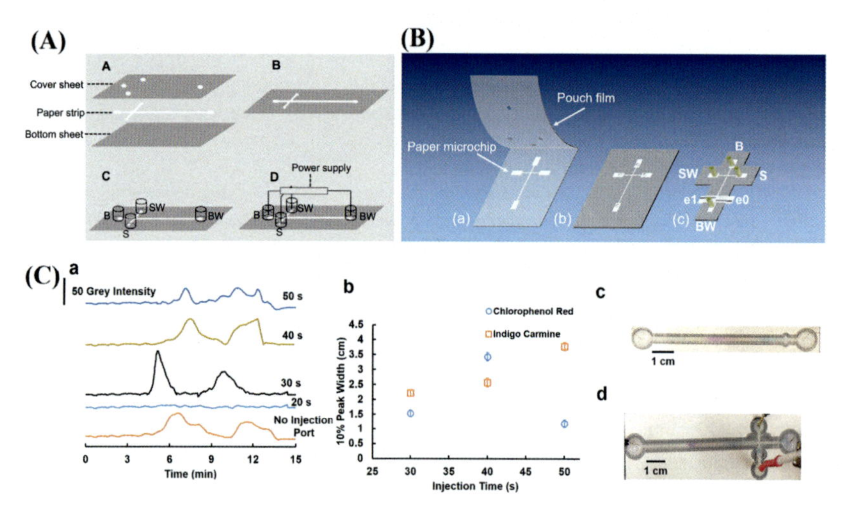

FIGURE 3.2 Representation of microchip manufacturing processes with thermolaminated paper channels for the separation of (A) dyes and (B) biomolecules. (C) Demonstration of the electrophoretic separation of two dyes, chlorophenol red and indigo carmine, in paper microchips with a single channel and cross channels. *(Figure 3.2A,B) Reproduced with permission from C. Xu, M. Zhong, L. Cai, Q. Zheng, X. Zhang, Sample injection and electrophoretic separation on a simple laminated paper based analytical device, Electrophoresis. 37 (2016) 476–481. C.L.S. Chagas, F.R. De Souza, T.M.G. Cardoso, R.C. Moreira, J.A.F. Da Silva, D.P. De Jesus, et al., A fully disposable paper-based electrophoresis microchip with integrated pencil-drawn electrodes for contactless conductivity detection, Anal. Methods. 8 (2016) 6682–6686. (Figure 3.2C) Reproduced with permission from Z.Y. Wu, B. Ma, S.F. Xie, K. Liu, F. Fang, Simultaneous electrokinetic concentration and separation of proteins on a paper-based analytical device, RSC Adv. 7 (2017) 4011–4016.*

When compared to the pioneering report [9], several improvements including the ability to create narrower channels (1 mm wide), the possibility of applying higher potentials (2.5 kV), the capacity to promote faster separations (250 s), and capacity to promote the integration of the device with a contactless conductivity detector are a few noticeable points to highlight the great potential of this emerging platform for analytical separations.

Mettakoonpitak and Henry developed a new methodology for manufacturing paper devices using Parafilm as a material to create hydrophobic barriers on filter and chromatographic papers using two different geometries, single and cross-shaped channels, the separation of two dyes, chlorophenol red, and indigo carmine, was successfully monitored (Fig. 3.2C). The cross-shaped chip provided a more efficient separation and offered a better resolution between the analytes, when compared to the single-channel chip. In addition, the authors demonstrated the electrophoretic separation of fluorescein isothiocyanate and L-glutamic acid labeled with fluorescein isothiocyanate, monitored using a fluorescence detector. Interestingly, the authors studied the main features that influence electrophoretic separations, such as Joule heating, electroosmotic flow, and electrophoretic mobility [21].

Although the geometry of the devices is a very important factor for getting successful separations, the optimization of the instrumental parameters is fundamental for the electrophoretic separation to be efficient. Wu et al. reported the use of the field amplified sample stacking technique to perform the concentration and separation of proteins in a paper analytical device. In this study, the authors generated an electric field gradient so that bovine hemoglobin and cytochrome C were concentrated and separated in sequence

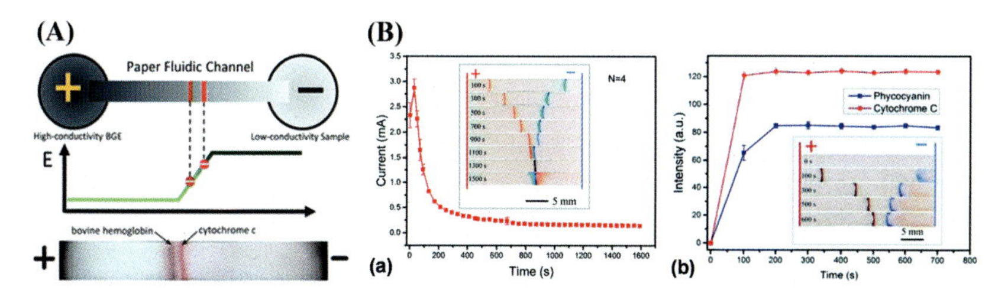

FIGURE 3.3 (A) Scheme of the electric field gradient generated during the electrophoretic separation of bovine hemoglobin and cytochrome C proteins in the paper device. (B) Demonstration of the electrophoretic separation of two ionic dyes, cationic probe of Rhodamine 6G and an anionic probe of Brilliant Blue, and two amphoteric ions-proteins, phycocyanin and cytochrome C, in devices manufactured with glass fiber paper channels. *(Figure 3.3A) Reproduced with permission from L. Liu, M.R. Xie, Y.Z. Chen, Z.Y. Wu, Simultaneous electrokinetic stacking and separation of anionic and cationic species on a paper fluidic channel, Lab. Chip. 19 (2019) 845–850. (Figure 3.3B) Reproduced with permission from L.D. Casto, J.A. Schuster, C.D. Neice, C.A. Baker, Characterization of low adsorption filter membranes for electrophoresis and electrokinetic sample manipulations in microfluidic paper-based analytical devices, Anal. Methods. 10 (2018) 3616–3623.*

in less than 30 s (Fig. 3.3A) [22]. Liu et al. demonstrated the separation of cationic and anionic species by means of electrokinetic stacking, as an electric field gradient, in an analytical device made with glass fiber paper channels. In this study, two different separations were monitored, the separation of two ionic dyes, cationic probe of Rhodamine 6G and an anionic probe of Brilliant Blue, and two amphoteric ions-proteins, phycocyanin and cytochrome C (Fig. 3.3B) [23]. In addition, Casto et al. studied the use of different filtration membranes (cellulose acetate, cellulose ester, and polyvinylidene fluoride) for the construction of analytical devices. The membranes were chosen due to the low adsorption of the analytes in their porous structures, so that these materials do not exert influence during the separation step. In this way, it was possible to perform the electrophoretic separations of three amino acids (arginine, glutamine, and glycine), without having a pronounced influence of the paper channels with the analytes present in the sample, thus simulating a zone electrophoretic separation [24].

Another possibility for conducting electrophoretic separations in paper devices refers to moving reaction boundary electrophoresis. In this mode, a solution containing a catholyte and another containing an anolyte are placed in different reservoirs. Thus, after the electric field is generated, the ions present in the solutions migrate through the paper with their respective mobilities, meeting and reacting after a certain time. As a consequence, it is possible to differentiate the analytes present in the sample by means of colorimetric reactions. Liang's research group explored this strategy in two separate reports [25,26]. In the first study, they managed to separate three metallic cations, Co^{2+}, Cu^{2+}, and Fe^{3+}, through complexation reactions with the anion $[EDTAH_2]^{2-}$, using a system composed of two reservoirs, made with centrifuge tubes, and a paper device, consisting of a channel made of chromatographic paper [25]. In a second study, the authors used a strategy similar to that of the first study. However, a sample concentration step, by field-amplified sample stacking, was combined with moving reaction boundary electrophoresis, using a similar paper analytical device, but with multiple channels. Thus, it was possible to show the concentration and electrophoretic separation through redox and complexation reactions [26].

3.2.2 Isoelectric focusing and isotachophoresis

The technique of isoelectric focusing consists in the separation of analytes based on their different isoelectric points through the application of an electric field through a pH gradient [27]. In paper devices, this technique is barely explored, and it is mainly used for preconcentration and protein separation. Gaspar et al. demonstrated the separation of two proteins, cytochrome C and myoglobin, applying isoelectric focusing on a paper device with silver electrodes printed on the surface with the aid of an inkjet printer (Fig. 3.4A). The microfluidic channel was delimited by hydrophobic wax barriers

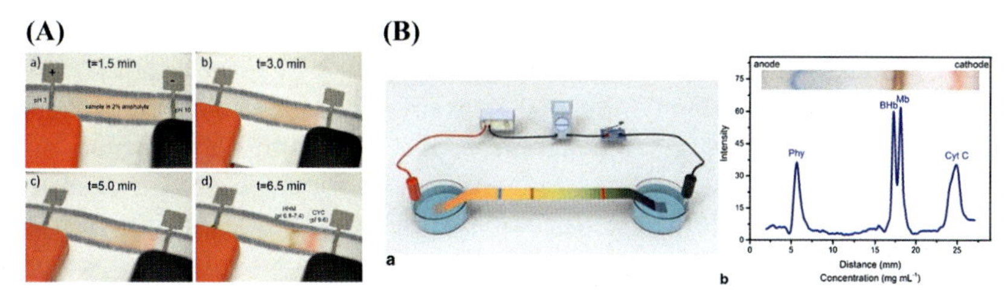

FIGURE 3.4 (A) Isoelectric focusing for the separation of cytochrome C and myoglobin, in devices made of paper with silver electrodes on the surface. (B) Demonstration of the separation of a complex sample containing four proteins using isoelectric focusing on paper strips. *(Figure 3.4A) Reproduced with permission from S.F. Xie, H. Gao, L.L. Niu, Z.S. Xie, F. Fang, Z. Y. Wu, et al., Carrier ampholyte-free isoelectric focusing on a paper-based analytical device for the fractionation of proteins, J. Sep. Sci. 41 (2018) 2085–2091. (Figure 3.4B) Reproduced with permission from S. Yu, C. Yan, X. Hu, B. He, Y. Jiang, Q. He, Isoelectric focusing on microfluidic paper-based chips, Anal. Bioanal. Chem. 411 (2019) 5415–5422.*

showing good print quality and resolution of the channels. However, the separation of proteins was identified only visually, thus comprising the possibility of quantifying these analytes [28].

The use of ampholytes, which are amphoteric electrolytes, is necessary for the creation of regions with different pHs, which are essential for isoelectric focusing to occur. Despite this, Xie et al. demonstrated the visual separation of two proteins, phycocyanin and bovine hemoglobin, without the use of an ampholyte in a paper analytical device. This method offered advantages such as low cost, low consumption of reagents and sample, in addition to compatibility with mass spectrometry, proving to be an excellent method for the pretreatment of complex protein samples [29]. In another study published by the same group, Niu et al. reported the use of the same paper microfluidic system to preconcentrate and separate four different proteins, phycocyanin, bovine hemoglobin, myoglobin, and cytochrome C (Fig. 3.4B) [30]. However, an ampholyte was used to improve the resolution of separation of these proteins, since it was a more complex sample. The studies reported by Xie et al. [29] and Niu et al. [30] evidenced the versatility of using isoelectric focusing in analytical paper-based devices composed of strips defined with 35 mm long and 3 mm wide for preconcentration and separation of proteins as well as enabled the integration with mass spectrometry.

The device design and the use of other reagents beyond the ampholytes are important factors for the success of the protein separation. In this sense, Yu et al. explored a novel manufacturing process based on the silanization of paper sheet followed by hydrophilization of the channel through a plasma treatment. In addition, the authors demonstrated the feasibility of using polyvinylpyrrolidone as a suppressor of the electroosmotic flow, thus helping in the separation of two proteins, myoglobin and cytochrome C [31].

Isoelectric focusing has been widely applied in other miniaturized platforms, fabricated often in polymeric platforms such as polycarbonate, poly (methyl methacrylate), poly(dimethyl siloxane) and cyclic olefin copolymer or even in glass substrates. Although paper is a very promising substrate, due to the ability to separate analytes without the need for an electric field, it still needs to be better explored for separations using isoelectric focusing.

Isotachophoresis is an analytical separation technique where the sample is placed between a discontinuous buffer system, where one buffer has ions with greater mobility and the other has ions with less electrophoretic mobility [32]. In this way, the sample, which has ions of intermediate mobility in relation to the buffers, is separated into zones due to the action of the electric field, and they move with equal speed. This technique has been widely explored in microfluidic systems, mainly those manufactured on polymeric substrates and glass [33]. However, few studies exploring isotachophoresis in μPADs have been reported in the literature, generally applied for sample preconcentration. Recently, Dong's group demonstrated the use of a paper analytical device for isotachophoretic separations [34,35]. In a first study, two important cardiac protein markers, acidic troponin T and basic troponin I, were simultaneously detected in human serum samples, in concentrations of the order of 10^{-12} mol L^{-1} [34]. In a second study, the authors were able to discriminate healthy exosomes, from exosomes derived from prostate cancer-causing cells, present in a human serum sample [35]. Both studies used the immobilization of antibodies as a strategy to detect the zones of each of the analytes, thus obtaining great selectivity. In addition, the cartridge used with support for the paper strip was reusable, inexpensive and portable, allowing analysis at the point-of-care (POC).

Another interesting strategy for performing isotachophoretic separations on a μPAD has been described by Schaumburg and coworkers. In their study, a sample containing a conjugate of the *Escherichia coli* bacterium with the enzyme β-galactosidase was incubated in the presence of chlorophenol red-β-D-galactopyranoside, which after reaction with the enzyme produced chlorophenol red in solution. After the incubation period, the sample was then separated by isotachophoresis, where it was possible to visually detect two color bands from chlorophenol red-β-D-galactopyranoside and chlorophenol red, indicating the presence of the bacteria in the sample. The tested method was able to detect both analytes within c. 90 min at concentrations of 9.2 CFU mL^{-1} and 920 CFU mL^{-1} in laboratory and apple juice samples, respectively [36].

3.3 Paper-based devices for chromatographic separations

In early 1970, the concept of lateral flow assays (LFA) was introduced and became an important tool for many applications, especially for bioanalytical studies [1]. This consists of a chromatographic system and an immunochemical reaction [37], in which the fluid is transported by capillary action without external pumping requirements [38]. It is well known that paper substrate has gained greater visibility as an LFA platform due to its ability to spontaneously transport fluids due to its porosity and hydrophilicity, which allows to perform pretreatment or sample separation. Chromatographic separation in paper-based devices generally occurs in microfluidic channels delimited by hydrophobic barriers. Different type of paper can be explored (such as chromatographic, qualitative, quantitative filter, and others) according to the purpose of the application, given that physicochemical properties such as thickness, surface area, capillary fluid rate, and ion exchange capacity may influence on the separation efficiency as well as analytical performance [39,40].

Chromatographic separations on paper have been explored for food, pharmaceutical, and mainly clinical applications. Generally, samples are complexes due to infinity constituent compounds, then a separation step is required before analyte detection. In summary, separation is performed by paper physicochemical properties and its interaction with the sample. Below are discussed some applications divided according to the detection techniques employed including electrochemistry, colorimetry, chemiluminescence, and fluorescence.

3.3.1 Electrochemical detection

Chromatographic separation on paper is widely associated to electrochemical techniques. Carvalhal and coworkers reported, in 2010, the use of Whatman chromatography paper the separation of uric acid (UA) and ascorbic acid (AA). The separation occurs due to different interactions between analytes and cellulose fibers, UA binds strongly due to greater polarity when compared to AA, which is easily ionizable and more soluble in the mobile phase. The device was constructed by pressing a narrow strip of paper onto a thin-layer gold electrochemical microcell made by photolithographic processes. The authors monitored the separation on channels defined with different lengths (5–20 mm) and observed that satisfactory separation was achieved using intermediary length values. In addition, the use of higher ionic strength has promoted peaks with greater intensity and better repeatability for both analytes [41].

The same group reported, in 2012, the use of a cation exchanger paper to separate paracetamol (PA) and 4-aminophenol (4-AP). They evaluated the use of Whatman Grade 1 chromatographic paper for this purpose, but did not observed any separation. So, chromatography paper Whatman P81 was chosen for separation due to its high ionic exchange capacity. The authors described that at pH 4.5, protonated 4-AP was easily exchanged between sodium ions from paper, which makes this analyte more retained to paper-based columns and PA presents lower interaction since it does not interact with the negatively charged surface. Acceptable separation with baseline resolution and good sensitivity was achieved using electrodes fabricated by the sputtering process [42].

Differently from the pioneering reports, Dossi and coworkers used commercial pencil for drawing electrodes on paper surface aiming to detect AA and sunset yellow in food matrices, especially soft drinks and juices that were separated by thin-layer chromatography on paper-based devices. The separation device was wax printed on paper in a geometry designed to contain two parallel wax strips forming longitudinal hydrophobic barriers and a hydrophilic channel where separation may be performed. Glass was used as support and it was denominated on-plate device. For separation, one extremity of the device was immersed in a buffer reservoir and on the other extremity, next to the electrodes, a piece of paper was placed to absorb the buffer residues. Therefore, AA and sunset yellow were separated based on their solubility in eluent and adsorption onto paper polar fibers [43].

The same group reported, in 2013, a similar device for separation of AA, PA, and dopamine (DA) in synthetic samples to demonstrate the system feasibility of the proposed device employing two electrodes positioned upstream and downstream (Fig. 3.5A). Separation occurred along the hydrophilic channel and amperometric detection was performed by applying upstream working electrodes with a potential for oxidation and reverse reduction was forced on the downstream working electrode [44].

Noiphung et al. reported for the first time the determination of glucose in whole blood samples using a paper-based device for plasma isolation coupled to electrochemical detection. The device was manufactured with a dumbbell layout to offer a uniform flow of plasma towards the detection zone created in the center of the device. Glass filter paper was used as zones for blood separation. The use of this paper type promoted a separation in 4 min without any conventional centrifugation step [45].

In 2015, Murphy et al. studied the effects of ion exchange capacity, porosity, and buffer pH on the separation performance. The authors investigated different types of cellulose papers to promote the separation of DA and AA on paper devices (Fig. 3.5B). In this study, the authors revealed that Whatman 5 filter paper has a slow flow rate, which makes the fluid not to reach the channel top. Then, the authors tested the separation of DA and AA in Whatman P81, Whatman 4, and VWR 413 papers, since the flow rate was reasonable. The authors observed that Whatman 4 and Whatman P81 were not capable of separating these analytes efficiently and VWR 413 paper performed the best separation of AA and DA due to the medium ion-exchange mechanism, it was capable to elute both analytes in different retention time [46].

In 2017, Ding et al. proposed a microfluidic sampling and separation system based on paper for potentiometric detection of chloride in the presence of salicylate ions. The device was fabricated using two pieces of filter paper with different pore sizes. For the detection zone, a piece of paper (Whatman

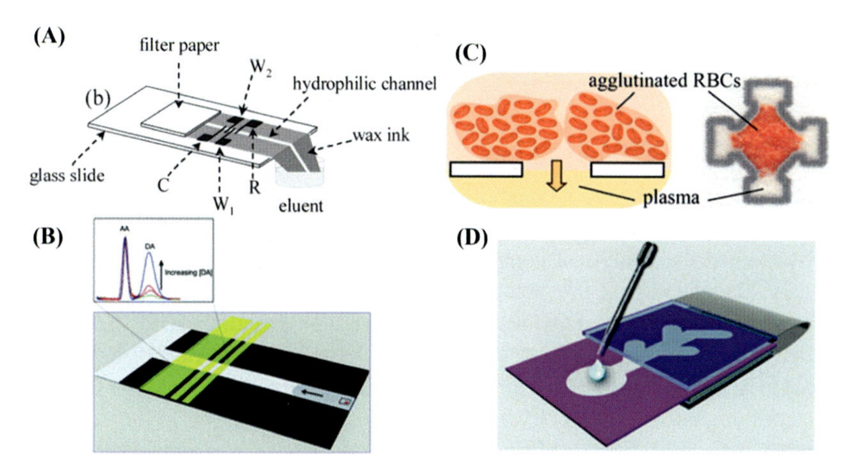

FIGURE 3.5 Paper-based devices for separation assays. In (A) dopamine and paracetamol in presence of Ascorbic acid is comigrated under on-plate separation conditions, whose electrochemical behavior displays different reversible characters. In (B) the μPAD proposed can detect dopamine in the presence of high concentrations of ascorbic acid selectively. In (C) μPAD was functionalized by antibodies which agglutinates RBC, allowing plasma separation for glucose determination. In (D) shows three-dimensional μPADs to detect four tumor markers by chemiluminescence. *(Figure 3.5A) Reproduced with permission from J. Noiphung, T. Songjaroen, W. Dungchai, C.S. Henry, O. Chailapakul, W. Laiwattanapaisal, Electrochemical detection of glucose from whole blood using paper-based microfluidic devices, Anal. Chim. Acta. 788 (2013) 39–45. (Figure 3.5B) Reproduced with permission from A. Murphy, B. Gorey, K. De Guzman, N. Kelly, E. P. Nesterenko, A. Morrin, Microfluidic paper analytical device for the chromatographic separation of ascorbic acid and dopamine, RSC Adv. 5 (2015) 93162–93169. (Figure 3.5C) Reproduced with permission from J. Ding, N. He, G. Lisak, W. Qin, J. Bobacka, Paper-based microfluidic sampling and separation of analytes for potentiometric ion sensing, Sens. Actuators B Chem. 243 (2017) 346–352. (Figure 3.5D) Reproduced with permission from X. Yang, O. Forouzan, T.P. Brown, S. S. Shevkoplyas, Integrated separation of blood plasma from whole blood for microfluidic paper-based analytical devices, Lab. Chip. 12 (2012) 274–280.*

Grade 589/1, 12–25 μm pore size) was cut in a "T" shape and, for the separation zone, a piece of paper (Whatman Grade 589/3, 2.0 μm pore size) was cut into a square shape with a small slot for the bottom. For the separation to occur, the separation zone was modified with a complexing agent, Fe (III), since Fe (III) presents intense interaction with cellulose due to electrostatic adsorption and hydrogen bonding and forms a stable complex with salicylate. The complex formation delays its transport to the detection region, while the analyte migrates faster toward the detection zone. Another important feature is the pore size of the paper, since the larger the pores, the higher the flow rate by capillarity and vice versa. Consequently, this parameter influences the contact time between the sample and the reagents. The authors selected the filter paper with the smallest pore size for the separation zone, thus aiming at a longer contact time for the complexation of Fe (III) with salicylate to occur efficiently. For the detection zone, the larger the pore size, the faster the sample transport. This would allow to obtain rapid analysis and a noticeable decrease in the water evaporation effect [47].

3.3.2 Colorimetric detection

Colorimetric detection consists of a reaction between analyte and a chromogenic reagent producing a colored product, and its color intensity is directly correlated to the analytical concentration [4]. This type of detection has been widely associated to LFA devices due to its simplicity, versatility, analysis speed, and potential for POC testing. Yang et al. proposed, in 2012, the use of paper as a separation platform coupled to colorimetric detection to determine glucose in whole blood plasma. Paper-based devices were manufactured using wax printing in a design containing a plasma separation zone in the center, a control zone, and three reading zones with colorimetric reagents (Fig. 3.5C). In this study, the separation zone was functionalized with agglutinating antibodies (anti-A, B) to promote the separation of plasma from whole blood by the aggregation of red blood cells (RBC), since free RBC can easily pass through pores of paper so much smaller than their diameter due to its deformability. Therefore, plasma separation was based on the agglutination of whole blood and the paper pore size enables the transport of plasma and blockage the passage of RBC aggregates (Fig. 3.5C). The plasma migrates toward the detection zones and reacts with the chromogenic agents generating color change, which is then related to the glucose concentration [48].

In 2015, Kar et al. [49] and Nilghaz et al. [50] also described the use of a LFA device for glucose colorimetric detection. Kar et al. proposed a device containing channels arranged in a H shape to promote the plasma separation without using expensive consumables and agglutination reagents. The proposed geometry was designed with four reservoirs, at the two same sides are added blood samples and phosphate buffer solution (PBS), respectively, these fluids by capillarity are carried through the separation channel where separation occurs by diffusion. Lighter molecules within the bloodstream diffuse into the PBS stream, when the separation is complete PBS stream and RBC stream are collected in different reservoirs. PBS reservoir is used to perform glucose colorimetric detection [49]. Nilghaz and coworkers introduced the use of a salt functionalized paper aiming to improve the separation performance. In a hypertonic medium, osmotic pressure leads to the deformation and crenation of RBCs, this increases contact with one to another and paper fibers, forming deflated RBC aggregate that is large to be separated chromatographically from plasma phase, which fills detection zone pretreated with chromogenic reagents for glucose determination [50].

Recently, Li et al. proposed the first instrument-free paper-based sensor using LFA paper-based device, which employs a smartphone and an automatic software detector system, what makes it portable and reliable due to the use of a box for image capture without light interference. This device was used to detect UA in whole blood, and the separation process was based on the capillary action, and aiming to improve its efficiency, the authors designed a sample zone, a background zone, and a detection zone [51].

3.3.3 Chemiluminescence detection

Clinical applications generally involve the whole blood separation, which RBC are separated from plasma and colorimetric detection is usually chosen for this analysis [48–50]. However, chemiluminescence detection is a promising analytical technique for paper-based analytical devices due to its high sensitivity and wide working range. Ge et al. developed a three-dimensional device based on the art of origami (Fig. 3.5D) to detect four tumor markers by chemiluminescence. RBC were successfully separated from pretreated whole blood with a binder, supplying only the plasma, which then traveled toward the test zone. This device presented several advantages including short analysis time (16 min), minimal experimental steps (e.g., the addition of reagent and washing steps for microzones), low cost, and potential for application at the POC [52].

3.3.4 Fluorescence detection

Fluorescence detection has been also employed to monitor separations on paper-based devices. Abbas and colleagues developed an analytical platform based on a paper strip offering a step of separating and preconditioning complex samples, which provided extremely low detection limits. The paper used in this work was cut out in a star shape, and all the five constituent fingers were coated with positively and negatively charged polyelectrolytes, such as poly(allylamine hydrochloride) and poly(sodium 4-styrene sulfonate), respectively. These polyelectrolytes promote the separation based on the chemical gradient formed on the surface of coated paper employing differentiated electrostatic interaction of the sample components with the different fingers of the star, since in each finger it was modified with different concentrations of polyelectrolyte [53].

Zhong and coworkers coated the paper by soaking in blocking solution overnight to avoid impregnating proteins on the paper surface. Fluorescein isothiocyanate was successfully separated from proteins based on size differences. Since proteins are larger molecules and consequently do not diffuse along the channel, fluorescein isothiocyanate presents greater diffusion, thus migrating to water flow along the channel in which it can be extracted [54].

3.4 Conclusions and perspectives

As presented in this chapter, paper substrate has been reinvented and emerged as powerful platform for microscale separations involving mainly electrokinetic and chromatographic techniques. This platform has offered several advantages as easiness for coupling different detection modes and ability to chemically modify the surface aiming to improve the separation performance. In comparison with other most popular platforms, the main disadvantage refers to the channel dimensions, which affects the separation resolution and efficiency. However, the low cost and possibility to create disposable platforms for simple and rapid separations enables paper substrate as potential candidate material for future investigations dedicated to fast screening, POC testing, or even previous separations of interfering agents, like red cells, on clinical or biomedical diagnostics. Due to the well-known chemistry, paper surface can be activated, modified, and widely explored for increasing the separation selectivity under electrokinetic and partition phenomena. Lastly, the simplicity and inherent global affordability of paper makes possible and quite attractive the integration of separation strips to replace capillary or microchannels on portable, mobile, and unmanned platforms in the near future.

Acknowledgments

The authors would like to thank CNPq (grants 426496/2018−3 and 308140/2016−8), CAPES, FAPEG, and INCTBio (grant 465389/2014−7) for the financial support and granted scholarships and researcher fellowships.

References

[1] C. Carrell, A. Kava, M. Nguyen, R. Menger, Z. Munshi, Z. Call, et al., Beyond the lateral flow assay: a review of paper-based microfluidics, Microelectron. Eng. 206 (2019) 45−54.

[2] D.M. Cate, J.A. Adkins, J. Mettakoonpitak, C.S. Henry, Recent developments in paper-based microfluidic devices, Anal. Chem. 87 (2015) 19−41.

[3] A.W. Martinez, S.T. Phillips, M.J. Butte, G.M. Whitesides, Patterned paper as a platform for inexpensive, low-volume, portable bioassays, Angew. Chem. Int. Ed. 46 (2007) 1318−1320.

[4] G. Sriram, M.P. Bhat, P. Patil, U.T. Uthappa, H.Y. Jung, T. Altalhi, et al., Paper-based microfluidic analytical devices for colorimetric detection of toxic ions: a review, Trends Anal. Chem 93 (2017) 212−227.

[5] J. Mettakoonpitak, K. Boehle, S. Nantaphol, P. Teengam, J.A. Adkins, M. Srisa-Art, et al., Electrochemistry on paper-based analytical devices: a review, Electroanalysis. 28 (2016) 1420−1436.

[6] E. Noviana, D.B. Carrão, R. Pratiwi, C.S. Henry, Emerging applications of paper-based analytical devices for drug analysis: a review, Anal. Chim. Acta. 1116 (2020) 70−90.

[7] N.A. Meredith, C. Quinn, D.M. Cate, T.H. Reilly, J. Volckens, C.S. Henry, Paper-based analytical devices for environmental analysis, Analyst. 141 (2016) 1874−1887.

[8] P. Nanthasurasak, J.M. Cabot, H.H. See, R.M. Guijt, M.C. Breadmore, Electrophoretic separations on paper: past, present, and future—a review, Anal. Chim. Acta 985 (2017) 7−23.

[9] H.G. Kunkel, A. Tiselius, Electrophoresis of proteins on filter paper, J. Gen. Physiol. 35 (1951) 89−118.

[10] W. Mejbaum-Katzenellenbogen, W.M. Dobryszycka, New method for quantitative determination of serum proteins separated by paper electrophoresis, Clin. Chim. Acta. 4 (1959) 515–522.
[11] W.P. Jencks, E.L. Durrum, Paper electrophoresis as a quantitative method: the staining of serum lipoproteins, J. Clin. Invest. 34 (1955) 1437–1448.
[12] D.R. Baker, Capillary Electrophoresis, Wiley, New York, 1995.
[13] A. Manz, D.J. Harrison, E.M.J. Verpoorte, J.C. Fettinger, A. Paulus, H. Lüdi, et al., Planar chips technology for miniaturization and integration of separation techniques into monitoring systems. Capillary electrophoresis on a chip, J. Chromatogr. A 593 (1992) 253–258.
[14] E.R. Castro, A. Manz, Present state of microchip electrophoresis: state of the art and routine applications, J. Chromatogr. A. 1382 (2015) 66–85.
[15] R.J. Block, E.L. Durrum, G. Zweig, A Manual of Paper Chromatography and Paper Electrophoresis, Academic Press Inc., 1955.
[16] C.L.S. Chagas, F.R. De Souza, T.M.G. Cardoso, R.C. Moreira, J.A.F. Da Silva, D.P. De Jesus, et al., A fully disposable paper-based electrophoresis microchip with integrated pencil-drawn electrodes for contactless conductivity detection, Anal. Methods. 8 (2016) 6682–6686.
[17] L. Ge, S. Wang, S. Ge, J. Yu, M. Yan, N. Li, et al., Electrophoretic separation in a microfluidic paper-based analytical device with an on-column wireless electrogenerated chemiluminescence detector, Chem. Commun. 50 (2014) 5699–5702.
[18] L. Luo, X. Li, R.M. Crooks, Low-voltage origami-paper-based electrophoretic device for rapid protein separation, Anal. Chem. 86 (2014) 12390–12397.
[19] Y. Matsuda, K. Sakai, H. Yamaguchi, T. Niimi, Electrophoretic separation on an origami paper-based analytical device using a portable power bank, Sensors 19 (2019) 1724.
[20] C. Xu, M. Zhong, L. Cai, Q. Zheng, X. Zhang, Sample injection and electrophoretic separation on a simple laminated paper based analytical device, Electrophoresis. 37 (2016) 476–481.
[21] J. Mettakoonpitak, C.S. Henry, Electrophoretic separations on parafilm-paper-based analytical devices, Sens. Actuators B Chem 273 (2018) 1022–1028.
[22] Z.Y. Wu, B. Ma, S.F. Xie, K. Liu, F. Fang, Simultaneous electrokinetic concentration and separation of proteins on a paper-based analytical device, RSC Adv 7 (2017) 4011–4016.
[23] L. Liu, M.R. Xie, Y.Z. Chen, Z.Y. Wu, Simultaneous electrokinetic stacking and separation of anionic and cationic species on a paper fluidic channel, Lab. Chip 19 (2019) 845–850.
[24] L.D. Casto, J.A. Schuster, C.D. Neice, C.A. Baker, Characterization of low adsorption filter membranes for electrophoresis and electrokinetic sample manipulations in microfluidic paper-based analytical devices, Anal. Methods. 10 (2018) 3616–3623.
[25] L. Ouyang, C. Wang, F. Du, T. Zheng, H. Liang, Electrochromatographic separations of multi-component metal complexes on a microfluidic paper-based device with a simplified photolithography, RSC Adv 4 (2014) 1093–1101.
[26] L. Ouyang, Q. Liu, H. Liang, Combining field-amplified sample stacking with moving reaction boundary electrophoresis on a paper chip for the preconcentration and separation of metal ions, J. Sep. Sci. 40 (2017) 789–797.
[27] R.C. Allen, C.A. Saravis, H.R. Maurer, Gel Electrophoresis and Isoelectric Focusing of Proteins: Selected Techniques. Walter de Gruyter, Walter de Gruyter (2019).
[28] C. Gaspar, T. Sikanen, S. Franssila, V. Jokinen, Inkjet printed silver electrodes on macroporous paper for a paper-based isoelectric focusing device, Biomicrofluidics. 10 (2016) 064120.

[29] S.F. Xie, H. Gao, L.L. Niu, Z.S. Xie, F. Fang, Z.Y. Wu, et al., Carrier ampholyte-free isoelectric focusing on a paper-based analytical device for the fractionation of proteins, J. Sep. Sci. 41 (2018) 2085–2091.

[30] J.C. Niu, T. Zhou, L.L. Niu, Z.S. Xie, F. Fang, F.Q. Yang, et al., Simultaneous preconcentration and separation on simple paper-based analytical device for protein analysis, Anal. Bioanal. Chem. 410 (2018) 1689–1695.

[31] S. Yu, C. Yan, X. Hu, B. He, Y. Jiang, Q. He, Isoelectric focusing on microfluidic paper-based chips, Anal. Bioanal. Chem. 411 (2019) 5415–5422.

[32] F.M. Everaerts, J.L. Beckers, T.P.E.M. Verheggen, Verheggen, Isotachophoresis: Theory, Instrumentation and Applications, Elsevier, 1976.

[33] P. Smejkal, D. Bottenus, M.C. Breadmore, R.M. Guijt, C.F. Ivory, F. Foret, et al., Microfluidic isotachophoresis: a review, Electrophoresis. 34 (2013) 1493–1509.

[34] S. Guo, W. Schlecht, L. Li, W.J. Dong, Paper-based cascade cationic isotachophoresis: multiplex detection of cardiac markers, Talanta. 205 (2019) 120112.

[35] S. Guo, J. Xu, A.P. Estell, C.F. Ivory, D. Du, Y. Lin, et al., Paper-based ITP technology: an application to specific cancer-derived exosome detection and analysis, Biosens. Bioelectron. 164 (2020) 112292.

[36] F. Schaumburg, C.S. Carrell, C.S. Henry, Rapid bacteria detection at low concentrations using sequential immunomagnetic separation and paper-based isotachophoresis, Anal. Chem. 91 (2019) 9623–9630.

[37] Z. Li, Y. Wang, J. Wang, Z. Tang, J.G. Pounds, Y. Lin, Rapid and sensitive detection of protein biomarker using a portable fluorescence biosensor based on quantum dots and a lateral flow test strip, Anal. Chem. 82 (2010) 7008–7014.

[38] E.B. Bahadır, M.K. Sezgintürk, Lateral flow assays: principles, designs and labels, Trends Anal. Chem 82 (2016) 286–306.

[39] A.K. Yetisen, M.S. Akram, C.R. Lowe, Paper-based microfluidic point-of-care diagnostic devices, Lab. Chip 13 (2013) 2210–2251.

[40] R.F. Carvalhal, E. Carrilho, L.T. Kubota, The potential and application of microfluidic paper-based separation devices, Bioanalysis. 2 (2010) 1663–1665.

[41] R.F. Carvalhal, M.S. Kfouri, M.H.O. De Piazetta, A.L. Gobbi, L.T. Kubota, Electrochemical detection in a paper-based separation device, Anal. Chem. 82 (2010) 1162–1165.

[42] L.Y. Shiroma, M. Santhiago, A.L. Gobbi, L.T. Kubota, Separation and electrochemical detection of paracetamol and 4-aminophenol in a paper-based microfluidic device, Anal. Chim. Acta. 725 (2012) 44–50.

[43] N. Dossi, R. Toniolo, A. Pizzariello, F. Impellizzieri, E. Piccin, G. Bontempelli, Pencil-drawn paper supported electrodes as simple electrochemical detectors for paper-based fluidic devices, Electrophoresis. 34 (2013) 2085–2091.

[44] N. Dossi, R. Toniolo, E. Piccin, S. Susmel, A. Pizzariello, G. Bontempelli, Pencil-drawn dual electrode detectors to discriminate between analytes comigrating on paper-based fluidic devices but undergoing electrochemical processes with different reversibility, Electroanalysis. 25 (2013) 2515–2522.

[45] J. Noiphung, T. Songjaroen, W. Dungchai, C.S. Henry, O. Chailapakul, W. Laiwattanapaisal, Electrochemical detection of glucose from whole blood using paper-based microfluidic devices, Anal. Chim. Acta. 788 (2013) 39–45.

[46] A. Murphy, B. Gorey, K. De Guzman, N. Kelly, E.P. Nesterenko, A. Morrin, Microfluidic paper analytical device for the chromatographic separation of ascorbic acid and dopamine, RSC Adv 5 (2015) 93162–93169.

[47] J. Ding, N. He, G. Lisak, W. Qin, J. Bobacka, Paper-based microfluidic sampling and separation of analytes for potentiometric ion sensing, Sens. Actuators, B Chem 243 (2017) 346–352.

[48] X. Yang, O. Forouzan, T.P. Brown, S.S. Shevkoplyas, Integrated separation of blood plasma from whole blood for microfluidic paper-based analytical devices, Lab. Chip 12 (2012) 274–280.

[49] S. Kar, T.K. Maiti, S. Chakraborty, Capillarity-driven blood plasma separation on paper-based devices, Analyst. 140 (2015) 6473–6476.

[50] A. Nilghaz, W. Shen, Low-cost blood plasma separation method using salt functionalized paper, RSC Adv 5 (2015) 53172–53179.

[51] N.S. Li, Y.T. Chen, Y.P. Hsu, H.H. Pang, C.Y. Huang, Y.L. Shiue, et al., Mobile health-care system based on the combination of a lateral flow pad and smartphone for rapid detection of uric acid in whole blood, Biosens. Bioelectron. 164 (2020) 112309.

[52] L. Ge, S. Wang, X. Song, S. Ge, J. Yu, 3D Origami-based multifunction-integrated immunodevice: Low-cost and multiplexed sandwich chemiluminescence immunoassay on microfluidic paper-based analytical device, Lab. Chip 12 (2012) 3150–3158.

[53] A. Abbas, A. Brimer, J.M. Slocik, L. Tian, R.R. Naik, S. Singamaneni, Multifunctional analytical platform on a paper strip: separation, preconcentration, and subattomolar detection, Anal. Chem. 85 (2013) 3977–3983.

[54] Z.W. Zhong, R.G. Wu, Z.P. Wang, H.L. Tan, An investigation of paper based microfluidic devices for size based separation and extraction applications, J. Chromatogr. B 1000 (2015) 41–48.

Chapter 4

Colorimetric paper-based analytical devices

Habdias A. Silva-Neto[1], Lucas R. Sousa[1] and Wendell K.T. Coltro[1,2]
[1]Instituto de Química, Universidade Federal de Goiás, Goiânia, Brazil, [2]Instituto Nacional de Ciência e Tecnologia de Bioanalítica (INCTBio), Campinas, Brazil

4.1 Introduction

The need for analytical platforms that meet portability and usability characteristics has received considerable attention for point-of-care testing (POCT) [1–3]. According to World Health Organization (WHO), an analytical device for monitoring human health in POCT, for example, must be Affordable, Sensitive, Specific, User-friendly, Rapid and Robust, Equipment free and Deliverable to end-users (ASSURED) [4]. Paper-based analytical devices (PADs) can be ideal for this purpose due to their main characteristics that include low cost, portability, ability to integrate multiple stages of analysis, being easy to handle, having a compatible substrate, etc. [5–10] However, the detection method associated to these devices is necessary also meets the ASSURED criteria [4,10].

Colorimetric detection is the most used detection method to quantify and identify the analytes of interest on PADs. The first detection technique reported for the use of PADs was colorimetric detection [11].

Although other detection techniques methods have been coupled to paper-based devices [12–14], colorimetric detection has been used mostly to ensure the portability and user-friendly aspects of PADs without losing sensitivity or specificity [8,15]. This technique offers a relatively lower cost than other methods such as electrochemical detection, for example [16]. Also, it makes use of much more straightforward and portable electronic devices for image capture that include smartphones [17–19], scanners [20,21], and digital cameras [22,23]. Besides, considering the global and commercial affordability of the mentioned electronic devices, colorimetric detection is quite attractive in association with paper platforms.

Even though previous studies have described the use of digital imaging for conventional analytical procedures [24], the pioneering report showing

Paper-Based Analytical Devices for Chemical Analysis and Diagnostics.
DOI: https://doi.org/10.1016/B978-0-12-820534-1.00009-8
© 2022 Elsevier Inc. All rights reserved.

colorimetric measurements on microfluidic paper-based analytical devices (μPADs) was successfully demonstrated in 2007 by the Whitesides group [11]. Since then, colorimetric detection on PADs has been moving toward applications in different areas of analytical chemistry such as bioanalysis [25], environmental chemistry [26,27], forensic chemistry [28,29], food control, and quality [30,31] and, especially, clinical diagnosis [32,33].

Colorimetric detection can be explored for qualitative and quantitative purposes. In a qualitative assay, the coloration development or change can be enough to provide YES/No answers, useful for rapid screening of potentially relevant markers. On the other hand, for quantitative assays, the colorimetric readout is based on quantifying an analyte through the correspondence between color intensity and its concentration. Depending on the reaction, the color intensity may increase or decrease [8,16]. How to explain the growing use of colorimetric detection on paper-based analytical devices? Most paper-based analytical devices, especially μPADs, have an extensive range of applications in colorimetric detection due to the material's physical and chemical properties [34]. Some studies have dedicated their efforts to studying the material's properties when adding a chromogenic agent and the behavior of the samples with their colorimetric reaction on that substrate, for example [35–37].

Despite the possibility of assessing color change with the naked eye, different methodologies for colorimetric analysis have been developed [38,39]. Because the analysis with the naked eye has a qualitative nature, the visual comparison of the color change after reaction can be distinguished differently by each user due to lighting, contrast levels, and even the user's reaction time [8,16]. In addition to the quantitative point of view, colorimetric detection can be explored using different analysis tools such as multivariate and univariate methods. Most of these tools are based on mathematical models and chemometric tools, which describe the color intensity in an analytical response, thus enabling the analysis of a series of different analytes in a single device [19].

The data acquisition related to a colorimetric assay depends on some factors such as luminosity, detection area, the gradient of color, etc. The software and app mechanism for color analysis fragment the image into simpler elements for data analysis in pixels or color intensity in the analyzed area. In general, graphical systems use color matching to decompose the image in terms of pixels or color intensity [18,34,40]. In general, primary colors of the additive system known as R (red), G (green), and B (blue) can be used as well as secondary colors of the subtractive system named C (cyan), M (magenta), Y (yellow), and K (black) [41]. The commonly used computational software for colorimetric analysis is Adobe Photo-Paint [42], Corel Photo-Paint [43], and Image J [44]. In portable systems, such as smartphones, we should highlight the app Color Grab [45], PhotoMetrix [46], etc.

The technology explored for capturing and evaluating the generated color intensity is compatible with portable electronic systems such as notebooks, tablets, colorimeters, and smartphones [18]. These systems do not require a skilled labor force to operate them, and, consequently, they can be easily transported for use in POCT. The assignments of colorimetric detection are similar to conventional spectrophotometric methods.

Perhaps one of the main advantages of using μPADs allied to colorimetric detection is the ability to perform multiple and simultaneous tests on single devices using the same sample [47,48]. However, some lateral flow consequences on paper-based devices can be detrimental to color uniformity inside the detection zone.

As it can be seen, the precursor work focused on the use of μPADs served as basis for the optimization of tools dedicated to colorimetric detection in different areas. This chapter is devoted to demonstrating the most usual colorimetric tools to be used in association with PADs. Also, this chapter discusses the challenges, perspectives, and main electronic devices for image capturing on μPADs. Initially, some fundamentals related to color formation in PADs and μPADs will be addressed, followed by a discussion about the alternatives for measuring the color and a summary about procedures dedicated for surface modifying aiming to improve the analytical performance. Lastly, an experimental data set was included to compare side-by-side the analytical performance and instrumental features of the most usual electronic devices for capturing images on paper-based devices,

4.2 Physical aspects of colorimetric detection

The geometry and layout of the PAD used are entirely dependent on user application. Although the fluid properties can significantly impact performance and fluidic transport, one of the biggest concerns for colorimetric reactions on paper is color uniformity [16]. The factors that attribute the color formation given by the colorimetric reaction between chromogen and analyte depend on the substrate–analyte chemistry interaction and drying time, reaction kinetics, absorption rate, etc. [36]. Capillary flow is a determinant property that affects the performance of colorimetric reactions on paper-based devices. When we are thinking about other devices that do not require lateral flow, such as spot tests, the user must pay attention to the uniformity of color and the substrate–analyte chemistry interaction to provide a reliable analytical response [21,34].

Since the pioneering reports, the lack of color uniformity inside the delimited detection or reaction zones has been an interesting point to be investigated. Since the color uniformity affects not only the analytical reliability but also the comprehensive understanding of the chemical assay, the knowledge of the phenomena involved is essential to ensure the analytical features toward commercialization.

The problem associated with the poor color uniformity is also known as the coffee ring or washing effect. It is most often observed for enzymatic assays. A noticeable coffee ring effect can be easily visualized. As displayed in Fig. 4.1A, the color uniformity inside detection zones is compromised, making it difficult to extract reliable information for quantitative purposes. The brown color appears as the result of the colorimetric reaction for detecting glucose in the presence of glucose oxidase and peroxidase. Nonuniformity is related to a movement of the reagents and products toward the zone borders [43]. Since 2014, different studies showing advances in colorimetric performance have been reported in the literature, including oxidation, use of nanomaterials, selection of paper substrates, and use of biopolymers.

Freitas et al. [49] successfully demonstrated that the washing away effect often observed for glucose enzymatic assay is promoted by the movement of iodide excess and triiodide (reaction product) by lateral flow toward the zone edge. This undesirable transport generates a color gradient, thus justifying the poor uniformity often visualized for enzymatic assays (see example in Fig. 4.1A). The authors used mass spectrometry imaging techniques like matrix-assisted laser desorption/ionization (MALDI) and desorption electrospray ionization (DESI) to understand the color gradient generation commonly seen on μPADs. Based on the images depicted in Fig. 4.1B, the color gradient has been confirmed on native and oxidized paper surfaces due to the transport of both iodide (excess) and triiodide (reaction product). On the other hand, MALDI experiments revealed that enzymes are kept fixed on the surface.

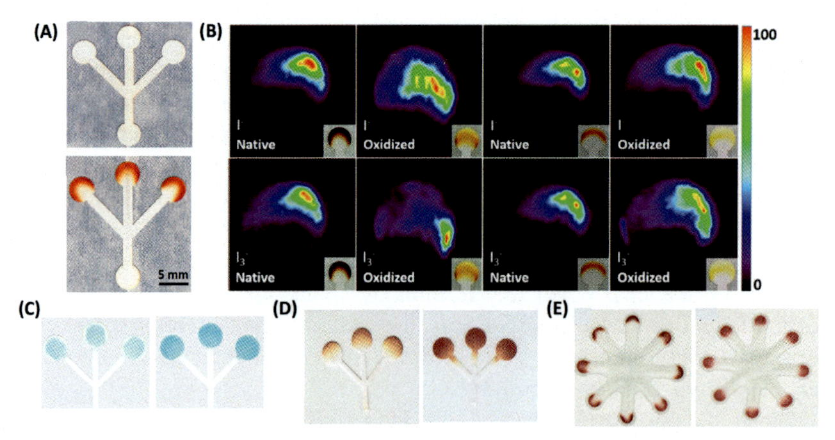

FIGURE 4.1 The images show examples of surface modifications for the colorimetric glucose assay. In (A) optical images showing the μPAD before and after colorimetric assays using KI as the chromogenic agent (A). In (B) DESI imaging showing the spatial distribution for the same glucose assay using KI on oxidized paper and native paper. Images showing the assays in paper modified with Fe_3O_4 nanoparticles (MNPs) (C), native paper and silica-modified (with SiO_2) papers (D) optical micrographs showing the analytical performance of μPADs without and with chitosan (E).

The substrate selection has been demonstrated to be an essential parameter for preparing μPADs dedicated to colorimetric assays. Due to the fact slew possibilities and types of the paper substrate, it is necessary to choose between material's properties, such as mechanical strength and stability, capillarity, hydrophilicity and hydrophobicity, thickness, fiber size, and porosity, all to make the lateral flow efficient and ensure the color uniformity more representative inside detection zones [16].

The most common types of paper commercially available to produce μPADs are those dedicated to chromatographic separation and qualitative filtration. Although the use of these paper substrates has been often reported, other types are also affordable for preparing analytical devices, including office paper, paper towels, etc. However, the porosity of these alternative platforms is not so useful for promoting fast lateral flow. In general, thicker substrates offer higher resistance to the lateral flow, and, consequently, it promotes low color development due to the slower flow rate, as reported by Evans et al. [36]. Nevertheless, these physical aspects are not sufficient to keep the uniformity of color in the detection zone.

For three-dimensional (3D) μPADs, some alternatives for improving the colorimetric signal through color homogeneity in the detection zone have also been reported. Morbioli et al. [50] reported an improvement in the homogeneous sample distribution in μPADs arranged in 3D format combining multiple layers. The addition of layers to the prototype proposed by them showed that a longer hydrophilic path for each channel present in the subsequent layer would increase the device hydrodynamic resistance. Also, they managed to show that for an enzymatic reaction, the prolonged hydration of the lower points of the devices allowed a more significant interaction between the enzymes and redox indicators.

Besides the mentioned features, other strategies using surface oxidation, the addition of nanoparticles, carbon-based nanomaterial, and biopolymers have also been reported for improving the colorimetric performance on μPADs.

4.2.1 Surface oxidation and addition of nanomaterials and biopolymers

The strategies for functionalizing the paper surface involve the chemical modification of hydroxyl groups present at cellulose structure naturally.

Other studies have shown that surface modification by cellulose oxidation has benefits for color formation for enzymatic colorimetric assays [51,52]. Garcia et al. reported the oxidation of hydroxyl groups present in the paper surface using sodium m-periodate [51]. The oxidation promoted hydroxyl conversion into aldehydes groups, allowing the enzymatic immobilization without requiring additional coupling agents. This strategy ensured better color uniformity minimizing the color gradient observed in native paper surfaces [53].

Despite the oxidation, the incorporation of materials for crosslinking reactions is proposed with or without previous oxidation [52]. One of the most explored approaches is the incorporation of nanomaterials [54]. Figueredo et al. [55] reported nanostructured materials to improve color uniformity on paper devices. They made use of spheres, tubes, and planar-shaped nanomaterials such as Fe_3O_4 nanoparticles (MNPs), multiwalled carbon nanotubes (MWCNT), and graphene oxide (GO) in an unprecedented way to treat μPADs detection zones (Fig. 4.1C). The objective of their work was to create a biocompatible layer with a high-resolution catalytic surface. The μPADs had the detection zone filled with a colloidal and stable solution of MNPs, MWCNT, and GO nanomaterial solution during 30 s in hydrophobic plastic film to be laminated. According to the achieved results, nanostructured materials ensure better analytical performance compared to native substrates.

Another reported approach includes modifying paper surfaces with functionalized silica nanoparticles (SiNPs) to improve color uniformity. SiNPs were previously modified with 3-aminopropyltriethoxysilane and then added to the paper devices to facilitate the adsorption of selected enzymes and prevent the washing away effect [52]. The μPADs were immersed in a suspension of SiNPs and dried at room temperature. They showed a balance between the amount of protein adsorbed to the surface of the nanoparticles (NPs) and the strength of the interaction. It must be emphasized the interaction of the redox indicator can also be associated with improved color uniformity, as can be seen in Fig. 4.1D.

Chitosan has been reported to reinforce the uniformity effects on paper [43]. Chitosan is a biopolymer that is well used for enzymatic assays due to its ability to modify the paper's surface to the point of favoring the mechanism of the enzymatic reaction [56]. Gabriel et al. [43] described chitosan use to improve the analytical performance of colorimetric measurements associated with enzymatic bioassays (Fig. 4.1E). In this method, a chitosan solution prepared in an acid medium works like a crosslinking agent without requiring chemical modifications or activations. The amino groups present in the chitosan structure benefit crosslinking with enzymes and biological structures in general [57,58].

The coupling of functional groups is also an alternative to improve color uniformity. The hydrophobic silanes groups, for example, were coupled to paper's cellulose fibers, becoming a zone area more hydrophilic [35,59]. Devadhasan et al. [60] described using a chemically patterned microfluidic paper-based analytical device (C-μPAD) by immobilizing silane groups terminating in amine, carboxyl and thiol through condensation so that the chromogens used by them to detect heavy metals are covalently coupled to these functional groups. The modification was valid since the cross-reaction with chelating agents was minimized. Consequently, they obtained relatively low detection limit values with those compared to the detection of the same metals in other works.

4.3 Colorimetric reaction methodology

The colorimetric reaction mechanism may vary according to the analyte nature. When μPADs were introduced around 15 years ago, the possibility of adapting bioassays on paper became one of the most attractive applications, especially those involving clinical or biomedical focus. A common approach for analyzing clinical biomarkers includes enzymatic reactions with well-known chemistry [61]. Although this reaction mechanism has been increasingly highlighted in bioanalytical applications, the use of redox indicators, acid–base indicators, NPs, and complexometric indicators have been reported in several studies [16]. These reaction mechanisms are used in addition to other tests to minimize side reactions and to improve the analytical performance associated with color uniformity and intensity.

4.3.1 Enzymatic reaction

Enzymes are a class of biocomposites, and their use in μPADs is frequently explored, especially for clinical diagnostic applications due to chemical catalyst properties in a high degree of specificity [11,56]. Colorimetric detection based on the enzymatic mechanism was the first example of colorimetric assay performed on μPADs [11]. In the pioneering work of Martinez et al. [11], enzymatic assay associated with colorimetric reaction for detecting glucose using the enzymes glucose oxidase (GOx) and horseradish peroxidase (HPR) was reported in an unprecedented way exploiting paper-based platforms. As well known, glucose is indirectly detected, that is, the enzymatic reaction product—hydrogen peroxide—can be colorimetrically measured by adding a redox chromogenic reagent, which is responsible for promoting changes in the coloration. Enzymes of the oxidase family (peroxidase enzymes) have been typically used in studies involving the analysis of more complex biomolecules [62]. Some compounds react with these enzymes in the presence of oxygen and water to form hydrogen peroxide. Peroxidase, in turn, helps in catalyzing the oxidation of the redox indicator and reducing the hydrogen peroxide by-product [63].

This mechanism is not limited to glucose alone. There are reports of peroxidase enzymes associated with a detection system for lactate, starch, uric acid, etc. [9,33,64,65]. In addition to oxidoreductase enzymes, other enzymatic families such as nucleases, proteases, transferases, hydrolases, etc. [57,66,67] can be used for similar purposes. In addition to enzyme-catalyzed reactions, enzyme-linked immunosorbent assays (ELISA) can be performed on a paper-based platform using the enzymes' chemical properties [68]. When it is necessary to use an antibody conjugated to an enzyme such as HRP, specific ligands, or oligonucleotides, the paper also represents a

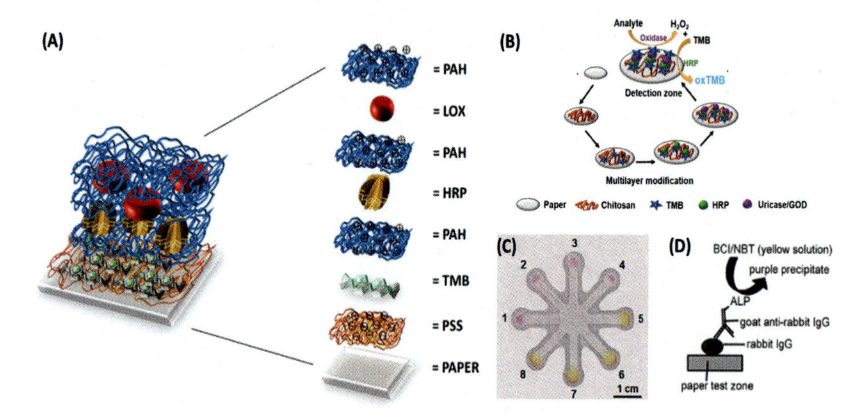

FIGURE 4.2 In (A), the representation of a "Wafer-like" system functionalized with some colorimetric reagents and enzymes for lactate detection. In (B), a surface modification with chitosan was performed to add the chromogenic reagents and the enzyme for detection of uric acid. In (C), the colorimetric response obtained for the glucose colorimetric assay using the enzymatic assay and two different chromogens to detect the by-product of the enzymatic reaction In (D), a scheme of the paper-based indirect ELISA for IgG.

versatile substrate for bioassays, which can occur spontaneous oxidation by photodynamic treatment, for example [69]. The color changes caused by these tests can be discriminated against with the naked eye and quantified by other data processing tools.

An important issue that should be highlighted in enzymatic assays is the reaction time and the time taken to acquire color intensity data in the region where the enzyme assay occurs. Each enzyme can have a specific activity time. Depending on the user's time to capture the image and analyze the data, it must be monitored together with the reaction kinetics. For this case, the use of smartphones for colorimetric analysis has a significant advantage over other colorimetric detection tools, as will be discussed later. Fig. 4.2 shows examples of analyzes made using the enzymatic mechanism.

4.3.2 Redox indicator

The reaction with redox chromogens involves a change in the chromogen's oxidation state when it comes in contact with an analyte in question. When this occurs, the color of the system changes [16]. Colorless potassium iodide, for example, is oxidized to iodine, which has a brown color in the presence of hydrogen peroxide. As mentioned before, usually, the reaction is performed in the presence of iodide in excess, thus promoting the formation of triiodide due to its reaction with iodine. Still following this example, it is possible to observe some characteristics that add advantages to the use of

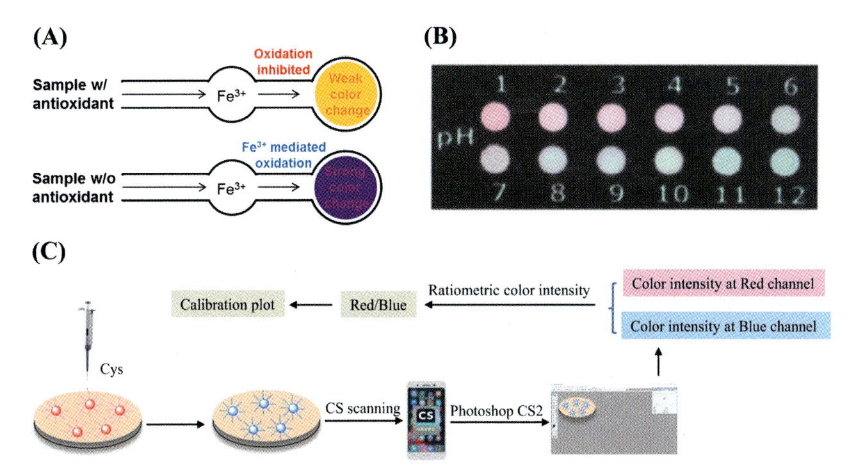

FIGURE 4.3 Examples of applications of the techniques involve the use of colorimetric reactions by a redox reaction (A), using specific chromogens for pH analysis (B) and detection mechanism, and using a nanoparticle (C).

this mechanism, such as the evident color change of the chromogen in the reduced form to the oxidized form and the medium being favorable so that the process is not reversible or that it does not occur considerable loss of color in a small fraction of time maintaining stability [49].

Most redox indicators respond to a large number of analytes. To keep the selectivity, it would be ideal for the user to use surface modifications or associate the technique with other mechanisms such as those involving NPs, for example [70]. This reaction mechanism is also widely used for the indirect detection of certain analytes. Some metals in their oxidized form are used in the zones along with some chromogenic reagents. Upon contact with the analyte, the metal can be reduced to another form that reacts directly with the added chromogen (Fig. 4.3A) [70,71].

4.3.3 pH indicator

pH indicators are a comprehensive class of colorimetric reagents that are substantially crucial for most analytical applications [22]. Because they have a simple mechanism that involves only the color change caused by differences in the acidity or basicity values, pH tests are often used as models to exemplify the proof of concept of different types of paper-based devices. In μPADs, it is common to observe an edge effect on pH tests, thus suggesting the requirement for adjusting the chromogen concentration and making surface changes. One of the alternatives to deflect the edge effect and improve this test's uniformity is the use of ammonium quaternary to act as an ion-pairing agent with the charged form of pH indicators [16].

The indicators can be used in conjunction with each other to align a specific pH range specificity. For this, one or more indicators are mixed with corresponding bands. Indicators can be obtained synthetically or naturally. Many natural products have in the structure of their flowers and anthocyanin fruits, which are compounds responsible for their color. Anthocyanins change color based on pH and have a slightly restricted pH range compared to synthetic indicators, as shown in Fig. 4.3B [72].

The applications of pH indicators are not limited to determining the pH of a sample of interest. They are essential for neutralization balances. About this fact, the pH indicators associated with μPADs can be used to perform volumetric measurements and be adjusted from the quantitative point of view using any of the data capture and analysis tools.

4.3.4 Nanoparticles

The NPs are materials ranging from 1 to 100 nm with properties easily changed according to another substance in the environment. The NPs are easily synthesized, functionalized, and impregnated on paper substrates for different applications, such as environmental and clinical [73]. Among the main NPs employed in colorimetric assays are the metallic such as SiNPs, gold nanoparticles (AuNPs), and silver nanoparticles (AgNPs). The NPs can offer excellent high contact surface, monodispersion, excellent adsorption and present changes in their optical properties. Electrical properties that correlate the analyte of interest with significative selectivity and detectability are then used as sensors [8,73]. Recently, Liu et al. [74] reported developing a PAD modified with chitosan and AuNPs for cysteine detection (Fig. 4.3C). The colorimetric response was observed due to the color changes associated with AuNPs, which ranges from red to blue in the presence of the cysteine sample.

4.3.5 Complexation reaction

The complexometric reactions on paper-based platforms are adaptations from classical quantitative methods using titration. The chemical reactions are carried out through the interaction between a chelator (monodentate or multidentate) and a central atom. Furthermore, the complexing compound (chromogen) reacts in a coordinated way with central atom forming a stable and colored product, such as metal-complexing [75].

One of the pioneering studies using titrations was the German chemist Justus Liebig in the 1850s, in which the cyanide ion was determined using silver nitrate [76]. Lately, Yamada et al. [77] developed an antibody-free μPAD for lactoferrin detection in human tear fluid samples. The assay on μPAD was based on the formation of the lactoferrin–terbium complexes and the image analysis via a digital camera and ImageJ software.

4.4 Detection and readout system

Different colorimetric reactions have been explored to measure chemical compounds, such as enzymatic, redox reactions, and pH indicators, with NPs and complexometric reaction, for example [16,78]. The pixel intensities are often collected by digitization using an office scanner and via smartphone cameras and instrument-free measurements based on lateral and radial distance. These different colorimetric detection approaches will be carefully presented in the following subsections.

4.4.1 Office scanner

Scanners are powerful image collection tools since they minimize the interference of external light during the measurement. Besides, scanners are easy to be operated, and the captured images offer suitable reproducibility [16,79]. The colorimetric responses are extracted from digitalized images through the pixel intensity analyzed in software, which is correlated with analyte concentration.

Aksorn et al. [80] reported the development of μPADs for performing enzymatic assays to detect sucrose, fructose and glucose in food samples colorimetrically. The sucrose measurement was realized by sequentially immobilization of three enzymes, invertase, glucose oxidase and peroxidase. The enzyme fructose dehydrogenase and the ferricyanide colored ion marker were impregnated on μPAD for colorimetric detection of fructose. In contrast, the colorimetric assay for glucose detection was realized by immobilizing glucose oxidase and peroxidase. The reaction zones were digitalized by an office scanner, and the pixels intensity calculated using ImageJ software, as described in the Fig. 4.4A.

4.4.2 Smartphone

Smartphones are small computers with high-quality camera and broad availability of media applications [19]. In the last years, smartphones have been standing out in sensing and biosensing areas due to accessibility, portability, low cost, free applications for image capture, and colorimetric analysis. Other remarkable features offered by smartphones include the multiple connection possibilities allowing the data transmission via USB port, Wi-Fi, and Bluetooth [84,85].

Recently, da Silva et al. [81] proposed a PAD combined with smartphone for colorimetric detection of phenacetin in cocaine samples (Fig. 4.4B). The sodium 1,2-naphthoquinone-4-sulfonate prepared in the basic medium was impregnated on circular paper zones (spot tests). The addition of phenacetin solution promoted changes from colorless to brown inside the spot tests. Afterward, the devices were photographed, and color intensity was analyzed using GIMP2 software.

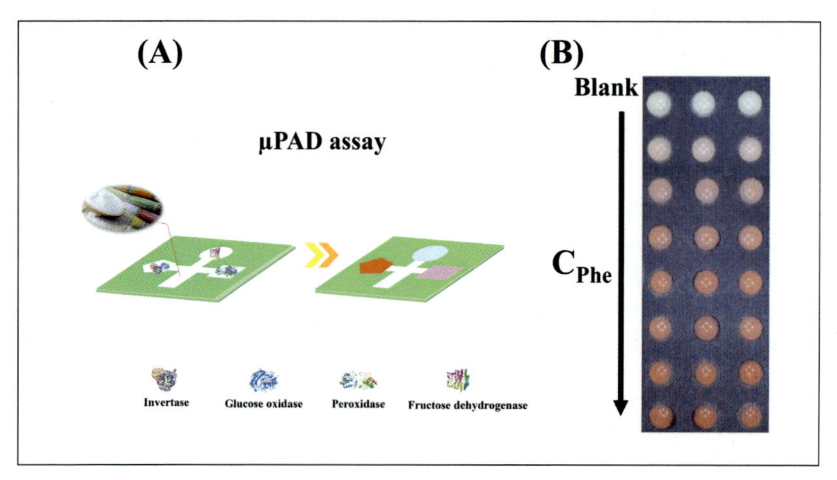

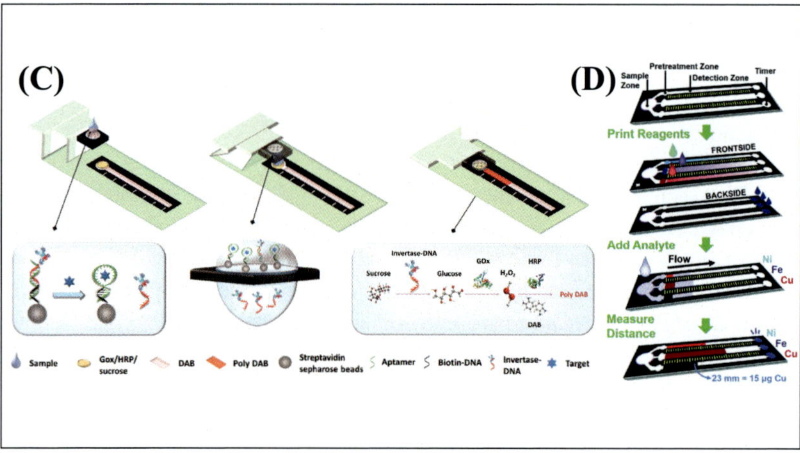

FIGURE 4.4 (A) Schematic of μPADs for colorimetric detection of sucrose, fructose, and glucose employed office scanner [79]. (B) PADs for colorimetric detection of phenacetin using a smartphone [81]. (C) μPAD for visual colorimetric detection of Fe, Ni, Cu [82]. (D) representation of the aptamer-μPAD for distance-based analysis [83]. *(Figure 4.4A) Reproduced and with permission from Yamada K., Henares T.G., Suzuki K., Citterio D. Paper-based inkjet-printed microfluidic analytical devices. Angew. Chem. Int. (Ed.). 2015;54(18):5294–5310. (Figure 4.4B) Reproduced and with permission from da Silva G.O., de Araujo W.R., Paixão T.R. L.C. Portable and low-cost colorimetric office paper-based device for phenacetin detection in seized cocaine samples. Talanta. 2018;176(July 2017):674–678. (Figure 4.4C) Reproduced and with permission from Cate D.M., Noblitt S.D., Volckens J., Henry C.S. Multiplexed paper analytical device for quantification of metals using distance-based detection. Lab Chip. 2015;15 (13):2808–2818. (Figure 4.4D) Reproduced and with permission from Tian T., An Y., Wu Y., Song Y., Zhu Z., Yang C. Integrated distance-based origami paper analytical device for one-step visualized analysis. ACS Appl. Mater. Interfaces. 2017;9(36):30480–30487.*

4.4.3 Visual detection

The current needs of the analytical methodologies for the POCT have probably driven the emergence of performing chemical analysis directly in the field without instrumentation requirements [86]. However, the first study using this simple approach was reported by Zuk et al. [87] in the 1985s. The authors used an immunoassay to determine theophylline in biological fluids. First, the paper strips were immobilized with antibodies, posteriorly with the enzyme glucose oxidase and peroxidase, and then a color developer solution containing glucose and 4-chloro-1-naphthol. In a general way, these colorimetric measurements are based on visual detection with the naked eye of color developed via lateral and vertical flow.

Recently, Cate et al. [82] proposed free-equipment μPAD for colorimetric detection of Fe, Ni, and Cu. The colorimetric assay was performed immobilizing the chromogenic agents dimethylglioxima (dmg), bathophenanthroline, and dithiooxamide on, respectively, microfluidic channels. When the sample is added, the complexation reactions are simultaneously initiated and the distance-based color development occurs proportionally to the metal concentration levels (Fig. 4.4C).

Tian et al. [83] fabricated aptamer-μPADs for separately visual analysis of invertase, cocaine, and adenosine. The aptamer sensor was based on enzyme embedded hydrogel for specific target detection. For instrument-free measurements, the microfluidic channel was incorporated with specific aptamers, and when the target sample is introduced on paper zones, the embedded enzyme is released, posteriorly these enzyme levels can be relational to target concentration. The principle of the aptamer-μPAD for distance-based detection is represented in Fig. 4.4D.

4.4.4 Comparison of image digitalization methods

It is well-known in the literature that the image analysis may be affected by external light interferences depending on measurement apparatus. In this way, a comparative study involving instrumental and free-instrumental colorimetric methods was performed using a well-established complexometric reaction as model. For this purpose, the chromogenic agent dmg was first impregnated on paper zones and channels. Then, a solution containing Ni(II) was added, and the color changed from colorless to purple due to the formation of $Ni(dmg)_2$ complex, as represented in Fig. 4.5.

After developing the color, the zone was scanned and photographed to compare the performance with the data obtained with a distance-based method. The achieved data versus the analyte concentration are displayed in

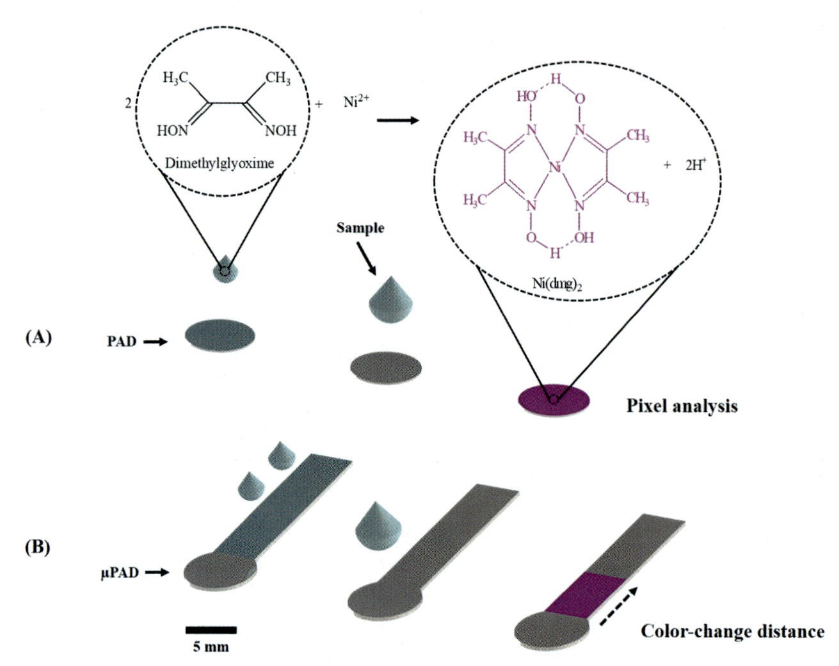

FIGURE 4.5 The colorimetric assay's representative scheme involving colorimetric detection based on pixels analysis (A) and distance (B).

Fig. 4.6. The analytical information extracted from the curves presented in Fig. 4.6 is summarized in Table 4.1. Based on the achieved data, Ni detection's sensitivity values were 0.13 A.U. (mg $L^{-1})^{-1}$, 0.0004 A.U. (mg $L^{-1})^{-1}$, and 0.03 mm (mg $L^{-1})^{-1}$ using office scanner, smartphone, and distance-based measurements, respectively.

As presented, an office scanner has ensured the highest sensitivity and the lowest limit of detection (LOD), most probably due to the reproducible conditions for image capture during the digitalization stage. The lowest standard deviation values also evidence this for scanner-based measurements. On the other hand, although the instrument-free and smartphone-based methods present lower sensitivity, they consume less energy and are easily integrated with PAD and μPAD. However, careful studies must be carried out regarding the uniformity of color between the images due to ambient light interference via a smartphone and the perception of human vision in distance-based measurements. Thus, both image analysis methods on paper-based sensors can be employed since they are low cost, employ low amount of reagents, and require minimum training time. Lastly, the use of smartphones and distance-based measurements are portable and can be easily controlled by analysts [88,89].

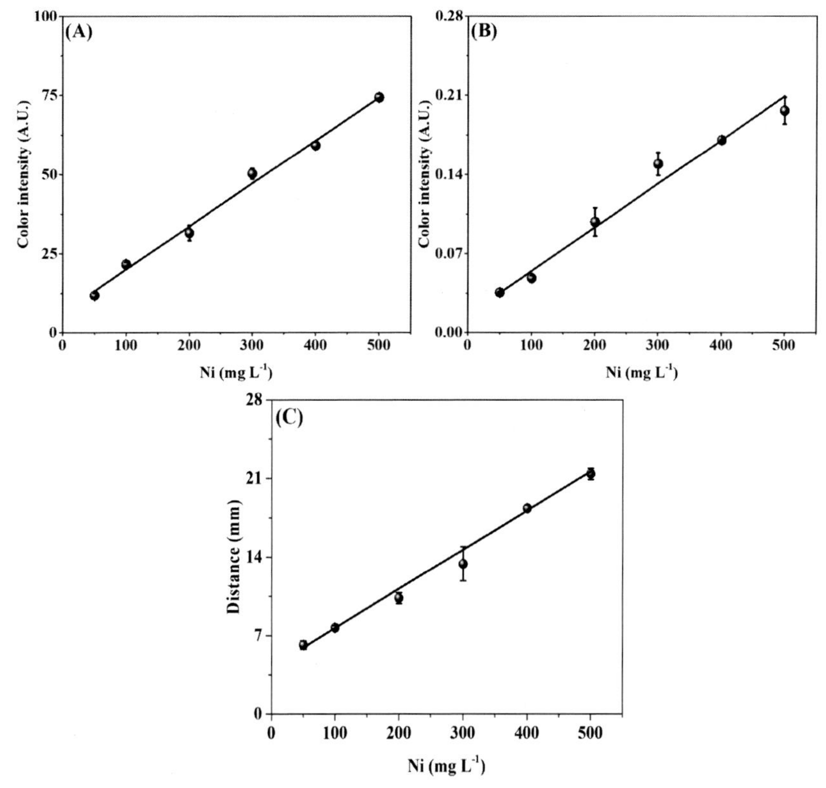

FIGURE 4.6 Analytical curves recorded for Ni using colorimetric readout based on (A) office scanner, (B) smartphone camera, and (C) distance-based information.

TABLE 4.1 Analytical data extracted from colorimetric measurements recorded on PADs using office scanner, smartphone and distance-based measurements.

Analytical parameters	Office scanner	Smartphone	Distance-based
Required instrument	Yes	Yes	No
Analysis time (min)	8	4	8
R^2	0.993	0.999	0.996
Sensitivity (a.u. mg^{-1} L)	0.13	0.0004	0.03[a]
LOD (mg L^{-1})	1.3	11.8	5.8
Inter-PAD (%)	3.3	4.8	3.6
Repeatability (%)	1.9	0.8	2.6

[a] *mm (mg $L^{-1})^{-1}$*

4.5 Conclusions and perspectives

This chapter has summarized the huge potential of colorimetric paper-based analytical devices for different applications using simple electronic devices for image capture or instrument-free measurements. As discussed, one of the main drawbacks refers to the color uniformity, which can be solved or minimized by promoting chemical oxidation of hydroxyl groups, adding nanostructured materials, selecting the paper type correctly or adjusting the reagent addition order. A few examples involving enzymatic assays, redox titrations, pH indicators, NPs, and complexometric reactions were presented to highlight these devices' general feasibility for applications directly in the field.

Based on the reported approaches, the perspectives for future investigations involve the integration of user-friendly and straightforward Apps on smartphones for widely boosting these portable platforms for point-of-care testing. The success of these advances may speed up the commercialization of newer paper-based products in addition to those already available in the market, like pregnancy tests and emerging kits for coronavirus testing. In the same way, instrument-free systems may also be useful for performing assays in remote areas since the color displacement can be easily measured using more straightforward metric tools, like a ruler.

References

[1] S. Solanki, C.M. Pandey, R.K. Gupta, B.D. Malhotra, Emerging trends in microfluidics based devices, Biotechnol. J. (2020) 1900279.

[2] A.T. Singh, D. Lantigua, A. Meka, S. Taing, M. Pandher, G. Camci-Unal, Paper-based sensors: emerging themes and applications, Sensors 18 (9) (2018) 1–22.

[3] S. Kasetsirikul, M.J.A. Shiddiky, N.T. Nguyen, Challenges and perspectives in the development of paper-based lateral flow assays, Microfluid. Nanofluid 24 (2) (2020) 1–18.

[4] S. Smith, J.G. Korvink, D. Mager, K. Land, The potential of paper-based diagnostics to meet the ASSURED criteria, RSC Adv 8 (59) (2018) 34012–34034.

[5] X. Weng, Y. Kang, Q. Guo, B. Peng, H. Jiang, Recent advances in thread-based microfluidics for diagnostic applications, Biosens. Bioelectron. 132 (2019) 171–185.

[6] J. Yang, K. Wang, H. Xu, W. Yan, Q. Jin, D. Cui, Detection platforms for point-of-care testing based on colorimetric, luminescent and magnetic assays: a review, Talanta. 202 (2019) 96–110.

[7] S. Ahmed, M.P.N. Bui, A. Abbas, Paper-based chemical and biological sensors: engineering aspects, Biosens. Bioelectron. 77 (2016) 249–263.

[8] D.M. Cate, J.A. Adkins, J. Mettakoonpitak, C.S. Henry, Recent developments in paper-based microfluidic devices, Anal. Chem. 87 (2015) 19–41.

[9] M. Park, B.H. Kang, K.H. Jeong, Paper-based biochip assays and recent developments: a review, Biochip J. 12 (1) (2018) 1–10.

[10] Z. Li, H. Liu, X. He, F. Xu, F. Li, Pen-on-paper strategies for point-of-care testing of human health, Trends Anal. Chem. 108 (2018) 50–64.

[11] A.W. Martinez, S.T. Phillips, M.J. Butte, G.M. Whitesides, Patterned paper as a platform for inexpensive, low-volume, portable, Angew. Chem. 46 (2007) 1318–1320.

[12] J. Liu, X. He, Y. He, M. Yu, L. Jiang, B. Chen, Development and application of paper spray ionization mass spectrometry, Prog. Chem. 29 (6) (2017) 659–666.

[13] J. Mettakoonpitak, K. Boehle, S. Nantaphol, P. Teengam, J.A. Adkins, M. Srisa-Art, et al., Electrochemistry on paper-based analytical devices: a review, Electroanalysis. 28 (7) (2016) 1420–1436.

[14] L. Ge, J. Yu, S. Ge, M. Yan, Lab-on-paper-based devices using chemiluminescence and electrogenerated chemiluminescence detection, Anal. Bioanal. Chem. 406 (23) (2014) 5613–5630.

[15] L.M. Fu, Y.N. Wang, Detection methods and applications of microfluidic paper-based analytical devices, Trends Anal. Chem. 107 (2018) 196–211.

[16] G.G. Morbioli, T. Mazzu-Nascimento, A.M. Stockton, E. Carrilho, Technical aspects and challenges of colorimetric detection with microfluidic paper-based analytical devices (μPADs) - a review, Anal. Chim. Acta 970 (2017) 1–22.

[17] N. Lopez-Ruiz, V.F. Curto, M.M. Erenas, F. Benito-Lopez, D. Diamond, A.J. Palma, et al., Smartphone-based simultaneous pH and nitrite colorimetric determination for paper microfluidic devices, Anal. Chem. 86 (19) (2014) 9554–9562.

[18] L. Hárendarčíková, J. Petr, Smartphones & microfluidics: marriage for the future, Electrophoresis. 39 (11) (2018) 1319–1328.

[19] A. Roda, E. Michelini, M. Zangheri, Di, M. Fusco, D. Calabria, et al., Smartphone-based biosensors: a critical review and perspectives, Trends Anal. Chem. 79 (2016) 317–325.

[20] S. Kim, D. Kim, S. Kim, Simultaneous quantification of multiple biomarkers on a self-calibrating microfluidic paper-based analytic device, Anal. Chim. Acta 1097 (2020) 120–126.

[21] A.W. Martinez, S.T. Phillips, G.M. Whitesides, Three-dimensional microfluidic devices fabricated in layered paper and tape, Proc. Natl Acad. Sci. 105 (2008) 19606–19611.

[22] L. Florea, C. Fay, E. Lahiff, T. Phelan, N.E. O'Connor, B. Corcoran, et al., Dynamic pH mapping in microfluidic devices by integrating adaptive coatings based on polyaniline with colorimetric imaging techniques, Lab. Chip 13 (6) (2013) 1079–1085.

[23] M. Ariza-Avidad, A. Salinas-Castillo, L.F. Capitán-Vallvey, A 3D μPAD based on a multi-enzyme organic-inorganic hybrid nanoflower reactor, Biosens. Bioelectron. 77 (2016) 51–55.

[24] L. Byrne, J. Barker, G. Pennarun-Thomas, D. Diamond, S. Edwards, Digital imaging as a detector for generic analytical measurements, TrAC. - Trends Anal. Chem. 19 (8) (2000) 517–522.

[25] M. Santhiago, E.W. Nery, G.P. Santos, L.T. Kubota, Microfluidic paper-based devices for bioanalytical applications, Bioanalysis. 6 (1) (2014) 89–106.

[26] G. Sriram, M.P. Bhat, P. Patil, U.T. Uthappa, H.Y. Jung, T. Altalhi, et al., Paper-based microfluidic analytical devices for colorimetric detection of toxic ions: a review, Trends Anal. Chem. 93 (2017) 212–227.

[27] M.I.G.S. Almeida, B.M. Jayawardane, S.D. Koleva, I.D. McKelvie, Developments of microfluidic paper-based analytical devices (μPADs) for water analysis: a review, Talanta. 177 (2018) 176–190.

[28] L. Wang, G. Musile, B.R. McCord, An aptamer-based paper microfluidic device for the colorimetric determination of cocaine, Electrophoresis. 39 (3) (2018) 470–475.

[29] W.R. de Araujo, T.M.G. Cardoso, R.G. da Rocha, M.H.P. Santana, R.A.A. Muñoz, E.M. Richter, et al., Portable analytical platforms for forensic chemistry: a review, Anal. Chim. Acta 1034 (2018) 1–21.

[30] D. Lin, B. Li, J. Qi, X. Ji, S. Yang, W. Wang, et al., Low cost fabrication of microfluidic paper-based analytical devices with water-based polyurethane acrylate and their application for bacterial detection, Sens. Actuators B Chem. 303 (2020)127213.

[31] E. Trofimchuk, Y. Hu, A. Nilghaz, M.Z. Hua, S. Sun, X. Lu, Development of paper-based microfluidic device for the determination of nitrite in meat, Food Chem. 316 (2020) 126396.

[32] C. Dincer, R. Bruch, E. Costa-Rama, M.T. Fernández-Abedul, A. Merkoçi, A. Manz, et al., Disposable sensors in diagnostics, food, and environmental monitoring, Adv. Mater. 31 (2019) 30.

[33] E. Lepowsky, F. Ghaderinezhad, S. Knowlton, S. Tasoglu, Paper-based assays for urine analysis, Biomicrofluidics. 11 (5) (2017) 051501.

[34] E.W. Nery, L.T. Kubota, Sensing approaches on paper-based devices: a review, Anal. Bioanal. Chem. 405 (24) (2013) 7573–7595.

[35] A. Shakeri, N.A. Jarad, A. Leung, L. Soleymani, T.F. Didar, Biofunctionalization of glass- and paper-based microfluidic devices: a review, Adv. Mater. Interfaces 6 (19) (2019) 1–16.

[36] E. Evans, E.F. Moreira Gabriel, W.K. Tomazelli Coltro, C.D. Garcia, Rational selection of substrates to improve color intensity and uniformity on microfluidic paper-based analytical devices, Analyst. 139 (9) (2014) 2127–2132.

[37] J. Credou, T. Berthelot, Cellulose: from biocompatible to bioactive material, J. Mater. Chem. B. 2 (30) (2014) 4767–4788.

[38] J. Narang, C. Singhal, A. Mathur, A.K. Dubey, P.N.A. Krishna, A. Anil, et al., Naked-eye quantitative assay on paper device for date rape drug sensing via smart phone APP, Vacuum. 153 (2018) 300–305.

[39] L.A. Santana-Jiménez, A. Márquez-Lucero, V. Osuna, I. Estrada-Moreno, R.B. Dominguez, Naked-eye detection of glucose in saliva with bienzymatic paper-based sensor, Sensors (Switz.) 18 (4) (2018) 1–12.

[40] A.W. Martinez, E. Carrilho, S.W. Thomas, S.T. Phillips, G.M. Whitesides, H. Sindi, Simple telemedicine for developing regions: camera phones and paper-based microfluidic devices for real-time, off-site diagnosis, Anal. Chem. 80 (10) (2008) 3699–3707.

[41] V. Hamedpour, P. Oliveri, R. Leardi, D. Citterio, Chemometric challenges in development of paper-based analytical devices: Optimization and image processing, Anal. Chim. Acta 1101 (2020) 1–8.

[42] S.A. Bhakta, R. Borba, M. Taba, C.D. Garcia, E. Carrilho, Determination of nitrite in saliva using microfluidic paper-based analytical devices, Anal. Chim. Acta 809 (2014) 117–122.

[43] E.F.M. Gabriel, P.T. Garcia, T.M.G. Cardoso, F.M. Lopes, F.T. Martins, W.K.T. Coltro, Highly sensitive colorimetric detection of glucose and uric acid in biological fluids using chitosan-modified paper microfluidic devices, Analyst 141 (2016) 4749–4756.

[44] K.L. Peters, I. Corbin, L.M. Kaufman, K. Zreibe, L. Blanes, B.R. McCord, Simultaneous colorimetric detection of improvised explosive compounds using microfluidic paper-based analytical devices (μPADs), Anal. Methods 7 (1) (2015) 63–70.

[45] G. Xiao, J. He, X. Chen, Y. Qiao, F. Wang, Q. Xia, et al., A wearable, cotton thread/ paper-based microfluidic device coupled with smartphone for sweat glucose sensing, Cellulose. (2019). 0123456789.

[46] L.R. Sousa, L.C. Duarte, W.K.T. Coltro, Instrument-free fabrication of microfluidic paper-based analytical devices through 3D pen drawing, Sens. Actuators B Chem. 312 (2020) 128018.

[47] T.M.G. Cardoso, F.R. de Souza, P.T. Garcia, D. Rabelo, C.S. Henry, W.K.T. Coltro, Versatile fabrication of paper-based microfluidic devices with high chemical resistance using scholar glue and magnetic masks, Anal. Chim. Acta 974 (2017) 63–68.

[48] W. Dungchai, O. Chailapakul, C.S. Henry, Use of multiple colorimetric indicators for paper-based microfluidic devices, Anal. Chim. Acta 674 (2) (2010) 227–233.

[49] De Freitas SV, De Souza R, J.C.R. Neto, V. Abdelnur, B.G. Vaz, C.S. Henry, et al., Uncovering the formation of color gradients for glucose colorimetric assays on micro fl uidic paper-based analytical devices by mass spectrometry imaging, Anal. Chem. 90 (2018) 11949–11954.

[50] G.G. Morbioli, T. Mazzu-Nascimento, L.A. Milan, A.M. Stockton, E. Carrilho, Improving sample distribution homogeneity in three-dimensional microfluidic paper-based analytical devices by rational device design, Anal. Chem. 89 (9) (2017) 4786–4792.

[51] P.T. Garcia, T.M.G. Cardoso, C.D. Garcia, E. Carrilho, W.K.T. Coltro, A handheld stamping process to fabricate microfluidic paper-based analytical devices with chemically modified surface for clinical assays, RSC Adv. 4 (71) (2014) 37637–37644.

[52] E. Evans, E.F. Moreira Gabriel, T.E. Benavidez, W.K. Tomazelli Coltro, C.D. Garcia, Modification of microfluidic paper-based devices with silica nanoparticles, Analyst. 139 (21) (2014) 5560–5567.

[53] E.B. Strong, C.W. Kirschbaum, A.W. Martinez, N.W. Martinez, Paper miniaturization via periodate oxidation of cellulose, Cellulose. 25 (6) (2018) 3211–3217.

[54] C. Jiang, B. Liu, M.-Y. Han, Z. Zhang, Fluorescent nanomaterials for color-multiplexing test papers toward qualitative/quantitative assays, Small Methods 2 (7) (2018) 1700379.

[55] F. Figueredo, P.T. Garcia, E. Cortón, W.K.T. Coltro, Enhanced analytical performance of paper microfluidic devices by using Fe_3O_4 nanoparticles, MWCNT, and graphene oxide, ACS Appl. Mater. Interfaces 8 (1) (2016) 11–15.

[56] B. Krajewska, Application of chitin- and chitosan-based materials for enzyme immobilizations: a review, Enzyme Microb. Technol. 35 (2–3) (2004) 126–139.

[57] E. Noviana, S. Jain, J. Hofstetter, B.J. Geiss, D.S. Dandy, C.S. Henry, Paper-based nuclease protection assay with on-chip sample pretreatment for point-of-need nucleic acid detection, Anal. Bioanal. Chem. 412 (13) (2020) 3051–3061.

[58] R.S. Juang, F.C. Wu, R.L. Tseng, Use of chemically modified chitosan beads for sorption and enzyme immobilization, Adv. Environ. Res. 6 (2) (2002) 171–177.

[59] Q. He, C. Ma, X. Hu, H. Chen, Method for fabrication of paper-based microfluidic devices by alkylsilane self-assembling and UV/O_3-patterning, Anal. Chem. 85 (3) (2013) 1327–1331.

[60] J.P. Devadhasan, J. Kim, A chemically functionalized paper-based microfluidic platform for multiplex heavy metal detection, Sens. Actuators B Chem. 273 (2018) 18–24.

[61] H. Karlsen, T. Dong, Biomarkers of urinary tract infections: State of the art, and promising applications for rapid strip-based chemical sensors, Anal. Methods 7 (19) (2015) 7961–7975.

[62] L. Mou, X. Jiang, Materials for microfluidic immunoassays: a review, Adv. Healthc. Mater. 6 (15) (2017) 1–20.

[63] L.S.A. Busa, M. Maeki, A. Ishida, H. Tani, M. Tokeshi, Simple and sensitive colorimetric assay system for horseradish peroxidase using microfluidic paper-based devices, Sens. Actuators B Chem. 236 (2016) 433–441.

[64] D. Calabria, C. Caliceti, M. Zangheri, M. Mirasoli, P. Simoni, A. Roda, Smartphone–based enzymatic biosensor for oral fluid L-lactate detection in one minute using confined multilayer paper reflectometry, Biosens. Bioelectron. 94 (2017) 124–130.

[65] X. Wang, F. Li, Z. Cai, K. Liu, J. Li, B. Zhang, et al., Sensitive colorimetric assay for uric acid and glucose detection based on multilayer-modified paper with smartphone as signal readout, Anal. Bioanal. Chem. 410 (10) (2018) 2647–2655.

[66] Z. Li, H. Yang, M. Hu, L. Zhang, S. Ge, K. Cui, et al., Cathode photoelectrochemical paper device for microRNA detection based on cascaded photoactive structures and

hemin/pt nanoparticle-decorated DNA dendrimers, ACS Appl. Mater. Interfaces 12 (15) (2020) 17177–17184.

[67] X. Wang, M. Mahoney, M.E. Meyerhoff, Inkjet-printed paper-based colorimetric polyion sensor using a smartphone as a detector, Anal. Chem. 89 (22) (2017) 12334–12341.

[68] C.M. Cheng, A.W. Martinez, J. Gong, C.R. Mace, S.T. Phillips, E. Carrilho, et al., Paper-based elisa, Angew. Chem. Int. (Ed.) 49 (28) (2010) 4771–4774.

[69] J. Elomaa, L. Gallegos, F.A. Gomez, Cord-based microfluidic chips as a platform for ELISA and glucose assays, Micromachines. 10 (9) (2019).

[70] S.H. Park, A. Maruniak, J. Kim, G.R. Yi, S.H. Lim, Disposable microfluidic sensor arrays for discrimination of antioxidants, Talanta. 153 (2016) 163–169.

[71] E.B. Strong, S.A. Schultz, A.W. Martinez, N.W. Martinez, Fabrication of miniaturized paper-based microfluidic devices (MicroPADs), Sci. Rep. 9 (1) (2019) 1–9.

[72] S.A. Nogueira, L.R. Sousa, N.K.L. Silva, P.H.F. Rodrigues, W.K.T. Coltro, Monitoring acid-base titrations on wax printed paper microzones using a smartphone, Micromachines. 8 (5) (2017) 139. Available from: https://doi.org/10.3390/mi8050139.

[73] X. Zhao, H. Zhao, L. Yan, N. Li, J. Shi, C. Jiang, Recent developments in detection using noble metal nanoparticles, Crit. Rev. Anal. Chem. 50 (2) (2020) 97–110.

[74] C. Liu, Y. Miao, X. Zhang, S. Zhang, X. Zhao, Colorimetric determination of cysteine by a paper-based assay system using aspartic acid modified gold nanoparticles, Microchim. Acta 187 (6) (2020).

[75] J. Zhai, E. Bakker, Complexometric titrations: new reagents and concepts to overcome old limitations, Analyst. 141 (14) (2016) 4252–4261.

[76] A. Mu, Verfahren zur Bestimmung des Blausauregehaltes der medicinischen Blausaure, des Bittermandel - und Hirechlorbeerwassers, J. Chem. Inf. Model. 53 (9) (2019) 1689–1699.

[77] K. Yamada, S. Takaki, N. Komuro, K. Suzuki, D. Citterio, An antibody-free microfluidic paper-based analytical device for the determination of tear fluid lactoferrin by fluorescence sensitization of Tb^{3+}, Analyst. 139 (7) (2014) 1637–1643.

[78] T. Akyazi, L. Basabe-Desmonts, F. Benito-Lopez, Review on microfluidic paper-based analytical devices towards commercialisation, Anal. Chim. Acta 1001 (2018) 1–17.

[79] K. Yamada, T.G. Henares, K. Suzuki, D. Citterio, Paper-based inkjet-printed microfluidic analytical devices, Angew. Chem. Int. (Ed.) 54 (18) (2015) 5294–5310.

[80] J. Aksorn, S. Teepoo, Development of the simultaneous colorimetric enzymatic detection of sucrose, fructose and glucose using a microfluidic paper-based analytical device, Talanta 207 (2020) 120302.

[81] G.O. da Silva, W.R. de Araujo, T.R.L.C. Paixão, Portable and low-cost colorimetric office paper-based device for phenacetin detection in seized cocaine samples, Talanta. 176 (2018) 674–678.

[82] D.M. Cate, S.D. Noblitt, J. Volckens, C.S. Henry, Multiplexed paper analytical device for quantification of metals using distance-based detection, Lab Chip 15 (13) (2015) 2808–2818.

[83] T. Tian, Y. An, Y. Wu, Y. Song, Z. Zhu, C. Yang, Integrated distance-based origami paper analytical device for one-step visualized analysis, ACS Appl. Mater. Interfaces 9 (36) (2017) 30480–30487.

[84] M.M. Calabretta, R. Álvarez-Diduk, E. Michelini, A. Roda, A. Merkoçi, Nano-lantern on paper for smartphone-based ATP detection, Biosens. Bioelectron. 150 (2020).

[85] D. Calabria, M. Mirasoli, M. Guardigli, P. Simoni, M. Zangheri, P. Severi, et al., Paper-based smartphone chemosensor for reflectometric on-site total polyphenols quantification in olive oil, Sens. Actuators B Chem. 305 (2020).

[86] T. Tian, J. Li, Y. Song, L. Zhou, Z. Zhu, C.J. Yang, Distance-based microfluidic quantitative detection methods for point-of-care testing, Lab Chip 16 (7) (2016) 1139–1151.
[87] I. Schelshorn, Enzyme immunochromatography-a quantitative immunoassay requinng no instrumentation, Clin. Chem. 31 (7) (2010) 1144–1150.
[88] T. Ozer, C. McMahon, C.S. Henry, Advances in paper-based analytical devices, Annu. Rev. Anal. Chem. 13 (1) (2020) 85–109.
[89] M.S. Verma, M.N. Tsaloglou, T. Sisley, D. Christodouleas, A. Chen, J. Milette, et al., Sliding-strip microfluidic device enables ELISA on paper, Biosens. Bioelectron. 99 (2018) 77–84.

Chapter 5

Electrochemical paper-based analytical devices

Iana V.S. Arantes[1], Juliana L.M. Gongoni[1], Letícia F. Mendes[1], Vanessa N. de Ataide[1], Wilson A. Ameku[1], Paulo T. Garcia[2], William R. de Araujo[3] and Thiago R.L.C. Paixão[4,5]

[1]Departamento de Química Fundamental, Instituto de Química, Universidade de São Paulo, São Paulo, Brazil, [2]Institute of Exact Sciences, Faculty of Chemistry, Federal University of South and Southeast of Pará (UNIFESSPA), Marabá, Brazil, [3]Portable Chemical Sensors Lab, Department of Analytical Chemistry, Institute of Chemistry, State University of Campinas—UNICAMP, Campinas, Brazil, [4]Department of Fundamental Chemistry, Institute of Chemistry, University of Sao Paulo, Sao Paulo, Brazil, [5]National Institute of Bioanalitical Science and Technology, Campinas, Brazil

5.1 Introduction

The search for analytical devices for in-field or/and point-of-need applications is not new in the Analytical Chemistry scenario [1]. The search for new sensing analytical methods, and new instruments to operate on-site without electrical power sources, consumables, and small size was still on the table in the analytical research field. However, since 2007 Whitesides group [2] put an old and well-known material on the table to fabricate low-cost and point-of-need analytical devices, the paper as a microfluidic device and colorimetric detection approach, creating a publication boom in the literature, as mentioned in the previous chapter. The paper, as discussed in chapter 2, Chemistry of Paper—Properties, Modification Strategies and Uses in Bioanalytical Chemistry, has a lot of exciting potential, highlighted one here is the nonnecessity of power source for pump solution, that is, one of the requirements for portability, and a lot of other characteristics related in this book for the sensing purpose.

Most of the paper devices developed started with colorimetric detection since the first work reported by Whitesides's group [2]. Still, since the pioneering work in 2009 from Henry's group coupling electrochemistry and paper substrate [3], the number of electrochemical paper-based analytical devices (ePADs) started to grow to overdo the number of visible colorimetric paper-based devices in 2012, Fig. 5.1.

Paper-Based Analytical Devices for Chemical Analysis and Diagnostics.
DOI: https://doi.org/10.1016/B978-0-12-820534-1.00011-6

© 2022 Elsevier Inc. All rights reserved.

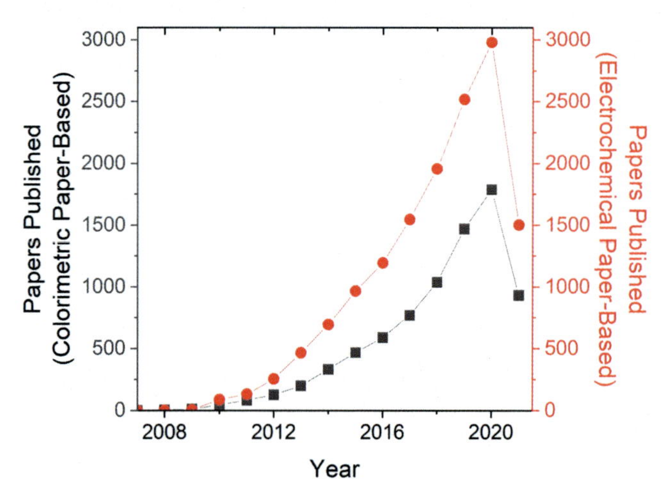

FIGURE 5.1 Paper published found in the Scopus database (http://www.scopus.com) using the keywords: *"colorimetric and paper-based and devices"* and *"electrochemical and paper-based and devices."*

The main advantages for enhancing the number of ePADs papers published are that these devices could increase the sensitivity and selectivity of the visible color detection on the paper-based devices without the necessity of eye detection or any bias related to that. ePADs are low-cost compared with other analytical instruments/methods, but not compared with eye color detection on paper. Additionally, the necessary sample volumes are low when compared with conventional electrochemical cells, and Arduini and collaborators made a comparison in a review article published in 2020 [4]. Commonly, these ePADs used less than 1,000 times sample solution volume than the conventional electrochemical detection, matching the requirements for the green analytical chemistry protocols and the necessities for clinical and point-of-need applications. Other advantages that ePADs try to circumvent are low linear concentration range, color saturation complications, as lightning limitations due to the sample or background [5,6] found on the color paper-based devices.

The last decade brought the old paper material back for the sensing area fostered new material development for sensing in different fields, especially in the electrochemical area. Hence, the chapter aims to relate the development in this field and clever protocols to fabricate the electrochemical paper-based devices for different applications and some perspectives for this field.

5.2 Principles of the electrochemical measurements using electrochemical paper-based analytical devices

The principle behind the ePADs detection is the same in conventional electrochemical cells, and it is based on electrochemical techniques. These techniques are based on potential, charge, or current information related to the analyte concentration. These properties are measured in the interface between the electrode surface and solution, except for conductometry techniques, where the conductivity measurements are based on the bulk solution. The usual lack of selectivity of conductometric measurements makes this technique mainly used as a detector in separation paper-based devices, as an example, to determine *Salmonella typhimurium* [7]. The electrochemical methods used in the ePADs detection are potentiometry, conductometry, coulometry, amperometry, and voltammetry. The distribution of the detection techniques used in the ePADs is summarized in Fig. 5.2.

As could be seen, the voltammetric techniques are the most common detection techniques used in the ePADs (52%), and it could be related to the possibility to detect multiples analytes using the same electrode and associated with the opportunity to achieve selectivity using the potential applied to the electrode surface used. Impedimetric measurements are the second one (28.9%), and potentiometry the third (9.5) used, followed by amperometry (6.8%), conductometry (1.5%), and coulometry (1.2%) techniques.

A brief overview of the interfacial techniques will be made here, and more details could be found in a fascinating free digital book published by Harvey [8]. Potentiometric detection is based on the measurement of the potential difference of an electrochemical cell under static conditions and the

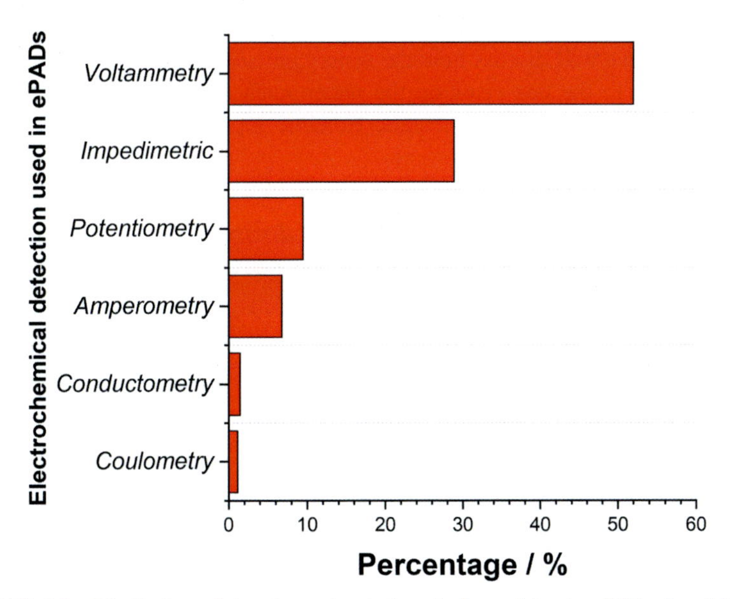

FIGURE 5.2 Distribution of the electrochemical methods used in the ePADs found in the Scopus database until May 4, 2021 (http://www.scopus.com). Keywords used: *"electrochemical and paper-based and device"* or *coulometry* or *conductometry* or *amperometry* or *potentiometry* or *impedimetric* or *voltammetry*.

absence of appreciable current flowing in the cell, which turns its technique into a nondestructive method. Usually, its detection is based on polymeric membrane ion-selective electrodes measuring the potential change at one electrode against a reference electrode. One example of potentiometric paper-based detection is the work published by Whitesides's group [9]. The other techniques, coulometry, amperometry, and voltammetry are based on potential applied to a working electrode, the sensor unit, versus a reference electrode, and the measured current results from electrochemical oxidation or reduction of the electroactive species flow between the working electrode and the counter-electrode, creating a three-electrode electrochemical cell. In the voltammetric technique, the potential is swept between the working and reference electrode, and the way how this potential is applied results in different types of voltammetric methods, such as cyclic voltammetry (CV), square wave voltammetry (SWV), differential pulse voltammetry (DPV), stripping voltammetry, polarography, and linear sweep voltammetry (LSV). The amperometry and potentiostatic coulometry work in a fixed potential between the working and reference electrode, and the measure is current and in charge of the electrochemical process.

Additionally, electrochemical impedance spectroscopy (EIS) detection monitors the response of a three-electrode electrochemical cell to an applied potential, in which the frequency dependence may reveal underlying chemical processes. Typically, the measurements provide information related to two different processes, as Faradaic or non-Faradaic mechanisms [10]. In the next section, we will discuss how the sensor units/electrochemical cells were manufactured in the last years on the paper surface to solve analytical problems using the ePADs, a schematic timeline representation can be found in Fig. 5.3.

5.3 Paper standardization and area definition

The ePADs were first reported by Henry's group in 2009 using screen-printed carbon electrodes [3] and then this type of sensor has received significant attention from different research groups. Briefly, for the fabrication of ePADs it is initially necessary to choose the paper type (chromatographic, filter paper, office paper, photographic paper, etc.) since the paper chosen will directly impact the properties of the device microfluidic features. Afterward different techniques can be employed to delimit the paper zone for analyses (electrochemical cell), and to fabricate the conductive tracks, *i.e.*, the electrodes on the paper surface, and then can be necessary an additional step to define the geometric area of the electrodes [11,12]. The fabrication of reaction zones and detection areas on paper requires the patterning of nonconductive materials. These standardizations of paper devices involve creating hydrophobic barriers or establishing physical boundaries by cutting the paper. On the other hand, the fabrication of electrodes on paper devices

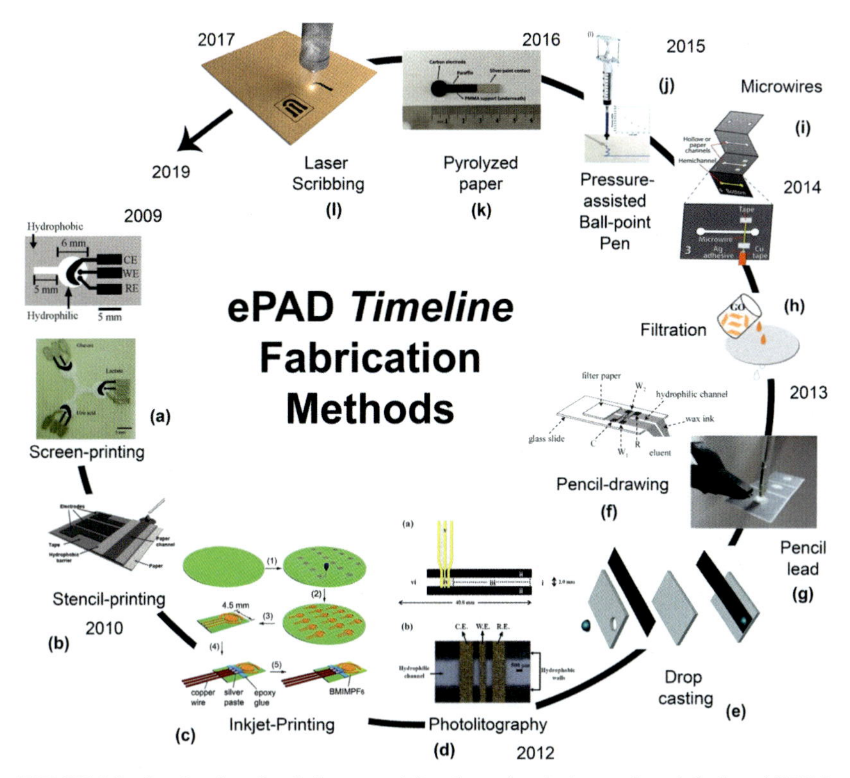

FIGURE 5.3 Landmarks of techniques used for electrochemical paper-based devices (ePADs) reported in the last years. [6] *Reproduced by permission of The Royal Society of Chemistry from V.N. Ataide, L.F. Mendes, L.I.L.M. Gama, W.R. de Araujo, T.R.L.C. Paixão, Electrochemical paper-based analytical devices: ten years of development, Analytical Methods 12(8) (2020) 1030–1054.*

can be achieved using different strategies to define conductive tracks on a paper surface [6,13]. In the following sections (sections 5.3.1 and 5.3.2), we will briefly discuss the most common procedures used to fabricate ePADs.

5.3.1 Patterning nonconductive materials on the paper surface

The definition of hydrophobic boundaries in paper-based devices denotes from the 19th century [14], as described in Chapter 1, Introduction Remarks for Paper-Based Analytical Devices and Timeline. Following, the resumption of paper used as a substrate for chemical testing by Whitesides's group in 2007 [15], where the authors reported for the first time the use of photolithography to create channels and detection zones in paper with high resolution and reproducibility. However, this technique requires expensive equipment, cleanroom facilities, and thus is not suitable for laboratories with limited resources.

Furthermore, another disadvantage is the high consumption of reagents to treat the entire paper [16]. Therefore printing techniques are the most commonly employed to pattern paper devices and decrease the consumption of hydrophobic materials. Several advantages encourage printing processes to obtain nonconductive patterns in the paper substrate, including low cost, instrumental simplicity, and good reproducibility [16,17]. Briefly, the printing techniques can be divided into template-based and nontemplate methods. The last one is the most used, once it transfers an image from computer to paper surface by deposition of nonconductive ink, it is digitally controlled without the use of masks or printing plates.

In this way, wax printing is the most representative of this kind of group, once it is simple, low cost, and provides adequate reproducibility. Carrilho and coauthors proposed this method in 2009 by using a commercial printer to define wax hydrophobic barriers on paper surfaces [18]. The mentioned method is nontoxic and cheaper than the other methods reported in the literature [11]. Furthermore, hydrophobic patterns can also be defined by others printing processes using various materials or solvents, such as alkyl/alkenyl ketene paraffin polydimethylsiloxane (PDMS), lacquer spraying, scholar glue, and others [19–36]. It is important to emphasize that the printing procedure presents high versatility and can define nonconductive barriers and conductive tracks on paper substrate by changing the ink/cartridges/masks, thus facilitating and integrating the entire fabrication process of ePADs.

Additionally, paper devices can be fabricated by cutting the paper, and this procedure consists of defining physical boundaries without the requirement of reagents as hydrophobic materials. In this way, this method is simple, once require only a cutting machine/system [37–39]. The channels can be obtained by cutting the paper substrate using a cutter plotter with X-Y axis control or a CO_2 laser cutter. It is essential to highlight that the device obtained by cutting is not mechanically rigid, and it is necessary to use solid supports to increase the mechanical resistance [40,41].

5.3.2 Electrodes fabrication by conductive patterning tracks on paper

The fabrication of electrodes on paper surfaces is the determinant step in the procedures to obtain useful ePADs. As previously mentioned, the first ePAD was proposed in 2009 using screen-printed technology [3]. Afterwards, several methods have been developed to define conductive tracks on paper surface and the main techniques are screen-printing [3,42,43], stencil-printing [44–46], inkjet-printing [47,48], pencil drawing [49–51], vacuum filtration [52–54], drop-casting [55,56], sputtering [57–59], wire placement [60,61], paper pyrolysis [62–64], and laser-scribing [65,66].

Screen-printing was the first method used to fabricate conductive tracks on the paper surface. It is a sophisticated stencil technique, once use more expensive masks compared to stencil-printing. Both screen/stencil printing techniques use masks to pattern conductive tracks on paper surfaces using conductive inks or paste [3,42–46]. It is essential to highlight that screen/stencil printing technology is highly accessible once it requires inexpensive materials. However, both techniques present moderate to high resolution (μm-range) and produce a significant amount of waste. Still, about printing process techniques, inkjet printing has become popular, once not requires templates, and the conductive inks can be printed onto the paper surface using a commercial inkjet printer [47,48].

Furthermore, inkjet-printed ePADs have been fabricated with innumerous conductive inks such as carbon, metallic nanoparticles, graphene, and others. However, the control of the rheological properties of the ink plays an essential role in the success of high-resolution printing and in avoiding nozzle clogging. Pencil graphite has attracted significant attention as carbon material for electrode fabrication on a paper surface [49–51,67]. This technique provides advantages that encourage their use, such as simplicity, low cost, high availability, and no waste generation. An obstacle for this method is the low reproducibility due to the manual nature of the procedure and the variation of commercial pencil types. In the same way, another technique to produce electrodes with carbon-based materials is vacuum filtration [52–54]. The mentioned method has become popular once it is fast and provides scalable electrode production. The main disadvantage is the low reproducibility associated with the need for uniform filling of paper fibers with the filtering material.

Nanomaterials can be deposited on the electrode surface, and for that, the drop-casting procedure is the most employed due to its instrumental simplicity [55,56]. Furthermore, paper-based working electrodes are widely fabricated by drop-casting. When this technique is used, some consideration should be given, like the amount and homogeneity of the nanomaterial suspension. On the other hand, metal films can be deposited onto the paper surface as electrodes, and the most traditional technique employed is sputtering [57–59]. Briefly, a mask is placed over the paper to define the electrode region, and metals are sputtered. This method offers high-quality films with excellent reproducibility but requires expensive instrumentation. An alternative technique to define conductive tracks on paper substrates is by using microwires as electrodes, once they have higher surface areas and lower electrical resistance when compared to carbon ink electrodes [60,61]. Besides, the wires can be cleaned and reused. However, one disadvantage that discourages their use is the fragility and difficulty to handle. Recently, the use of pyrolyzed paper and laser-scribing technology were reported to fabricate conductive carbon materials using tube furnace and CO_2 laser, respectively [62–66]. When a laser is used to controlled carbonize the surface of paper-based material to produce electrodes, some attractive advantages encourage their implementation, such as large-scale production and does not require masks and reagents.

Last but not least, in the final step of the ePADs fabrication process is necessary to use procedures to delimit the geometric area, isolate the detection zones and protect the electrical contacts. Some previously mentioned methods can be used for this purpose, manually changing the conductive material by a dielectric material, and thus easily/quickly delimiting the ePADs electrochemical cell area. The following sections will discuss in more detail the main fabrication methods of ePADs.

5.4 The fabrication processes of electrochemical paper-based analytical devices and their applications

This section will be subdivided into different processes reported in Fig. 5.3.

5.4.1 Screen-printing and stencil printing

The screen-printing technique was the first one introduced for ePAD fabrication, and since then, it has been the most popular and used technique for fabricating carbon electrodes on paper [68]. The screen-printed electrodes (SPE) come from the same approach used with mashes or screens and were incorporated into electrode production [13]. The process consists of designing the negative layout, or mask, with the desired shape, which can be made in computer software [69], and cut them into the form, then is used to create a barrier so that the conductive ink can be spread in the desired configuration with the aid of a squeegee; hence it is left to dry so that the ink can cure, therefore the ePAD is ready to use.

The first screen-printed paper-based electrode was made in 2009, when Dungchai and Henry's group [3,13] putting together the knowledge from barriers, in this case, photolithography from Muller [70] and paper microfluidics from Whiteside's group [2] with the μPADs, creating an ePAD. This was an innovation because it used the good features of the paper, such as capillarity, porosity, easy modification, low cost, among others, and added an enhancement in the detection when compared with color detection used in the first paper-based device.

An alternative and cheaper way to print electrodes on paper is using stencil or mask technique, generally called for this feature as stencil printed electrodes. In this process, the conductive ink is spread on the paper and stencil set, then the stencil is removed from the top of the paper. Hence, the electrode, the set of electrodes, is fabricated with the desired shape of the mask, working in the same way that we used to paint a desired drawing or shape in a piece of fabric or wall. This technique can be used with carbon ink and its variations and to manufacture silver ink pseudo-reference electrodes [71].

The stencil printing technique is necessary to make the mask first, the design can be made like the screen printing, but using less expensive materials. Plastic or tape [72] can be used to produce the stencil and be cut into a frame with a CO_2 laser or with a commercial cutting printer, such as Silhouette. When the mask is ready, it can be affixed onto the paper, and it can be painted, when the ink is cured enough the mask is taken off, therefore the electrode is ready. The advantage of this method is that stencils can be made for different purposes, painting the conductive track for the electrode fabrication and also can be used to create hydrophobic barriers with a laminating pouch that can be put around the detection zone to prevent leaking [69].

Usually, for both stencil printing and screen-printing, the application of multiple inks coats is necessary to create genuinely conductive tracks and measure an analyte, which is a disadvantage because it is time-consuming and needs to be optimized to generate a functional electrode. However, the screen-printing is not only designing and fabricating the electrodes, but also the ink composition needs to attend the desired purposes and matches the paper properties. The central part is the screen that receives the ink, it must be compatible with the ink and accepts it, so one can be attached to the other and not peel off. Thus, it is essential to study what kind of paper is suitable to receive the ink and there are fewer modifiers that could interfere with the measurements in the final sensor, the most common one used is Whatman filter paper.

The other important part of the printing process is the ink properties, as composition, stretching force, viscosity, etc. [73]. One way to change its composition to obtain conductive tracks is to mixture more conductive components like carbon powder, carbon black, carbon nanotubes, graphite to the commercially available carbon inks, and even change the mix ratios and the solvents as well [11]. Besides these materials, other (nano)materials, like gold nanoparticles, inorganic modifiers (Prussian blue), enzymes, and antibodies, can be used, depending on the purpose of the analytical application. It is essential to be tailored to the target molecule so the detection is accurate, and some discussion using paper as a substrate will be made here.

One application of electrochemical detection in our life is the glucometer, where a screen-printed electrode modified with an enzyme, glucose oxidase, is responsible for monitoring the hydrogen peroxide that came from the enzymatic conversion of the glucose to H_2O_2 [74]. In the same way, the paper-based sensor could be used to detect other vital analytes like the device to detect the influenza virus at the point of need. The device was fabricated on the Whatman paper surface using a mask to spray PDMS onto the paper to create a hydrophobic barrier. A stencil is applied to paint the conductive tracks for the electrode's fabrication. A third step is the immobilization of the biological recognition element by a drop-casting process on the surface of the dried working electrode [75].

Another innovative application of screen/stencil-printed electrodes is their integration with colorimetric detection [76]. In the aforementioned work, the authors fabricated the device in a Whatman number 4 paper, where in one side is a wax barrier that will place the colorimetric detection and the other one the electrochemical detection to prevent cross-contamination with the colorimetric reagents. The electrodes are produced outside the barrier paper with stencil printing technique and put together to produce the entire sensor. This sensor was used for multiple catecholamine neurotransmitters detection using both detection approaches [76].

There are not only modifications with biological substrates but there are other modifications for the conductive track surface, like this one to detect capsaicin in pepper [77]. The ePADs were fabricated using a screen-printed electrode, which the commercial graphite ink was modified with carbon black to achieve a higher sensitivity to capsaicin. The geometric area of the working electrode was delimitated by a wax printing step and the device was tested for capsaicin in real pepper samples.

Screen-printing using commercial or homemade inks is a vast field with many point-of-need applications using paper as a substrate for the detection of different analytes [6]. It is inexpensive because it can be made with simple materials and does not necessarily require fancy materials or expensive machinery. Due to its simplicity, it can be incorporated into various types of paper-based devices, even complex designs, like 3D origami, in a reproducible way [78]. But commercialization of these ePADs is still complicated, even though the screen-printing technique is fast, low cost, and presents high sensitivity and repeatability. The fabrication in the lab is primarily handmade, and there are imprecisions on the method, and in higher scale production, the storage of this devices could be a problem especially with biosensors, where enzymes and antibodies require special storage conditions. On the other hand, stencil printing, is a vast field of interest, following with the advantages and disadvantages of screen-printing, but here there is also the tape or mask, and this cause to consider more factors into the ink formulation because it can peel off when the stencil is removed. Besides, the electrodes may not retain its shape when the mask is taken off or even the mask itself may not works properly, nevertheless is an easy way to produce devices without great resources [68]. Even though it is still a field of interest, there is an excellent possibility for growth in industrial production and enhancing its fabrication reproducibility.

5.4.2 Digital printing and printer fabrication process without the use of masks

Printing processes are primarily used in many research laboratories for ePAD fabrication. They are practical, simple to use, cost-effectively, and reproducible, apart from meeting the requirements for large-scale production, solving some reproducibility problems discussed previously. Another essential characteristic of the printing methods is their versatility, especially considering ePADs fabrication. By only making changes in the ink, cartridges, and mask layout, it is possible to pattern both conductive tracks and hydrophobic barriers to delimitate the electrode area on the paper surface using the same equipment. These methods can also be used to modify the electrode surface, incorporating reagents or sensing materials on the final ePAD [6].

The challenge in fabricating printing-based devices is how to pattern electrodes faster and precisely. In this scenario, digital printing, such as inkjet printing and automatic craft cutter/plotter machines, are the main representative techniques because of their ease of use and commercial availability [6,79]. Unlike other popular methods, digital printing does not require predeposition or templates. A computer image can be directly transferred onto a paper substrate by digitally controlled deposition of the ink, saving both time and costs, eliminating some of the waste involved in screen and stencil printing, and providing better resolution [6,13].

In this context, inkjet printing is a well-known technology showing a disruptive potential as a deposition technique for ePAD fabrication due to its prevalence in home and office environments, allowing many research groups to implement this technique in their laboratories using commercially available home inkjet printers. However, depending on the properties of the ink and the presence of organic solvents, the internal system needs to be adapted to support the ink composition without damage the printer system. These printers work by precisely depositing tiny ink droplets row by row to form the desired two-dimensional shape in the paper [80,81]. Essentially, the difference between printing an electrode and an ordinary document using an inkjet printer relies on the ink properties [6]. For ePAD fabrication, the printer cartridges are filled with different conductive inks to be used as the electrode material, such as carbon [82,83], multiwalled carbon nanotubes (CNT) [84,85], silver nanoparticles [80,86], gold nanoparticles [87,88], conducting polymers like poly(3,4-ethylenedioxythiophene) doped with polystyrene sulfonate (PEDOT:PSS) [48], and graphene nanocomposites [89], which are printed in the desired patterns onto the paper substrates.

Furthermore, these inks require adjustments in the surface tension and viscosity to prevent clogging the printing system and promote an adequate printability of the sensing materials [13]. Inkjet usually requires low viscosity inks, from 1 to 40 $mPa \cdot s$. Some of these inks are commercially available or synthesized in the laboratory [90].

As mentioned, in low-budget circumstances, it is possible to modify regular office printers with customized inks for ePAD's fabrication [81]. For example, in the work of da Costa et al., an office HP Deskjet D4260 printer was used to print a custom-made multiwalled carbon nanotube sodium n-dodecyl sulfate ink (MWCNT-SDS). The formulated ink was transferred to an empty, cleaned cartridge and the three-electrode sensor was printed onto a paper substrate. The SDS acted as a surfactant, optimizing the ink dispensing by the cartridge's nozzle. As the minimum spacing between lines, a 0.7 mm value was achieved. It was observed that the sheet resistance of the CNT network depends on the number of sequential prints, thus requiring over 20 prints to yield electrodes with a lower sheet resistance. Additionally, mineral oil was printed on the electrodes to act as a hydrophobic barrier during sensing. The electrochemical experiments were carried out using a homemade potentiostat by applying a potential step voltammetry technique for dopamine detection. [85]

On the other hand, when greater precision and control of the printing process are required, research-grade printers are available, allowing for precise adjustment of ink droplet volume, ejection speed, and spacing. Moreover, in these sophisticated machines, multiple ink cartridges can be loaded with various functional materials aiming to detect distinct target analytes [81].

Another example of an inkjet-printed electrochemical sensing platform, this time using a piezoelectric Dimatix DMP-2800 materials printer, was presented by Määttänen et al. In this work, the three-electrode configuration was printed onto a multilayer recyclable paper, and consisted of inkjet-printed custom-made gold nanoparticles ink as working (WE) and counter electrodes and a commercially available silver nanoparticle ink, on which an Ag/AgCl layer was deposited electrochemically, resulting in a quasireference electrode. A single nozzle and a drop spacing of 20 and 30 μm were used to print the silver and gold electrodes, respectively. To increase the conductivity of the inkjet-printed gold electrodes, an additional sintering step had to be carried out using an infrared dryer. Lastly, hydrophobic PDMS based ink was printed to define the sensing area of the electrodes. The authors presented different applications for the fabricated sensors through modification steps, including electropolymerization of a polyaniline film on the working electrode surface using cyclic voltammetry, to develop a potentiometric pH sensor; as well as the development of a glucose biosensor with amperometric detection by modifying the electrode surface with PEDOT/glucose oxidase enzyme (PEDOT/GOx) [47].

As seen so far, while inkjet printing automates different ink's deposition processes, it requires several layers of printing as well as a significant drying time between print cycles or sintering steps to create conductive films [68]. Besides, sophisticated inkjet printers are expensive, while the use of standard inkjet printers requires specialized skills to promote appropriate changes in the cartridge materials to better resist the solvents and formulated inks [6].

Recently, the use of automatic cutter/plotter printers in the fabrication process of electrochemical paper-based devices has increased due to the popularization of those printers, which are inexpensive ($150–300) and extremely easy to use [72,91]. These machines can be used with either a thin cutting blade or an engraving tip. They are usually designed for personal use with decorative purposes, scrapbooking, card making, paper crafting, etc. [91,92].

Regarding ePADs manufacture, in most of the reported works, only the cutting feature of the cutter/plotters has been implemented rather than the plotting [93], since by cutting transparency or adhesive films it is possible to prepare inexpensive stencils and masks for the construction of screen-printed electrodes with moderate reproducibility areas [72]. The plotting feature has been used only a few times, most for creating hydrophobic barriers on paper by deposition of permanent marker inks [94]. This feature works by installing pens that can be filled with hydrophobic or conductive inks [79], or even commercial mechanical pencils [50], into the computer-controlled mechanical plotter, which enables both vertical pressure and writing speed to be controlled for transferring an easily modifiable digital pattern onto the paper substrate to fabricate precise and reproducible printed electrodes [95,96].

About this topic, Whitesides's group developed a device for electrochemical enzyme-linked immunosorbent assay (ELISA) using a Silhouette Craft Robo Pro electronic printer to print three-electrode systems in a hydrophobic chromatography paper, produced by silanization with decyl trichlorosilane. The modified paper can be easily printed with conductive inks, and it facilitates the physical adsorption of biomolecules on its surface through hydrophobic interactions. Moreover, to minimize the fouling of the electrodes by ELISA's typical compounds, the device was divided into two zones: an embossed well made from 3D printed molds, on which surface the antigen/antibody immobilization and recognition events occur; and a detection zone, where the electrodes are printed. A graphite ink fabricated in the lab was loaded into a 960-μm rollerball pen, which was installed in the blade-set of the cutter/plotter machine, and the electrodes were printed according to a digital file. After the enzymatic reactions in the embossed well occurred, the device was folded, allowing the electrodes to contact the solution. The electrochemical signal of the species produced in the enzymatic amplification step of ELISA was then recorded using the square-wave voltammetry technique to detect rabbit IgG and malarial histidine-rich protein from *Plasmodium falciparum* in spiked human serum [95].

Another approach of the cutter/plotter machines for ePAD development was presented by Orzari et al. [97]. In this work, a conductive ink made of graphite and automotive varnish mixture was applied on a self-adhesive paper sheet with the aid of a squeegee. After drying, a Silhouette Cameo 3 printer was used to cut the format of the devices, consisting of a three-electrode system. After that, to execute the electrochemical measurements, the cut electrodes adhered to a polyethylene terephthalate (PET) substrate. The proposed device was then applied for the electrochemical determination of dopamine and serotonin by square-wave voltammetry and differential pulse voltammetry techniques. The WE was also modified with GOx and dihexadecyl phosphate film to obtain a biosensor to detect glucose by cyclic voltammetry, demonstrating acceptable analytical efficiency and reproducibility [97].

In conclusion, it was shown that cutting using an electronic craft cutter equipped with standard blades offers much more reproducible fabrication than hand-cutting and is less costly than laser cutting [93]. Furthermore, the plotting feature of those machines, although not yet widely explored in the literature, enables automated and reproducible deposition of reagents and conductive materials on paper, as well as the inkjet printing technique, both being suitable alternatives to ePAD fabrication with terrific possibility for the commercialization of these devices.

5.4.3 Photolithography and sputtering

Photolithography is often applied to produce well-defined and reproducible microfluidic channels [99], and is convenient for prototyping devices and applicable to micropatterns [2]. Henry's group in 2009 work reported a method to pattern hydrophobic channels on ePAD, as discussed in section 5.4.1 [3]. They produced microfluidic channels on the filter paper with cured photoresist exposing to ultraviolet (UV) light using a photomask with the desired decal to percolate the electrolyte solution toward the detection zone [3].

However, this method is expensive because it requires specialized equipment and high-cost materials [23,98]. Other related problems are the yield of rigid hydrophobic walls by the use of photoresists like SU-8 and PMMA that hinder PADs from being folded without damaging channels [23], needing multiple steps during fabrication and usage of organic solvents to remove the unpolymerized photoresist [99]. The problem of hydrophobic-wall inflexibility can be overcome by photo-induced silanization of cellulose hydroxyls using the lithographic method of making foldable microfluidic channels [101].

Employing photolithography, Cao et al. hydrophobized an aldehyde-functionalized Whatman No. 1 baking negative photoresist SU-8 2007 using photomask under the UV irradiation [105]. The electrodes were fabricated by screen printing using carbon and Ag/AgCl inks. In the hydrophilic region between working and counter/reference electrodes, the capture antibodies were chemically adsorbed. The ePAD was built assembling the adhesive backing layer, blotting paper, screen-printed paper electrode, two slices of glass fibers (one used as a sampling zone containing Tween-20/Bovine Serum Albumin (BSA) and the other as a conjugated pad with Au nanoparticles (AuNPs) functionalized with primary antibodies), and antacrylonitrile butadiene styrene layer with injection hole. The ePAD was applied to the detection of the human chorionic gonadotropin biomarker.

Unlike the work previously described, photolithography was used to manufacture the three-electrode configuration on polymers to be integrated into a functional paper to fabricate ePADs. The substrates were covered with a positive photoresist spread by the spinner. Then, using the photomask-aid method and light exposition, the three-electrode configuration was defined. After that, the revelation step was made, putting the photoresist in contact with the developer solution. Thus, Lee et al. fabricated single-wall carbon nanotube (SWCNT) electrodes on PET substrate [106]. The three-electrode configuration prepatterned on PET covered with SWCNT layer was treated with O_2 plasma. After removing the photoresist with ethanol, on the working electrode, AuNPs and polyglutamic acid were codeposited. Stacking Nafion-modified nitrocellulose paper on top of the SWCNT array, a vertical-flow ePAD was developed. Thus the ascorbic acid interference was suppressed from acetaminophen determination by DPV. In another work, Carvalhal et al. produced Au electrodes on polyester film [107]. After patterning the microcell with photoresist on polyester, it was covered with thin layers of Ti and Au by electron-beam evaporation. All the remaining photoresist was lift-off producing Au three-electrode cell. The microcell was coupled with paper-based separation devices for ascorbic and uric acids made by Whatman cellulose chromatographic paper (grade 1Chr).

As photolithography, sputtering is an expensive technique. However, it allows large-scale electrode production even at a lab-scale [102] and can achieve highly reproducible geometrical area and strictly controlled surface film morphology on almost any surface (hydrophilic or hydrophobic) [6,58,100]. The produced thin metal layer does not hinder the microfluidic solution flow, as happened in some cases when other techniques make dense hydrophobic structures [100]. Despite the high-cost drawback, low-cost devices can be fabricated by miniaturizing the sensor and using a cost-effective platform like paper [57,58,103,104].

The sputtering method can be done using a physical mask to shape the electrodes. Adhesive plastic [100] polyester [102], or copper mask [57] with a slit can cover the paper allowing metal atom deposition. Nunez-Bajo et al. [100] made an Au layer on Whatman grade 1 for glucose enzymatic biosensing using an adhesive plastic mask with a 1 mm × 20 mm slit. Previously to Au deposition, hydrophobic barriers of 10 mm × 10 mm were drawn on both sides of the paper with a permanent marker to define the working zone. The Au working electrode was modified with GOx and horseradish peroxidase enzymes, and the ferrocyanide was used as a mediator to yield analytical signals measured by chronoamperometry. The electrolyte solution could be placed on the side where Au was deposited or on the back side of paper. In both cases, the three electrodes can be connected because of paper capillarity. Shiroma et al. fabricated the three electrodes depositing Au film on the end of the hydrophilic channel patterned with a melted wax barrier made on cation exchanger Whatman P81. The 200-nm-thick Au film was designed using a copper mask, and its deposition was controlled by quartz balance. In that way, a mixture of paracetamol and 4-aminophenol was separated by paper chromatography and detected by amperometry [57].

Kokkinos et al. [58] used three different polyester masks to fabricate Sn, Pt, and Ag tracks on the paper platform to develop low-cost ePADs. The authors printed wax barriers on Whatman No. 1, parallel metal layers were made on the paper backside. Following, the wax was melted in an oven to form the microfluidic channel. The method allowed 72 devices to be produced simultaneously in every fabrication run. They were applied to direct and simultaneously Cd(II) and Zn(II) determination by anodic stripping voltammetry (ASV) [58]. Devices prepared following a similar fabrication procedure were applied as biosensing for the subpicomolar detection of biotinylated DNA using streptavidin-conjugated CdSe/ZnS quantum dots (STV-QDs) as labeling [102]. The Sn working electrode was modified with capture DNA strands, which hybridized with target biotinylated DNA. The STV-QDs conjugated with it were dissolved in an HCl solution generating Cd(II) ions, which were quantified by ASV.

Also, the metal layer can be sputtered without the aid of a mask. Parrila et al. [103] and Borràs-Brull et al. [104] sputtered a Pt layer on Whatman grade 5, Núnez-Bajo et al. [100] have been fabricated Au layer on Whatman grade 1 and 3MM Chr paper for glucose determination, and Li et al. [108] sputtered Au layer on art paper for bisphenol-A determination. After that, the papers were cut in the desired format. So, the paper with Pt was cut into strips (20×5 mm) and then sandwiched between two acrylic adhesives, and the top one had a 3-mm-diameter circular window to delimit the Pt-layer exposed area to the electrolyte solution. Afterward, it was coated with glucose oxidase (GOx) enzyme entrapped between Nafion layers, which allowed potentiometric detection of H_2O_2 (one of the GOx reaction by-products) in serum [103] and saliva [104]. Before sputtering Au, Núnez-Bajo et al. [109] printed a circular wax barrier on the paper and melted it. Then, the hydrophobic area was coated with adhesive spray. After that, the paper with sputtered Au was cut with a paper punch and attached to the working electrode of carbon SPE. AuNPs were electrogenerated in a reproducible way by galvanostatic procedure to a nonenzymatic determination of glucose at the low potentials in beverages [109]. Finally, Li et al. [108] cut the art paper with an Au layer into small pieces (8×16 mm); the procedure resulted in 50 electrodes in a single sputtering process. Then, the geometrical area was delimited with adhesive tape with a 5-mm-diameter window and modified with multiwalled carbon nanotubes to determine bisphenol-A in real plastic samples.

5.4.4 Drop-casting

Drop-casting is a manual and straightforward method and needs no sophisticated equipment to deposit conductive materials [56] to fabricate electrodes or nanomaterials/enzymes/membranes [110,111] to modify the working electrodes [6,98]. Concern related to nonuniformity of the electrodic surface causing irreproducibility of the devices should be aware, phenomena of the coffee ring and cracking are frequently observed [98]. The first happens when static droplets of volatile and low-viscous solvents containing nonvolatile compounds evaporate faster at the edge resulting in ring-like solid stain [98,112]. The cracking is caused by reducing large volume during the thermal annealing when both the stabilizer agents are removed and the structure starts getting denser because of the sintering process [112]. Some modifiers such as poly (ethylene oxide), poly(vinyl alcohol), and poly(vinylpyrrolidone) were applied to improve the uniformity [98]. Additionally, when nanomaterials have been employed some considerations should be taken about homogeneity and amount of nanomaterial suspension, and the ratio between nanomaterial and additives [6]. Thus, this section focused on gathering different approaches to fabricate electrodes utilizing drop-casting.

Interestingly, Costa-Rama et al. [113] fabricated ePAD for diclofenac drug detection, the device could preconcentrate the analyte for ultrasensitive determination achieving a LOD of 70 nM. The working electrode was fabricated depositing 2 μL of conductive carbon ink [40% of carbon paste in Dimethylformamide (DMF)] on Whatman grade 1 containing circular printed wax barriers with 4 mm of diameter melted in a heating plate, which was left to dry for 12 h. Thus without the need for adhesives or cables, the paper was inserted among three bent pins of the gold-plated connector header, one of them was in physical contact with the carbon layer served as a connector. The other two, positioned on the other side of the paper, are employed as reference and auxiliary electrodes. The electrolyte solution was placed upon the paper on the opposite side of the ink, closing the electrochemical circuit, while the connector of the working electrode was not in contact with the solution. The method allowed to fabricate a multiplexed platform with eight electrochemical cells that work independently.

In another work, Ding et al. [114] developed an origami paper device for potentiometric biosensing of organophosphate pesticide by inhibiting the butyrylcholinesterase enzyme activity. They applied carbon and Ag/AgCl inks by drop-casting on Whatman No. 1 paper with previously patterned wax barriers to design the chip and confer high mechanical strength when wet. The inks were used as transducers of indicator and reference electrodes both had 3 mm of diameter and were separated by a 1.5 mm gap. Three folding tabs with hydrophilic regions surrounded the test pad. Two of them served as a sample and enzyme-load zone. The third served as a zone to load substrates to initiate enzymatic hydrolysis. Both electrodes were covered with cocktail solutions to generate solid-state membranes for selective detection of butyrylcholine by the indicator electrode and the other one contained KCl salt to confer stability to the reference electrode.

Nanomaterials can be used as well to fabricate conductive tracks. Ameku et al. [55] fabricated a paper-based analytical device (PAD) with dual electrochemical/surface-enhanced Raman scattering detections, reinforcing the possibility to combine different detection techniques on PAD to identify pharmaceutical compounds in seized cocaine samples. They applied 1 μL of AuNP concentrated suspension on the office paper patterned with wax mold melted in a heating plate. After the curing step of AuNP suspension under an infrared lamp, this process was repeated two times to prepare Au-porous conductive tracks for the electrical contact of the working electrode and this process was made again to fabricate the working electrode. The electrode contact was reinforced using silver ink to avoid damage on the Au layer by the potentiostat alligator, and pseudo-reference and auxiliary electrodes were drawn. To control the exposed area to the electrolyte solution and protect electrical contacts, an adhesive was used.

5.4.5 Pencil-drawing

Pencil-drawing is a simple technique to transfer conductive carbon tracks onto cellulosic substrates, such as paper, cardboard, and plastic materials using a pencil or a mechanical pencil [50,51,115,116]. Graphite transfer mechanism occurs due to the exfoliation of its particles, which get stuck on the substrate surface, creating interconnected layers. This interconnection of the graphite layers is necessary for the production of a carbon conductive film. An important characteristic of the substrate to receive the graphite tracks is that it should be rough enough because it facilitates the exfoliation of the material [117].

The commercial graphite lead matrix comprises graphite, clay, and binding agents (wax, polymers, or resins) [118]. The hardness or softness of the pencil/mechanical pencil depends on the proportion between graphite and clay. The more graphite present in the matrix, the softer the pencil is. Hence, these writing materials are classified according to their hardness/softness, defined by a scale ranging from 9H to 8B. The letters H and B that accompany the numbering of the pencils stand for Hardness (low graphite content) and Blackness (high graphite content) [119,120].

In addition to simplicity, this technique offers several advantages for fabricating ePADs, such as pencils are widely available, easy to use, portable, cheap, different patterns can be created, fast fabrication (few minutes) [6,51,115]. One of the disadvantages and a challenge of the pencil-drawing technique is to guarantee high reproducibility in the transfer of the graphite layers and, consequently, in the conductivity of the films. The conditions of exfoliation of graphite in a manual way (hand drawing), such as the pressure applied, depend from person to person. Besides, the variation in commercial graphite leads from a different batch can hinder reproducibility [6]. For this reason, positioning systems, such as machines with pen/pencil holders, controlled by the software, are being used for the fabrication of electrodes by pencil-drawing to ensure greater control of the applied pressure, the number of graphite layers transferred, and the writing speed [50,96].

The electrodes fabricated using the pencil-drawing technique (PDEs) can be applied to detect a variety of analytes, such as ascorbic acid, catechol, salicylic acid, B6 vitamin, dopamine, and uric acid, p-nitrophenol, furosemide [50,115,121–125]; coupled with microfluidic devices [96,116], biosensors [51,126], ion-selective electrodes [127], energy storage devices [128,129], among others. The following will discuss some approaches used for PDEs fabrication.

Orzari et al. [115] applied the pencil-drawing technique to develop a disposable electrochemical device onto a corrugated fiberboard substrate.

They used a cutter printer to fabricate a mask using adhesive paper. This mask was fixed onto the substrate. Next, the three-electrode system was hand-drawn using a 6B pencil, in which several graphite layers were transferred until suitable electrical resistance was obtained. The resistance control of the graphite film was done using a multimeter. After, the corrugated fiberboard was cut to fit in the coupler, and the mask was removed. The electrical contacts were made through three banana-type connectors integrated into the coupler. The working, pseudo-reference, and counter electrodes were made of graphite, and the electrochemical measurements were recorded after 30 s to overcome the water absorption. The volume of solution used in the detection zone of this device was 50 μL. As a proof-of-concept, the device was used to detect catechol in environmental samples. The PDE linear concentration ranged from 0.05 to 1.1 mmol L^{-1}, and the detection limit was 0.01 mmol L^{-1}.

Another interesting approach is to combine the fabrication of the electrodes using the pencil-drawing technique with biological elements to produce paper-based biosensors. In this sense, Li et al. [126] described an origami paper-based device that coupled PDE with GOx for glucose sensing. The device was fabricated using a Whatman chromatographic paper (grade 1) in A4 size. The paper-based sensor layout included both detection and enzyme zones. The areas to be painted by the pencil to create the working, reference, and counter electrodes were previously printed in the detection zone by a laser printer. Following, the electrodes were painted by hand drawing with a 6B pencil in the detection zone. The quality of the formed graphite films was measured through electrical resistance measurements. The authors reported that adequate resistances were obtained to perform the electrochemical measurements. Finally, the enzyme zone on the paper device was functionalized with GOx, and the detection zone with the ferrocenecarboxylic acid using 5 μL of each reagent. After the reagents dried, the device was folded, making contact between the two regions. The glucose electrochemical detection was performed by chronoamperometry. The sensor presented a limit of detection of 0.05 mmol L^{-1} and a linear concentration range of 1–12 mmol L^{-1}. The results of the PDE glucose biosensor were compared with a commercial glucometer in human blood samples.

Graphite has heterogeneous characteristics, and its electrochemical properties and reactivity are influenced by the presence of two central surface regions—the basal plane and edges. These two regions have different structures; the basal plane contains C-C bonds with sp^2 hybridization. The edge region comprises dangling carbon bonds and oxygenated functional groups (hydroxyls, carboxyl, and carbonyls) [130,131]. Therefore the role played by the graphite surface is essential for better comprehension concerning the electron transfer process between this surface and different redox probes. In this context, some works in the literature have proposed activations/treatments of graphite surfaces to improve their electrochemical performance.

Santhiago et al. [51] developed a method to improve the PDE electrochemical properties by oxidizing its surface and reducing it. Initially, the paper-based device was fabricated using an office paper patterned with wax, which was thermally treated on a hot plate at 120°C for 120 s. Before the electrodes were assembled on the patterned paper it was allowed to cool at room temperature for 5 min. The reference and counter electrodes were fabricated on opposite sides on the same patterned paper. A gold film thermally deposited on one side of the waxed paper worked as a counter electrode. On the other side, the reference electrode was fabricated using silver ink. Subsequently, 200 μL of a commercial bleach solution containing NaClO 2.5% (wt.:vol.) was dropped on the silver layer to form an Ag/AgCl electrode. Another waxed paper was used to fabricate the working electrode, drawn with a 4B pencil. The electrical contact of the PDE was made with silver ink.

Next, a single-sided tape was punched (with a punch hole of 1 mm diameter) and fixed onto the substrate. Then, a double-sided tape was used to assemble the three-electrode configuration. After the fabrication steps, the surface of the PDE was treated electrochemically. The PDE was used as the working electrode, the saturated calomel electrode (SCE) as the reference, and a Pt wire as the counter. The following procedure was carried out: (1) oxidation was performed by applying 1.8 V versus SCE for 180 s and (2) the reduction used a cathodic potential window in a range from 0 V to −1.5 V, 1 cycle and with a scan rate of 50 mV s^{-1}. The electrolyte used for both surface treatment processes was 0.1 mol L^{-1} phosphate buffer (pH = 7.0). Following that, the PDE surface was modified with the redox mediator Meldola's Blue, and the paper-based device was applied to detect nicotinamide adenine dinucleotide.

Another approach to treat/activate the graphite's surface was demonstrated by Ataide et al. [125]. In this work, a CO_2 laser machine was used to irradiate the electrodic surface to enhance the PDE electrochemical performance. The PDE was fabricated by hand-drawing (6B commercial pencil) squares (4 × 4 cm) previously printed in an A4 office paper. Next, the graphite layer was photo-thermally treated using a CO_2 laser under the following optimized experimental conditions: laser power of 6.6% (500 mW), Z-distance of 10 mm, and a scan rate of 25 mm s^{-1}. Then, the working and counter electrode patterns were cut using the CO_2 laser (laser power of 10%, Z-distance of 10 mm, and a scan rate of 20 mm s^{-1}). Subsequently, an office paper was patterned using a wax printer, creating hydrophobic barriers onto its surface to avoid leakage. The wax barriers were melted using a thermal press at 120°C for 30 s. After that, a reference electrode was painted using commercial silver ink. The working and counter electrodes were attached to the paper platform with double-sided tape as a final step. The CO_2 laser-treated PDE was applied to detect furosemide in synthetic urine samples.

The pencil leads composition can be tuned by making them in the laboratory, which was the strategy developed by Dossi et al. [132]. The authors assembled a pencil lead doped with desired modifiers. They used a stainless steel extruder in a syringe shape, which consisted of a cylindrical tube with a moving plunger inside it—the cylindrical tube outlet with a circular orifice of 3 mm diameter. The pencil lead composition was graphite powder (80% wt./wt.), sodium bentonite (8% wt./wt.), and potassium silicate (12% wt./wt.). Each one of these worked as a conductive material, binding agent, and hardening agent, respectively. This mixture was loaded inside the tube and pressed by the plunger. Next, the pencil lead was extruded by pushing the plunger and cut into the desired size (approximately 30 mm). The resulting pencil leads were left to dry at room temperature for 24 h before use. Finally, they were placed in the lead holders. The doped pencil lead fabrication involved the same steps, only adding to the graphite mixture (aforementioned) 1% decamethylferrocene or 8% cobalt(II) phthalocyanine. A filter paper was patterned with wax-based ink to delimit the PDE detection zone in a subsequent step. The ink was melted by heating at 120°C for 10 min. Next, the paper sheet was cut, and the backside of the patterned paper was thermally laminated using a PET layer to avoid leakage while the analysis was performed. Then, a PET mask was placed onto the patterned paper to draw the electrodes. The working, reference, and counter electrodes were drawn by applying the lab-made pencil leads, passing them four times with a width graphite track of 1 mm and a distance between one another of 1 mm. The electrochemical behavior of the undoped pencil leads with different compositions was evaluated by cyclic voltammogram measurements recorded using hexacyanoferrate(II). The electrochemical response of the doped pencil leads, decamethylferrocene, and cobalt(II) phthalocyanine were studied towards cysteine and hydrogen peroxide detection, respectively.

As previously discussed, hand drawing has the disadvantage of providing low reproducibility in paper-based device fabrication. Therefore Dossi et al. [50] developed a simple method of transferring graphite films to paper substrates by coupling a mechanical pencil to the positioning system of a commercial plotter/cutter. The fabrication of the PDEs involved the following steps: (1) the filter paper (20 $\times$ 20 cm) was placed onto a plastic platform that was moved by an X-Y device and patterned with multiple ring shapes using a marker filled with a hydrophobic ink (wax-based ink) coupled to the plotter/cutter by a pen holder; (2) the hydrophobic ink was allowed to cool at room temperature; (3) a 4B commercial mechanical pencil replaced the marker in the pen holder, and a three-electrode system (working, reference, and counter) was drawn under the optimized conditions preestablished in the software that regulates the parameters, such as vertical pressure ($\sim$ 1.9 N), writing speed (3 cm s^{-1}), and the number of graphite layers (4 deposition steps with double passage); and (4) a cutting blade was inserted in the holder and the PDEs were cut into the desired size. The electrodes were electrically connected using miniature crocodile connectors. The voltammetric experiments were recorded with the PDE cell (detection zone) placed onto a closed box with water to decrease the electrolyte evaporation. The PDE was applied in the analysis of vitamin B_6 in food-supplement tablets.

The pencil-drawing is a promising method for electrochemical paper-based sensors concerning its simplicity and the use of low-cost materials and available around the world. Especially for developing countries, this technique may be suitable for the production of electrochemical sensors. However, one of the challenges to overcome is the fabrication reproducibility of these devices. In this sense, machines with commercial or homemade/lab-made positioning systems controlled by software, in which it is possible to manage the writing parameters, can be applied.

5.4.6 Graphite leads and wire materials as electrodes on paper

Besides the pencil drawing, pencil leads were also used to fabricate conductive tracks on paper surfaces. In 2013, Santhiago et al. [133] reported for the first time the use of graphite pencil, as electrodes, on paper platforms. In this work, they fabricated an electrochemical paper-based device for glucose biosensing using a two-electrodes system. Wax-printing was used in the chromatographic paper as a standard procedure to create a hydrophobic barrier to delimitate reaction microzones. Three microzones were fabricated, one for a filtration step, one for the enzymatic reaction, and then for the final electrochemical detection. The graphite electrodes were fabricated using a glass capillary tube and epoxy resin to attach the commercial cylindrical graphite leads within the tube to be used as working electrodes. This electrode was positioned vertically on the electrochemical zone for glucose detection, and the reference electrode was fabricated using silver ink. A low-cost electrode material was used to fabricate the device in a paper-based platform. The analyses were performed a short time and without the necessity of electrode modification.

Another way to use the graphite leads as electrodes on paper platforms is to attach them parallel to paper using tape, to produce conductive working and counter electrodes tracks on paper [67]. The advantages to using graphite pencil leads as electrodes on paper-based platforms are the low cost, simplicity, and the facility to find worldwide in different radius and containing different graphite proportions. However, the drawbacks are the lack of flexibility, besides the compromised reproducibility due to the necessity to apply mechanical polishing or different treatments to reuse the electrode between the electrochemical measurements, which could influence the electrode active surface area and consequently on the reproducibility of the measurements. Although this last drawback cited could be circumvented using disposable graphite leads [118].

Another method applied to construct conductive tracks on paper, using the same procedure, is to couple a conductive wire on the platform. The technique was reported in 2014 when Fosdick and coauthors presented Au wires and carbon fibers on the paper platform to provide the working electrodes conductive tracks [60]. This work used wax printing to fabricate paper channels with a hydrophobic barrier on the chromatography paper. The wire was parallel attached using a tape, and the device was finally origami folded. Electrochemical measurements were recorded, and the results obtained were compatible with all the simulations performed, showing the tremendous electrochemical performance of the fabricated devices.

This method of fabrication presents advantages when compared with screen or stencil-printed ones, for example. Incorporating the wires on the paper is very simple, and they can easily be modified before being integrated on the platform. This modification can be performed using a considerable wire size, which could be cut and then used to fabricate a significant amount of these origami paper-based devices at once. Moreover, the wire can also be cleaned using different treatments, and the material can be reused to fabricate other ePAD [60,68]. Although only Au and carbon were used in this work, different metal wires can also be applied. Adkins and Henry studied the use of platinum, copper, silver, platinum with tungsten, and platinum with iridium microwires as an alternative way to fabricate ePADs [61]. The devices were fabricated using filter paper, and the electrochemical measurements were carried out using a redox probe to confront the theoretical predicted electrochemical behavior. These ePADs presented excellent electrochemical performance compared with carbon screen-printed electrodes, besides the microwires being easy to incorporate on paper platforms. Furthermore, in this work, after the electrochemical characterization, as proof-of-concept, the Cu microwire was modified before being attached to the platform, and nonenzymatic devices were obtained for glucose, fructose, and sucrose detection in beverages samples.

Due to the advantages of this fabrication technique, in another work, Au wire was modified to be integrated into a paper platform to obtain a biosensor responsible for trace detection of virus particles [134]. Additionally, graphite leads could be modified with gold to work as an electrochemical sensor on paper for chemometric discrimination of drugs [135] together with colorimetric detection. Moreover, this kind of system using microwires was also used to apply flow injection analysis on paper [136]. However, the disadvantage of this presented method is the decrease of the active surface area when the wire is attached on the paper platform [61], besides the use of noble metals being, in general, expensive [6,68].

5.4.7 Filtration

The filtration of conductive materials directly on the paper platform to fabricate electrochemical devices was also explored as an alternative method to develop the ePADs. This method consists of the retention of different conductive materials directly on the device substrate, such as on a filter paper platform, to create the conductive tracks and finally obtain the ePADs. The advantages of this technique are simplicity, low cost, and large-scale production [137,138]. In 2011, Zhe Weng and collaborators demonstrated the development of a graphene-cellulose paper using vacuum filtration to retain the graphene dispersed in N-methyl pyrrolidone directly on the filter paper substrate [53]. The device was applied as flexible supercapacitors. Posteriorly, the technique was also applied to obtain electrochemical paper-based devices using materials, such as carbon nanotubes [138], graphene oxide [139], nanoparticles [54], or a mix of reduced graphene oxide and metal nanoparticles [140,141] to construct different devices.

Wang and coworkers developed an ePAD to measure adenosine triphosphate (ATP) [139]. This work used a vacuum filtration system to fabricate a conductive film on mixed cellulose ester (MCE) filter paper. Reduced graphene oxide dispersed on deionized water was used as a conductive material, and after the filtration, the film was vacuum-dried. Then a template was used to cut the paper-based working electrodes (graphene/MCE), which were connected with copper wire before the use. An improvement in ATP electrochemical behavior was observed for graphene/MCE electrodes compared with commercial glassy carbon electrodes or commercial glassy carbon electrodes functionalized with reduced graphene oxide. Moreover, the paper-based system presents disposability, besides being simple and easy to fabricate in a scalable way.

In another work, Bhargav Guntupalli et al. [54] fabricated a paper-based porous gold film for electrochemical biosensing. In this case, a solution of metallic single-walled carbon nanotubes (M-SWCNTs) was added to MCE filter paper in a vacuum filtration system. The film obtained was vacuum-dried, and a citrate-capped AuNP solution was used on the top of the M-SWCNTs layer to create a hybrid gold film. After, the electrodes were cut using a template and prepared to be applied using copper wire. The ePADs presented results similar to the observed in commercial gold electrodes. The paper-based porous gold electrodes were successfully applied for serotonin and dopamine detection as proof of concept. The demonstrated method represented a fast way to fabricate the devices (all the process takes 20 min), which requires a simple vacuum filtration system.

Another approach using the vacuum filtration method was reported by Lei et al. [142]. The authors used a metal mask during the vacuum filtration as a template for the electrodes to define the geometry of the sensor directly on the filter paper. CNT solution was used to construct the conductive tracks. The CNTs were functionalized with biotin, and the device was used to quantify avidin by binding interaction using the direct current measurement as a fast immunoassay analysis.

Like other techniques, vacuum filtration also has disadvantages. In this method, the material penetrates the paper matrix, which is made of cellulose fibers. For this reason, it is difficult to control the uniformity of the formed film. Furthermore, the patterning of the layout of the electrodes directly on the filter paper is still challenging.

5.4.8 Pyrolysis process

The thermal process has also emerged as an alternative method to fabricate carbon conductive material with different structures depending on the precursor substrate [143,144]. The technique consists of pyrolyzing organic matrices, which leads to the carbonization of the material and, consequently, the obtention of carbon-based materials.

In terms of ePAD development, Giuliani and collaborators used the thermal process to develop carbon-based electrodes using paper as a substrate [62]. In this work, a furnace was used to pyrolyze papers with different properties. The pyrolysis was realized under a temperature of 1000 °C using an Ar-controlled atmosphere with H_2 to provide a reducing environment. Then, a commercial CO_2 laser machine was used to cut the electrodes into the desired layout to pattern them. A poly(methyl methacrylate) (PMMA) was used as a platform under carbonized paper-based material and parafilm was employed to delimitate the reaction area. To finalize, the electrical contact was painted with silver ink. The authors observed the obtention of different materials according to the distinct precursor papers. Although the characterization of the final carbon-based electrodes showed most of the domains with disorder carbon, the material was successfully applied to uric acid detection. In 2016, the same research group demonstrated the possibility of incorporates metallic nanoparticles in the pyrolyzed carbon material in a single-step approach [64]. The method is based on the pyrolysis (1000 °C under a mixture of 95% Ar/5% H_2 for 1 h) of paper strips modified with a saturated solution of $CuSO_4$ and yields to the formation of abundant copper nanoparticles on the surface of carbonized cellulose fibers.

The pyrolysis method showed an alternative technique to fabricate carbon electrodes directly from a low-cost and abundant platform. However, the disadvantages are the need for controlled conditions and multiple steps that the paper pyrolysis requires, besides the demanded time to finalize the process.

5.4.9 Laser-scribing

Another thermal-related method is the laser-scribing technique, which also consists of the carbonization of organic platforms to generate conductive carbon materials [145]. In this case, the carbonization is localized and controlled by laser radiation, with wavelength varying from UV to infrared [146,147]. Initially, the laser accidentally reached the polymeric material during laser irradiation used to reduce graphene oxide under a polyimide platform, and a carbonized track was observed. Studies revealed this material as a conductive carbon-based material with domains similar to graphene, containing organized carbon structures [148]. First, the material was electrochemically applied in the development of supercapacitors. After, electrochemical sensors were also fabricated using the laser-scribing material derived from polyimide platforms. Other platforms were also employed, such as other polymers [145,149], clothes, food, cellulosic materials [150,151], phenolic resin [152], and phenolic paper [153]. However, the carbonization of paper platforms using this technique was still a challenge due to the fragility of the platform.

In 2017, for the first time, an electrochemical paper-based device was fabricated using paperboard and an one-step laser-scribing technique [65]. de Araujo et al. used a CO_2 laser machine coupled with software to provide the carbonization of the three electrodes system on the paperboard with optimized parameters. The laser machine was also used to cut each ePAD, and a nonconductive glue was applied to delimitate the electrochemical reaction area. Finally, a silver ink was added to the electrical contacts and reference electrode, to provide mechanical stability to contact the devices with the electrochemical workstation and ensure the obtention of a stable pseudo-reference electrode. The developed devices presented carbon-based electrodes with graphene-like domains and excellent electrochemical behavior compared with conventional carbon electrodes. As a proof-of-concept, the material was successfully applied to electrochemical detection in different areas.

The laser-scribing method provided a simple and easy technique to fabricate conductive tracks directly on paper. It could be applied to a scalable production, besides being a maskless and reagentless method and demanding low-time consumption during the fabrication [65]. It was also applied in the development of laser-scribed electrochemical paper-based devices (LS-ePADs) for sulfite analysis [154] and lead analysis using square-wave anodic stripping voltammetry [155]. Recently, other work was carried out to investigate some unique aspects of the electrochemical behavior of the LS-ePADs [156]. Another work reported using carbon fiber paper as a precursor material to fabricate laser-scribed electrodes using a laser machine with a wavelength of 450 nm [66]. The device was decorated with nickel nanoparticles and demonstrated ultrasensitive glucose detection.

However, it is a recent technique and the lack of studies in the influence of the parameters applied during the laser-scribing on the final material obtained is still a challenge to overcome to fabricate devices with better characteristics and high reproducibility. Moreover, due to the laser machine's power, the method is not compatible with most paper types, requiring higher grammage paper platforms to support the photo-thermal process without carbonizing the entire substrate.

5.5 Conclusions and perspectives

In this chapter, we presented an overview of the main techniques and approaches commonly used to manufacture electrochemical paper-based devices and demonstrate the versatility of these portable (bio) sensors for a wide range of applications. In the last decade, we have noticed a boom in new creative manufacturing approaches aimed at reducing costs, improving analytical parameters (sensitivity and selectivity), or providing new functionality or process integration. It is worth noting that given many works in the area, inevitably, several excellent and innovative contributions using ePADs were left out of this chapter. Although the remarkable advantages in associate paper-based devices with electrochemistry techniques, following the literature reported in this chapter on existing paper-based (bio)sensors, it is clear that extensive developments should still be reached. In our vision, further research needs to be considered to overcome some issues associated with:

1. Enhance the reproducibility of the fabrication methods and the incorporation of functional materials on the paper surface;
2. Improve the strategies for controlling sample evaporation and retention, which can cause inefficiencies in sample delivery and compromised analytical results;
3. Perform multiple steps and premixing of samples/solutions, which may hinder point of care applications;
4. Whole-genome/transcriptome amplification, which may suffer from bias and nonspecific products;
5. Evaluate the robustness of the analytical results provided by ePADs in biological or complex matrices, comparing the results with standard methods;
6. Develop handheld readout devices, printing the electrical potentiostat circuit on paper to fabricate an entirely paper-based sensing tool easy to dispose of, that is, incineration.
7. Development of user-friendly interface that provides the analytical result without the user's intervention, like the commercial glucometers, which facilitates the use by the nonskilled person and accelerates the commercialization of these devices.

Acknowledgments

Financial support for this project has been provided by Brazilian agencies FAPESP (Grant Numbers: 2019/15065−7, 2018/08782−1, 2018/14462−0 and 2018/16250−0), CAPES, CNPq (grant numbers: 302839/2020−8, and 438828/2018−6), and INCTBio (573672/2008−3).

References

[1] Portable Analytical Instruments, Anal. Chem. 63(11) (1991) 641A-644A.

[2] A.W. Martinez, S.T. Phillips, M.J. Butte, G.M. Whitesides, Patterned paper as a platform for inexpensive, low-volume, portable bioassays, Angew. Chem. Int. (Ed.) Engl. 46 (8) (2007) 1318−1320.

[3] W. Dungchai, O. Chailapakul, C.S. Henry, Electrochemical detection for paper-based microfluidics, Anal. Chem. 81 (14) (2009) 5821−5826.

[4] A. Antonacci, V. Scognamiglio, V. Mazzaracchio, V. Caratelli, L. Fiore, D. Moscone, et al., Paper-based electrochemical devices for the pharmaceutical field: state of the art and perspectives, Front. Bioeng. Biotechnol. 8 (2020) 339.

[5] F. Arduini, S. Cinti, V. Scognamiglio, D. Moscone, Paper-based electrochemical devices in biomedical field, in: I. Palchetti, P.-D. Hansen, D. Barceló (Eds.), Past, Present and Future Challenges of Biosensors and Bioanalytical Tools in Analytical Chemistry: A Tribute to Professor Marco Mascini, Elsevier, 2017, pp. 385–413.

[6] V.N. Ataide, L.F. Mendes, L.I.L.M. Gama, W.R. de Araujo, T.R.L.C. Paixão, Electrochemical paper-based analytical devices: ten years of development, Anal. Methods 12 (8) (2020) 1030–1054.

[7] W. Wonsawat, S. Limvongjaroen, S. Supromma, W. Panphut, N. Ruecha, N. Ratnarathorn, et al., A paper-based conductive immunosensor for the determination of *Salmonella typhimurium*, Analyst 145 (13) (2020) 4637–4645.

[8] D. Harvey, Analytical chemistry 2.0—an open-access digital textbook, Anal. Bioanal. Chem. 399 (1) (2011) 149–152.

[9] W.J. Lan, X.U. Zou, M.M. Hamedi, J. Hu, C. Parolo, E.J. Maxwell, et al., Paper-based potentiometric ion sensing, Anal. Chem. 86 (19) (2014) 9548–9553.

[10] E.B. Bahadir, M.K. Sezginturk, A review on impedimetric biosensors, Artif. Cell Nanomed. Biotechnol. 44 (1) (2016) 248–262.

[11] W.J. Paschoalino, S. Kogikoski, J.T.C. Barragan, J.F. Giarola, L. Cantelli, T.M. Rabelo, et al., Emerging considerations for the future development of electrochemical paper-based analytical devices, ChemElectroChem 6 (1) (2019) 10–30.

[12] T. Gebretsadik, T. Belayneh, S. Gebremichael, W. Linert, M. Thomas, T. Berhanu, Recent advances in and potential utilities of paper-based electrochemical sensors: beyond qualitative analysis, Analyst 144 (8) (2019) 2467–2479.

[13] E. Noviana, C.P. McCord, K.M. Clark, I. Jang, C.S. Henry, Electrochemical paper-based devices: sensing approaches and progress toward practical applications, Lab. Chip 20 (1) (2020) 9–34.

[14] H. Yagoda, Applications of confined spot tests in analytical chemistry: preliminary paper, Ind. & Eng. Chem. Anal. (Ed.) 9 (2) (2002) 79–82.

[15] A.W. Martinez, S.T. Phillips, M.J. Butte, G.M. Whitesides, Patterned paper as a platform for inexpensive, low-volume, portable bioassays, Angew. Chem. 119 (8) (2007) 1340–1342.

[16] K. Yamada, T.G. Henares, K. Suzuki, D. Citterio, Paper-based inkjet-printed microfluidic analytical devices, Angew. Chem. Int. (Ed.) Engl. 54 (18) (2015) 5294–5310.

[17] X. Jiang, Z.H. Fan, Fabrication and operation of paper-based analytical devices, Annu. Rev. Anal. Chem. (Palo Alto Calif.) 9 (1) (2016) 203–222.

[18] E. Carrilho, A.W. Martinez, G.M. Whitesides, Understanding wax printing: a simple micropatterning process for paper-based microfluidics, Anal. Chem. 81 (16) (2009) 7091–7095.

[19] K. Abe, K. Suzuki, D. Citterio, Inkjet-printed microfluidic multianalyte chemical sensing paper, Anal. Chem. 80 (18) (2008) 6928–6934.

[20] K. Maejima, S. Tomikawa, K. Suzuki, D. Citterio, Inkjet printing: an integrated and green chemical approach to microfluidic paper-based analytical devices, RSC Adv. 3 (24) (2013) 9258–9263.

[21] T. Songjaroen, W. Dungchai, O. Chailapakul, W. Laiwattanapaisal, Novel, simple and low-cost alternative method for fabrication of paper-based microfluidics by wax dipping, Talanta 85 (5) (2011) 2587–2593.

[22] W. Dungchai, O. Chailapakul, C.S. Henry, A low-cost, simple, and rapid fabrication method for paper-based microfluidics using wax screen-printing, Analyst 136 (1) (2011) 77–82.
[23] D.A. Bruzewicz, M. Reches, G.M. Whitesides, Low-cost printing of poly(dimethylsiloxane) barriers to define microchannels in paper, Anal. Chem. 80 (9) (2008) 3387–3392.
[24] Y. Sameenoi, P.N. Nongkai, S. Nouanthavong, C.S. Henry, D. Nacapricha, One-step polymer screen-printing for microfluidic paper-based analytical device (muPAD) fabrication, Analyst 139 (24) (2014) 6580–6588.
[25] J. Olkkonen, K. Lehtinen, T. Erho, Flexographically printed fluidic structures in paper, Anal. Chem. 82 (24) (2010) 10246–10250.
[26] T. Nurak, N. Praphairaksit, O. Chailapakul, Fabrication of paper-based devices by lacquer spraying method for the determination of nickel (II) ion in waste water, Talanta 114 (2013) 291–296.
[27] T.M.G. Cardoso, F.R. de Souza, P.T. Garcia, D. Rabelo, C.S. Henry, W.K.T. Coltro, Versatile fabrication of paper-based microfluidic devices with high chemical resistance using scholar glue and magnetic masks, Anal. Chim. Acta 974 (2017) 63–68.
[28] P. de Tarso Garcia, T.M. Garcia Cardoso, C.D. Garcia, E. Carrilho, W.K. Tomazelli Coltro, A handheld stamping process to fabricate microfluidic paper-based analytical devices with chemically modified surface for clinical assays, RSC Adv. 4 (71) (2014) 37637–37644.
[29] K.L. Dornelas, N. Dossi, E. Piccin, A simple method for patterning poly(dimethylsiloxane) barriers in paper using contact-printing with low-cost rubber stamps, Anal. Chim. Acta 858 (2015) 82–90.
[30] N.N. Hamidon, Y. Hong, G.I. Salentijn, E. Verpoorte, Water-based alkyl ketene dimer ink for user-friendly patterning in paper microfluidics, Anal. Chim. Acta 1000 (2018) 180–190.
[31] Q. He, C. Ma, X. Hu, H. Chen, Method for fabrication of paper-based microfluidic devices by alkylsilane self-assembling and UV/O3-patterning, Anal. Chem. 85 (3) (2013) 1327–1331.
[32] X. Li, J. Tian, T. Nguyen, W. Shen, Paper-based microfluidic devices by plasma treatment, Anal. Chem. 80 (23) (2008) 9131–9134.
[33] Y. Zhang, J. Liu, H. Wang, Y. Fan, Laser-induced selective wax reflow for paper-based microfluidics, RSC Adv. 9 (20) (2019) 11460–11464.
[34] N.K. Mani, A. Prabhu, S.K. Biswas, S. Chakraborty, Fabricating paper based devices using correction pens, Sci. Rep. 9 (1) (2019) 1752.
[35] S. Oyola-Reynoso, A.P. Heim, J. Halbertsma-Black, C. Zhao, I.D. Tevis, S. Cinar, et al., Draw your assay: fabrication of low-cost paper-based diagnostic and multi-well test zones by drawing on a paper, Talanta 144 (2015) 289–293.
[36] M.A. Ostad, A. Hajinia, T. Heidari, A novel direct and cost effective method for fabricating paper-based microfluidic device by commercial eye pencil and its application for determining simultaneous calcium and magnesium, Microchemical J. 133 (2017) 545–550.
[37] C. Renault, X. Li, S.E. Fosdick, R.M. Crooks, Hollow-channel paper analytical devices, Anal. Chem. 85 (16) (2013) 7976–7979.
[38] E.M. Fenton, M.R. Mascarenas, G.P. Lopez, S.S. Sibbett, Multiplex lateral-flow test strips fabricated by two-dimensional shaping, ACS Appl. Mater. Interfaces 1 (1) (2009) 124–129.

[39] J. Nie, Y. Liang, Y. Zhang, S. Le, D. Li, S. Zhang, One-step patterning of hollow microstructures in paper by laser cutting to create microfluidic analytical devices, Analyst 138 (2) (2013) 671–676.

[40] C.L. Cassano, Z.H. Fan, Laminated paper-based analytical devices (LPAD): fabrication, characterization, and assays, Microfluid Nanofluid 15 (2) (2013) 173–181.

[41] A.K. Yetisen, M.S. Akram, C.R. Lowe, Paper-based microfluidic point-of-care diagnostic devices, Lab. Chip 13 (12) (2013) 2210–2251.

[42] E.L. Fava, T.A. Silva, T.M.D. Prado, F.C. Moraes, R.C. Faria, O. Fatibello-Filho, Electrochemical paper-based microfluidic device for high throughput multiplexed analysis, Talanta 203 (2019) 280–286.

[43] F. Arduini, S. Cinti, V. Caratelli, L. Amendola, G. Palleschi, D. Moscone, Origami multiple paper-based electrochemical biosensors for pesticide detection, Biosens. Bioelectron. 126 (2019) 346–354.

[44] C.C. Wang, J.W. Hennek, A. Ainla, A.A. Kumar, W.J. Lan, J. Im, et al., A Paper-based "pop-up" electrochemical device for analysis of beta-hydroxybutyrate, Anal. Chem. 88 (12) (2016) 6326–6333.

[45] K. Scida, J.C. Cunningham, C. Renault, I. Richards, R.M. Crooks, Simple, sensitive, and quantitative electrochemical detection method for paper analytical devices, Anal. Chem. 86 (13) (2014) 6501–6507.

[46] J.A. Adkins, K. Boehle, C. Friend, B. Chamberlain, B. Bisha, C.S. Henry, Colorimetric and electrochemical bacteria detection using printed paper- and transparency-based analytic devices, Anal. Chem. 89 (6) (2017) 3613–3621.

[47] A. Määttänen, U. Vanamo, P. Ihalainen, P. Pulkkinen, H. Tenhu, J. Bobacka, et al., A low-cost paper-based inkjet-printed platform for electrochemical analyses, Sens. Actuators B: Chem. 177 (2013) 153–162.

[48] E. Bihar, S. Wustoni, A.M. Pappa, K.N. Salama, D. Baran, S. Inal, A fully inkjet-printed disposable glucose sensor on paper, npj Flex. Electron. 2 (1) (2018).

[49] N. Dossi, R. Toniolo, A. Pizzariello, F. Impellizzieri, E. Piccin, G. Bontempelli, Pencil-drawn paper supported electrodes as simple electrochemical detectors for paper-based fluidic devices, Electrophoresis 34 (14) (2013) 2085–2091.

[50] N. Dossi, S. Petrazzi, R. Toniolo, F. Tubaro, F. Terzi, E. Piccin, et al., Digitally controlled procedure for assembling fully drawn paper-based electroanalytical platforms, Anal. Chem. 89 (19) (2017) 10454–10460.

[51] M. Santhiago, M. Strauss, M.P. Pereira, A.S. Chagas, C.C. Bufon, Direct drawing method of graphite onto paper for high-performance flexible electrochemical sensors, ACS Appl. Mater. Interfaces 9 (13) (2017) 11959–11966.

[52] C.-Y. Lee, K.F. Lei, S.-W. Tsai, N.-M. Tsang, Development of graphene-based sensors on paper substrate for the measurement of pH value of analyte, BioChip J. 10 (3) (2016) 182–188.

[53] Z. Weng, Y. Su, D.-W. Wang, F. Li, J. Du, H.-M. Cheng, Graphene-cellulose paper flexible supercapacitors, advanced energy, Materials 1 (5) (2011) 917–922.

[54] B. Guntupalli, P. Liang, J.H. Lee, Y. Yang, H. Yu, J. Canoura, et al., Ambient filtration method to rapidly prepare highly conductive, paper-based porous gold films for electrochemical biosensing, ACS Appl. Mater. Interfaces 7 (49) (2015) 27049–27058.

[55] W.A. Ameku, W.R. de Araujo, C.J. Rangel, R.A. Ando, T.R.L.C. Paixão, Gold nanoparticle paper-based dual-detection device for forensics applications, ACS Appl. Nano Mater. 2 (9) (2019) 5460–5468.

[56] E. Nunez-Bajo, M.C. Blanco-Lopez, A. Costa-Garcia, M.T. Fernandez-Abedul, Electrogeneration of gold nanoparticles on porous-carbon paper-based electrodes and application to inorganic arsenic analysis in white wines by chronoamperometric stripping, Anal. Chem. 89 (12) (2017) 6415–6423.
[57] L.Y. Shiroma, M. Santhiago, A.L. Gobbi, L.T. Kubota, Separation and electrochemical detection of paracetamol and 4-aminophenol in a paper-based microfluidic device, Anal. Chim. Acta 725 (2012) 44–50.
[58] C. Kokkinos, A. Economou, D. Giokas, Paper-based device with a sputtered tin-film electrode for the voltammetric determination of Cd(II) and Zn(II), Sens. Actuators B: Chem. 260 (2018) 223–226.
[59] X.-M. Bi, H.-R. Wang, L.-Q. Ge, D.-M. Zhou, J.-Z. Xu, H.-Y. Gu, et al., Gold-coated nanostructured carbon tape for rapid electrochemical detection of cadmium in rice with in situ electrodeposition of bismuth in paper-based analytical devices, Sens. Actuators B: Chem. 260 (2018) 475–479.
[60] S.E. Fosdick, M.J. Anderson, C. Renault, P.R. DeGregory, J.A. Loussaert, R.M. Crooks, Wire, mesh, and fiber electrodes for paper-based electroanalytical devices, Anal. Chem. 86 (7) (2014) 3659–3666.
[61] J.A. Adkins, C.S. Henry, Electrochemical detection in paper-based analytical devices using microwire electrodes, Anal. Chim. Acta 891 (2015) 247–254.
[62] J.G. Giuliani, T.E. Benavidez, G.M. Duran, E. Vinogradova, A. Rios, C.D. Garcia, Development and characterization of carbon based electrodes from pyrolyzed paper for biosensing applications, J. Electroanal. Chem. (Lausanne) 765 (2016) 8–15.
[63] L.A.J. Silva, W.P. da Silva, J.G. Giuliani, S.C. Canobre, C.D. Garcia, R.A.A. Munoz, et al., Use of pyrolyzed paper as disposable substrates for voltammetric determination of trace metals, Talanta 165 (2017) 33–38.
[64] G.M. Duran, T.E. Benavidez, J.G. Giuliani, A. Rios, C.D. Garcia, Synthesis of CuNP-modified carbon electrodes obtained by pyrolysis of paper, Sens. Actuators B Chem. 227 (2016) 626–633.
[65] W.R. de Araujo, C.M.R. Frasson, W.A. Ameku, J.R. Silva, L. Angnes, T. Paixao, Single-step reagentless laser scribing fabrication of electrochemical paper-based analytical devices, Angew. Chem. Int. (Ed.) Engl. 56 (47) (2017) 15113–15117.
[66] L. Hou, S. Bi, B. Lan, H. Zhao, L. Zhu, Y. Xu, et al., A novel and ultrasensitive nonenzymatic glucose sensor based on pulsed laser scribed carbon paper decorated with nanoporous nickel network, Anal. Chim. Acta 1082 (2019) 165–175.
[67] R.F. e Silva, T.R. Longo Cesar Paixão, M. Der Torossian Torres, W.R. de Araujo, Simple and inexpensive electrochemical paper-based analytical device for sensitive detection of Pseudomonas aeruginosa, Sens. Actuators B: Chem. 308 (2020).
[68] J. Adkins, K. Boehle, C. Henry, Electrochemical paper-based microfluidic devices, Electrophoresis 36 (16) (2015) 1811–1824.
[69] D. Martin-Yerga, I. Alvarez-Martos, M.C. Blanco-Lopez, C.S. Henry, M.T. Fernandez-Abedul, Point-of-need simultaneous electrochemical detection of lead and cadmium using low-cost stencil-printed transparency electrodes, Anal. Chim. Acta 981 (2017) 24–33.
[70] R.H. Müller, D.L. Clegg, Automatic paper chromatography, Anal. Chem. 21 (9) (2002) 1123–1125.
[71] J. Mettakoonpitak, K. Boehle, S. Nantaphol, P. Teengam, J.A. Adkins, M. Srisa-Art, et al., Electrochemistry on paper-based analytical devices: a review, Electroanalysis 28 (7) (2016) 1420–1436.

[72] A.S. Afonso, C.V. Uliana, D.H. Martucci, R.C. Faria, Simple and rapid fabrication of disposable carbon-based electrochemical cells using an electronic craft cutter for sensor and biosensor applications, Talanta 146 (2016) 381–387.
[73] Z. Chu, J. Peng, W. Jin, Advanced nanomaterial inks for screen-printed chemical sensors, Sens. Actuators B: Chem. 243 (2017) 919–926.
[74] M. Pohanka, Screen printed electrodes in biosensors and bioassays, A Review, Int. J. Electrochem. Sci. (2020) 11024–11035.
[75] S. Devarakonda, R. Singh, J. Bhardwaj, J. Jang, Cost-effective and handmade paper-based immunosensing device for electrochemical detection of influenza virus, Sens. (Basel) 17 (11) (2017).
[76] S. Nantaphol, A.A. Kava, R.B. Channon, T. Kondo, W. Siangproh, O. Chailapakul, et al., Janus electrochemistry: simultaneous electrochemical detection at multiple working conditions in a paper-based analytical device, Anal. Chim. Acta 1056 (2019) 88–95.
[77] P.B. Deroco, O. Fatibello-Filho, F. Arduini, D. Moscone, Electrochemical determination of capsaicin in pepper samples using sustainable paper-based screen-printed bulk modified with carbon black, Electrochim. Acta 354 (2020).
[78] E. Costa-Rama, M.T. Fernandez-Abedul, Paper-based screen-printed electrodes: a new generation of low-cost electroanalytical platforms, Biosens. (Basel) 11 (2) (2021).
[79] V. Soum, H. Cheong, K. Kim, Y. Kim, M. Chuong, S.R. Ryu, et al., Programmable contact printing using ballpoint pens with a digital plotter for patterning electrodes on paper, ACS Omega 3 (12) (2018) 16866–16873.
[80] G. Rosati, M. Ravarotto, M. Sanavia, M. Scaramuzza, A. De Toni, A. Paccagnella, Inkjet sensors produced by consumer printers with smartphone impedance readout, Sens. Bio-Sensing Res. 26 (2019).
[81] R. Tortorich, H. Shamkhalichenar, J.-W. Choi, Inkjet-printed and paper-based electrochemical sensors, Appl. Sci. 8 (2) (2018).
[82] J. Lessing, A.C. Glavan, S.B. Walker, C. Keplinger, J.A. Lewis, G.M. Whitesides, Inkjet printing of conductive inks with high lateral resolution on omniphobic "R(F) paper" for paper-based electronics and MEMS, Adv. Mater. 26 (27) (2014) 4677–4682.
[83] S. Cinti, N. Colozza, I. Cacciotti, D. Moscone, M. Polomoshnov, E. Sowade, et al., Electroanalysis moves towards paper-based printed electronics: carbon black nanomodified inkjet-printed sensor for ascorbic acid detection as a case study, Sens. Actuators B: Chem. 265 (2018) 155–160.
[84] O.-S. Kwon, H. Kim, H. Ko, J. Lee, B. Lee, C.-H. Jung, et al., Fabrication and characterization of inkjet-printed carbon nanotube electrode patterns on paper, Carbon 58 (2013) 116–127.
[85] T.H. da Costa, E. Song, R.P. Tortorich, J.-W. Choi, A. Paper-Based, Electrochemical sensor using inkjet-printed carbon nanotube electrodes, ECS J. Solid. State Sci. Technol. 4 (10) (2015) S3044–S3047.
[86] G. Yang, L. Xie, M. Mantysalo, J. Chen, H. Tenhunen, L.R. Zheng, Bio-patch design and implementation based on a low-power system-on-chip and paper-based inkjet printing technology, IEEE Trans. Inf. Technol. Biomed. 16 (6) (2012) 1043–1050.
[87] C. Hu, X. Bai, Y. Wang, W. Jin, X. Zhang, S. Hu, Inkjet printing of nanoporous gold electrode arrays on cellulose membranes for high-sensitive paper-like electrochemical oxygen sensors using ionic liquid electrolytes, Anal. Chem. 84 (8) (2012) 3745–3750.
[88] A. Maattanen, P. Ihalainen, P. Pulkkinen, S. Wang, H. Tenhu, J. Peltonen, Inkjet-printed gold electrodes on paper: characterization and functionalization, ACS Appl. Mater. Interfaces 4 (2) (2012) 955–964.

[89] N. Ruecha, O. Chailapakul, K. Suzuki, D. Citterio, Fully inkjet-printed paper-based potentiometric ion-sensing devices, Anal. Chem. 89 (19) (2017) 10608–10616.

[90] C. Gaspar, S. Passoja, J. Olkkonen, M. Smolander, IR-sintering efficiency on inkjet-printed conductive structures on paper substrates, Microelectronic Eng. 149 (2016) 135–140.

[91] A.C. Glavan, R.V. Martinez, E.J. Maxwell, A.B. Subramaniam, R.M. Nunes, S. Soh, et al., Rapid fabrication of pressure-driven open-channel microfluidic devices in omniphobic R(F) paper, Lab. Chip 13 (15) (2013) 2922–2930.

[92] P.K. Yuen, V.N. Goral, Low-cost rapid prototyping of flexible microfluidic devices using a desktop digital craft cutter, Lab. Chip 10 (3) (2010) 384–387.

[93] M. Rahbar, P.N. Nesterenko, B. Paull, M. Macka, Geometrical alignment of multiple fabrication steps for rapid prototyping of microfluidic paper-based analytical devices, Anal. Chem. 89 (22) (2017) 11918–11923.

[94] N. Nuchtavorn, M. Macka, A novel highly flexible, simple, rapid and low-cost fabrication tool for paper-based microfluidic devices (muPADs) using technical drawing pens and in-house formulated aqueous inks, Anal. Chim. Acta 919 (2016) 70–77.

[95] A.C. Glavan, D.C. Christodouleas, B. Mosadegh, H.D. Yu, B.S. Smith, J. Lessing, et al., Folding analytical devices for electrochemical ELISA in hydrophobic R(H) paper, Anal. Chem. 86 (24) (2014) 11999–12007.

[96] N. Dossi, R. Toniolo, F. Terzi, N. Sdrigotti, F. Tubaro, G. Bontempelli, A cotton thread fluidic device with a wall-jet pencil-drawn paper based dual electrode detector, Anal. Chim. Acta 1040 (2018) 74–80.

[97] L.O. Orzari, R. Cristina de Freitas, I. Aparecida de Araujo Andreotti, A. Gatti, B.C. Janegitz, A novel disposable self-adhesive inked paper device for electrochemical sensing of dopamine and serotonin neurotransmitters and biosensing of glucose, Biosens. Bioelectron. 138 (2019) 111310.

[98] E. Noviana, D.B. Carrao, R. Pratiwi, C.S. Henry, Emerging applications of paper-based analytical devices for drug analysis: a review, Anal. Chim. Acta 1116 (2020) 70–90.

[99] P. Batista Deroco, Jd.F. Giarola, D. Wachholz Júnior, G. Arantes Lorga, L. Tatsuo Kubota, Chapter four—paper-based electrochemical sensing devices, in: A. Merkoçi (Ed.), Comprehensive Analytical Chemistry, Elsevier, 2020, pp. 91–137.

[100] E. Nunez-Bajo, M. Carmen Blanco-Lopez, A. Costa-Garcia, M. Teresa Fernandez-Abedul, Integration of gold-sputtered electrofluidic paper on wire-included analytical platforms for glucose biosensing, Biosens. Bioelectron. 91 (2017) 824–832.

[101] T.M. Nargang, R. Dierkes, J. Bruchmann, N. Keller, K. Sachsenheimer, C. Lee-Thedieck, et al., Photolithographic structuring of soft, extremely foldable and autoclavable hydrophobic barriers in paper, Anal. Methods 10 (33) (2018) 4028–4035.

[102] C.T. Kokkinos, D.L. Giokas, A.S. Economou, P.S. Petrou, S.E. Kakabakos, Paper-based microfluidic device with integrated sputtered electrodes for stripping voltammetric determination of dna via quantum dot labeling, Anal. Chem. 90 (2) (2018) 1092–1097.

[103] M. Parrilla, R. Canovas, F.J. Andrade, Paper-based enzymatic electrode with enhanced potentiometric response for monitoring glucose in biological fluids, Biosens. Bioelectron. 90 (2017) 110–116.

[104] M. Borràs-Brull, P. Blondeau, J. Riu, Characterization and validation of a platinum paper-based potentiometric sensor for glucose detection in saliva, Electroanalysis 33 (1) (2020) 181–187.

[105] L. Cao, C. Fang, R. Zeng, X. Zhao, Y. Jiang, Z. Chen, Paper-based microfluidic devices for electrochemical immunofiltration analysis of human chorionic gonadotropin, Biosens. Bioelectron. 92 (2017) 87–94.

[106] S.H. Lee, J.H. Lee, V.-K. Tran, E. Ko, C.H. Park, W.S. Chung, et al., Determination of acetaminophen using functional paper-based electrochemical devices, Sens. Actuators B: Chem. 232 (2016) 514–522.

[107] R.F. Carvalhal, M.S. Kfouri, M.H. Piazetta, A.L. Gobbi, L.T. Kubota, Electrochemical detection in a paper-based separation device, Anal. Chem. 82 (3) (2010) 1162–1165.

[108] H. Li, W. Wang, Q. Lv, G. Xi, H. Bai, Q. Zhang, Disposable paper-based electrochemical sensor based on stacked gold nanoparticles supported carbon nanotubes for the determination of bisphenol A, Electrochem. Commun. 68 (2016) 104–107.

[109] E. Nunez-Bajo, M.C. Blanco-Lopez, A. Costa-Garcia, M.T. Fernandez-Abedul, In situ gold-nanoparticle electrogeneration on gold films deposited on paper for non-enzymatic electrochemical determination of glucose, Talanta 178 (2018) 160–165.

[110] M. Moccia, V. Caratelli, S. Cinti, B. Pede, C. Avitabile, M. Saviano, et al., Paper-based electrochemical peptide nucleic acid (PNA) biosensor for detection of miRNA-492: a pancreatic ductal adenocarcinoma biomarker, Biosens. Bioelectron. 165 (2020) 112371.

[111] R. Canovas, M. Parrilla, P. Blondeau, F.J. Andrade, A novel wireless paper-based potentiometric platform for monitoring glucose in blood, Lab. Chip 17 (14) (2017) 2500–2507.

[112] D. Tobjörk, H. Aarnio, P. Pulkkinen, R. Bollström, A. Määttänen, P. Ihalainen, et al., IR-sintering of ink-jet printed metal-nanoparticles on paper, Thin Solid. Films 520 (7) (2012) 2949–2955.

[113] E. Costa-Rama, H.P.A. Nouws, C. Delerue-Matos, M.C. Blanco-Lopez, M.T. Fernandez-Abedul, Preconcentration and sensitive determination of the anti-inflammatory drug diclofenac on a paper-based electroanalytical platform, Anal. Chim. Acta 1074 (2019) 89–97.

[114] J. Ding, B. Li, L. Chen, W. Qin, A. Three-Dimensional, Origami paper-based device for potentiometric biosensing, Angew. Chem. Int. (Ed.) Engl. 55 (42) (2016) 13033–13037.

[115] L.O. Orzari, I.A. de Araujo Andreotti, M.F. Bergamini, L.H. Marcolino, B.C. Janegitz, Disposable electrode obtained by pencil drawing on corrugated fiberboard substrate, Sens. Actuators B: Chem. 264 (2018) 20–26.

[116] N. Dossi, S. Petrazzi, F. Terzi, R. Toniolo, G. Bontempelli, Electroanalytical cells pencil drawn on PVC supports and their use for the detection in flexible microfluidic devices, Talanta 199 (2019) 14–20.

[117] N. Kurra, G.U. Kulkarni, Pencil-on-paper: electronic devices, Lab. Chip 13 (15) (2013) 2866–2873.

[118] I.G. David, D.E. Popa, M. Buleandra, Pencil graphite electrodes: a versatile tool in electroanalysis, J. Anal. Methods Chem. 2017 (2017) 1905968.

[119] M.C. Sousa, J.W. Buchanan, Observational models of graphite pencil materials, Computer Graph. Forum 19 (1) (2000) 27–49.

[120] P.H.C.P. Tavares, P.J.S. Barbeira, Influence of pencil lead hardness on voltammetric response of graphite reinforcement carbon electrodes, J. Appl. Electrochem. 38 (6) (2008) 827–832.

[121] V.X.G. Oliveira, A.A. Dias, L.L. Carvalho, T.M.G. Cardoso, F. Colmati, W.K.T. Coltro, Determination of ascorbic acid in commercial tablets using pencil drawn electrochemical paper-based analytical devices, Anal. Sci. 34 (1) (2018) 91–95.

[122] H.-R. Wang, X.-M. Bi, Z.-J. Fang, H. Yang, H.-Y. Gu, L.-J. Sun, et al., Real time sensing of salicylic acid in infected tomato leaves using carbon tape electrodes modified with handed pencil trace, Sens. Actuators B: Chem. 286 (2019) 104–110.

[123] E. Alipour, M.R. Majidi, A. Saadatirad, S.M. Golabi, A.M. Alizadeh, Simultaneous determination of dopamine and uric acid in biological samples on the pretreated pencil graphite electrode, Electrochim. Acta 91 (2013) 36–42.
[124] M. Santhiago, C.S. Henry, L.T. Kubota, Low cost, simple three dimensional electrochemical paper-based analytical device for determination of p-nitrophenol, Electrochim. Acta 130 (2014) 771–777.
[125] V.N. Ataide, W.A. Ameku, R.P. Bacil, L. Angnes, W.R. de Araujo, T.R.L.C. Paixão, Enhanced performance of pencil-drawn paper-based electrodes by laser-scribing treatment, RSC Adv. 11 (3) (2021) 1644–1653.
[126] W. Li, D. Qian, Q. Wang, Y. Li, N. Bao, H. Gu, et al., Fully-drawn origami paper analytical device for electrochemical detection of glucose, Sens. Actuators B: Chem. 231 (2016) 230–238.
[127] T. Fayose, L. Mendecki, S. Ullah, A. Radu, Single strip solid contact ion selective electrodes on a pencil-drawn electrode substrate, Anal. Methods 9 (7) (2017) 1213–1220.
[128] S. Zhu, Y. Li, H. Zhu, J. Ni, Y. Li, Pencil-drawing skin-mountable micro-supercapacitors, Small 15 (3) (2019) e1804037.
[129] Z. Li, L. Wu, S. Dong, T. Xu, S. Li, Y. An, et al., Pencil drawing stable interface for reversible and durable aqueous zinc-ion batteries, Adv. Funct. Mater. 31 (4) (2020).
[130] J.K. Kariuki, An electrochemical and spectroscopic characterization of pencil graphite electrodes, J. Electrochem. Soc. 159 (9) (2012) H747–H751.
[131] R.L. McCreery, Advanced carbon electrode materials for molecular electrochemistry, Chem. Rev. 108 (7) (2008) 2646–2687.
[132] N. Dossi, R. Toniolo, F. Impellizzieri, G. Bontempelli, Doped pencil leads for drawing modified electrodes on paper-based electrochemical devices, J. Electroanalytical Chem. 722–723 (2014) 90–94.
[133] M. Santhiago, L.T. Kubota, A new approach for paper-based analytical devices with electrochemical detection based on graphite pencil electrodes, Sens. Actuators B: Chem. 177 (2013) 224–230.
[134] R.B. Channon, Y. Yang, K.M. Feibelman, B.J. Geiss, D.S. Dandy, C.S. Henry, Development of an electrochemical paper-based analytical device for trace detection of virus particles, Anal. Chem. 90 (12) (2018) 7777–7783.
[135] W.A. Ameku, J.M. Goncalves, V.N. Ataide, M.S. Ferreira Santos, I.G.R. Gutz, K. Araki, et al., Combined colorimetric and electrochemical measurement paper-based device for chemometric proof-of-concept analysis of cocaine samples, ACS Omega 6 (1) (2021) 594–605.
[136] J.A. Adkins, E. Noviana, C.S. Henry, Development of a quasi-steady flow electrochemical paper-based analytical device, Anal. Chem. 88 (21) (2016) 10639–10647.
[137] K.F. Lei, S.I. Yang, Bundled carbon nanotube-based sensor on paper-based microfluidic device, J. Nanosci. Nanotechnol. 13 (10) (2013) 6917–6923.
[138] K.F. Lei, K.-F. Lee, S.-I. Yang, Fabrication of carbon nanotube-based pH sensor for paper-based microfluidics, Microelectronic Eng. 100 (2012) 1–5.
[139] P. Wang, Z. Cheng, Q. Chen, L. Qu, X. Miao, Q. Feng, Construction of a paper-based electrochemical biosensing platform for rapid and accurate detection of adenosine triphosphate (ATP), Sens. Actuators B: Chem. 256 (2018) 931–937.
[140] Y. Wang, J. Ping, Z. Ye, J. Wu, Y. Ying, Impedimetric immunosensor based on gold nanoparticles modified graphene paper for label-free detection of Escherichia coli O157: H7, Biosens. Bioelectron. 49 (2013) 492–498.

[141] K. Dagci, M. Alanyalioglu, Preparation of free-standing and flexible graphene/ag nanoparticles/poly(pyronin y) hybrid paper electrode for amperometric determination of nitrite, ACS Appl. Mater. Interfaces 8 (4) (2016) 2713–2722.

[142] K.F. Lei, S.I. Yang, S.W. Tsai, H.T. Hsu, Paper-based microfluidic sensing device for label-free immunoassay demonstrated by biotin-avidin binding interaction, Talanta 134 (2015) 264–270.

[143] Y.-R. Rhim, D. Zhang, D.H. Fairbrother, K.A. Wepasnick, K.J. Livi, R.J. Bodnar, et al., Changes in electrical and microstructural properties of microcrystalline cellulose as function of carbonization temperature, Carbon 48 (4) (2010) 1012–1024.

[144] W. Zhang, S. Zhu, R. Luque, S. Han, L. Hu, G. Xu, Recent development of carbon electrode materials and their bioanalytical and environmental applications, Chem. Soc. Rev. 45 (3) (2016) 715–752.

[145] J. Lin, Z. Peng, Y. Liu, F. Ruiz-Zepeda, R. Ye, E.L. Samuel, et al., Laser-induced porous graphene films from commercial polymers, Nat. Commun. 5 (2014) 5714.

[146] L. Huang, J. Su, Y. Song, R. Ye, Laser-induced graphene: en route to smart sensing, Nanomicro Lett. 12 (1) (2020) 157.

[147] A.K. Dubey, V. Yadava, Laser beam machining—a review, Int. J. Mach. Tools Manufacture 48 (6) (2008) 609–628.

[148] R. Ye, D.K. James, J.M. Tour, Laser-induced graphene: from discovery to translation, Adv. Mater. 31 (1) (2019) e1803621.

[149] Y. Zhu, H. Cai, H. Ding, N. Pan, X. Wang, Fabrication of low-cost and highly sensitive graphene-based pressure sensors by direct laser scribing polydimethylsiloxane, ACS Appl. Mater. Interfaces 11 (6) (2019) 6195–6200.

[150] Y. Chyan, R. Ye, Y. Li, S.P. Singh, C.J. Arnusch, J.M. Tour, Laser-induced graphene by multiple lasing: toward electronics on cloth, paper, and food, ACS Nano 12 (3) (2018) 2176–2183.

[151] R. Ye, Y. Chyan, J. Zhang, Y. Li, X. Han, C. Kittrell, et al., Laser-induced graphene formation on wood, Adv. Mater. 29 (37) (2017).

[152] Z. Zhang, M. Song, J. Hao, K. Wu, C. Li, C. Hu, Visible light laser-induced graphene from phenolic resin: A new approach for directly writing graphene-based electrochemical devices on various substrates, Carbon 127 (2018) 287–296.

[153] L.F. Mendes, A. de Siervo, W. Reis de Araujo, T.R. Longo Cesar Paixão, Reagentless fabrication of a porous graphene-like electrochemical device from phenolic paper using laser-scribing, Carbon 159 (2020) 110–118.

[154] A. Bezerra Martins, A. Lobato, N. Tasić, F.J. Perez-Sanz, P. Vidinha, T.R.L.C. Paixão, et al., Laser-pyrolyzed electrochemical paper-based analytical sensor for sulphite analysis, Electrochem. Commun. 107 (2019).

[155] N. Tasić, L. Sousa de Oliveira, T.R.L.C. Paixão, L. Moreira Gonçalves, Laser-pyrolysed paper electrodes for the square-wave anodic stripping voltammetric detection of lead, Med. Devices & Sens. 3 (6) (2020).

[156] N. Tasić, A. Bezerra Martins, X. Yifei, M. Sousa Góes, D. Martín-Yerga, L. Mao, et al., Insights into electrochemical behavior in laser-scribed electrochemical paper-based analytical devices, Electrochem. Commun. 121 (2020).

Chapter 6

Surface-enhanced Raman scattering paper-based analytical devices

Ana Carolina Marques[1,2,3], Hugo Águas[1], Rodrigo Martins[1], Bruno Costa-Silva[2], Maria Goreti Sales[3,4,5] and Elvira Fortunato[1]
[1]i3N/CENIMAT, Department of Materials Science, Faculty of Science and Technology, Universidade NOVA de Lisboa and CEMOP/UNINOVA, Campus de Caparica, Caparica, Portugal, [2]Champalimaud Research, Champalimaud Centre for the Unknown, Lisbon, Portugal, [3]BioMark@UC, Department of Chemical Engineering, Faculty of Science and Technology, Coimbra University, Coimbra, Portugal, [4]BioMark@ISEP, School of Engineering, Polytechnic Institute Porto, Porto, Portugal, [5]CEB—Centre of Biological Engineering, University of Minho, Braga, Portugal

6.1 Introduction

6.1.1 Raman spectroscopy: a historical perspective

"*I propose this evening to speak to you on a new kind of radiation or light emission from atoms and molecules.*" [1] These were the words, from Sir Chandrasekhara Venkata Raman, that anticipated one of the most impacting announcements on analytical methods. C. V. Raman, born in 1888, Indian physicist and a 1930' Nobel laureate "*for his work on the scattering of light and for the discovery of the effect named after him*" [2] (Fig. 6.1).

After discovering a new phenomenon on the scattering of light, C. V. Raman made the announcement to the press on the same day, February 28, 1928, which read in the Associated Press of India in the next day as "*Prof. C. V. Raman, F.R.S., of the Calcutta University, has made a discovery which promises to be of fundamental significance to physics... The new phenomenon exhibits features even more startling than those discovered by Prof. Compton with X-rays. The principal feature observed is that when matter is excited by light of one colour, the atoms contained in it emit light of two colours, one of which is different from the exciting colour and is lower down the spectrum. The astonishing thing is that the altered colour is quite independent of the nature of the substance used.*" [3]

Paper-Based Analytical Devices for Chemical Analysis and Diagnostics.
DOI: https://doi.org/10.1016/B978-0-12-820534-1.00001-3
© 2022 Elsevier Inc. All rights reserved.

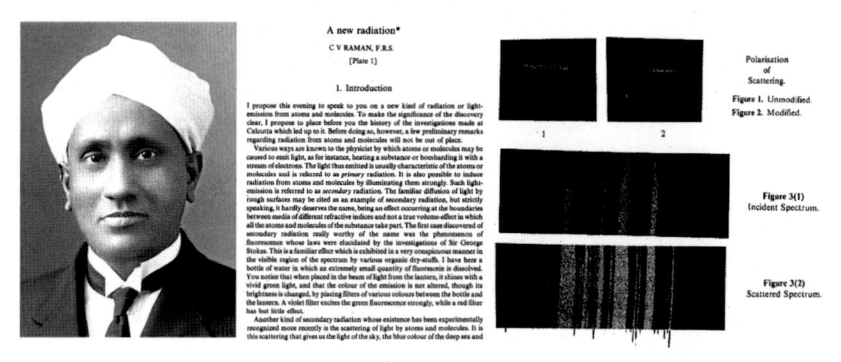

A new radiation*

C V RAMAN, F.R.S.

[Plate 1]

1. Introduction

I propose this evening to speak to you on a new kind of radiation or light-emission from atoms and molecules. To make the significance of the discovery clear, I propose to place before you the history of the investigations made at Calcutta which led up to it. Before doing so, however, a few preliminary remarks regarding radiation from atoms and molecules will not be out of place.

Various ways are known to the physicist by which atoms or molecules may be caused to emit light, as for instance, heating a substance or bombarding it with a stream of electrons. The light thus emitted is usually characteristic of the atoms or molecules and is referred to as *primary* radiation. It is also possible to induce radiation from atoms and molecules by illuminating them strongly. Such light-emission is referred to as *secondary* radiation. The familiar diffusion of light by rough surfaces may be cited as an example of secondary radiation, but strictly speaking, it hardly deserves the name, being an effect occurring at the boundaries between media of different refractive indices and not a true volume-effect in which all the atoms and molecules of the substance take part. The first case discovered of secondary radiation really worthy of the name was the phenomenon of fluorescence whose laws were elucidated by the investigations of Sir George Stokes. This is a familiar effect which is exhibited in a very conspicuous manner in the visible region of the spectrum by various organic dry-stuffs. I have here a bottle of water in which an extremely small quantity of fluorescein is dissolved. You notice that when placed in the beam of light from the lantern, it shines with a vivid green light, and that the colour of the emission is not altered, though its brightness is changed, by placing filters of various colours between the bottle and the lantern. A violet filter excites the green fluorescence strongly, while a red filter has but little effect.

Another kind of secondary radiation whose existence has been experimentally recognized more recently is the scattering of light by atoms and molecules. It is this scattering that gives us the light of the sky, the blue colour of the deep sea and

FIGURE 6.1 Sir Chandrasekhara Venkata Raman photograph, front page of his inaugural address detailing the Raman effect and relevant figures. *Reprinted with permission from Raman, C.V., Indian Journal of Physics, 2, 387-398, 1928.*

The events that led to his findings together with his coworker K. S. Krishnan were first detailed in an inaugural address delivered to the South Indian Science Association, on March 16, 1928 [1] (Fig. 6.1). Later a paper was published in Nature on March 31, entitled "A new type of secondary radiation." [4]

The experiment that allowed to find the Raman effect (Fig. 6.2) was based on the passage of sun's violet light, isolated with a violet filter, through a liquid sample. This led to the observation that although most of the emerging scattered light had the same wavelength as the incident violet beam—known as Rayleigh scattering—a small portion of the emitted light had a different wavelength.

This was possible by placing a green filter between the sample and the observer. The phenomenon was observed, at the time, by C. V. Raman and his coworker K. S. Krishnan, in more than 60 different liquid samples. [1,4]

From 1928 onward, more than 300,000 contributions (Web of science database search in October 2020) have used the Raman effect. The improvement on instrumentations, especially the invention of charge-coupled devices detectors and light amplification by stimulated emission of radiation (LASER) devices, has sparked an exponential increase on Raman spectroscopy–associated experiments. For its high significance as analytical technique, it was designated in 1998 as an International Historic Chemical Landmark by the American Chemical Society. [5]

6.1.2 Theoretical fundamentals of Raman spectroscopy and SERS

The Raman effect describes the phenomenon of the inelastic scattering of light. Briefly, when light passes through matter, part of the light is scattered in different directions: most of it with the same wavelength—elastic scattering (Rayleigh scattering)—but a small portion is scattered with a different

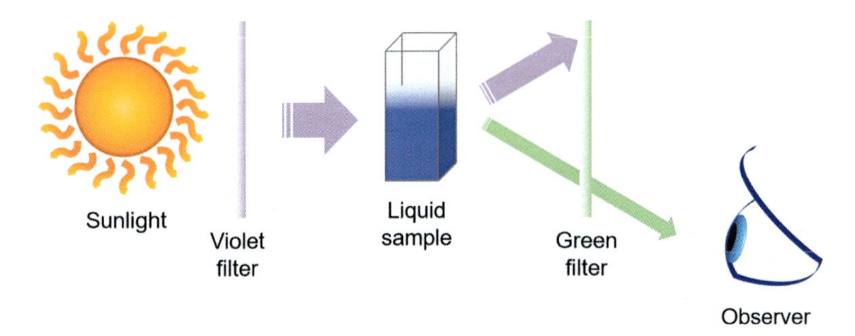

FIGURE 6.2 Schematic representation of the experiment that led to the finding of the Raman effect.

wavelength—inelastic scattering (Raman scattering). The Raman scattering is divided into two processes: Anti-Stokes and Stokes scattering. In an anti-Stokes scattering process, the scattered photon has a higher energy than the incident photon, which only happens if the molecule is already in an excited vibrational state. This event is very unlikely to succeed leading to the weakest signals. The Stokes scattering is the energy transference from light to a vibration of a molecule where the scattered photon has a lower energy, a higher wavelength than the incident photon. Since the energy transferred is the precise energy to excite one of the permitted molecular vibrations, the resulting scattered light is entirely dependent on the type of molecule, conferring a high selectivity to Raman spectroscopy. [6]

Over the years, Raman spectroscopy—which measures and analyzes the photons arising from the Raman effect—has in fact proven to be a powerful analytical technique due to its versatility and selectivity on the identification of analytes, through the distinctive vibration of molecules. [7] The energy loss on a Raman scattering event is called the Raman shift and corresponds to the wavenumber of the vibrational mode involved in the scattering. [6]

Despite its selectivity, the Raman scattering is an event with a very low probability; in 10 million scattered photons only one is inelastically scattered, leading to extremely weak signals. [7] To surpass this drawback, Surface-Enhanced Raman Spectroscopy or Surface-Enhanced Raman Scattering (SERS) has emerged as a valuable solution to enhance the highly specific but weak Raman signals. [6–8] The implementation of SERS has been exponentially increasing in the past decades (Fig. 6.3) since the reported enhancement of Raman signals of pyridine adsorbed on a rough silver electrode, by M. Fleischmann, in 1974. [9]

Three years later it was suggested that the signal enhancement was due to, not only, the presence of rough surfaces that increases the surface area but also to an increase in the electromagnetic field [10] and charge transferences between the analyte and silver. [11] The increase in the number of publications was sparked with the boom of nanotechnology and

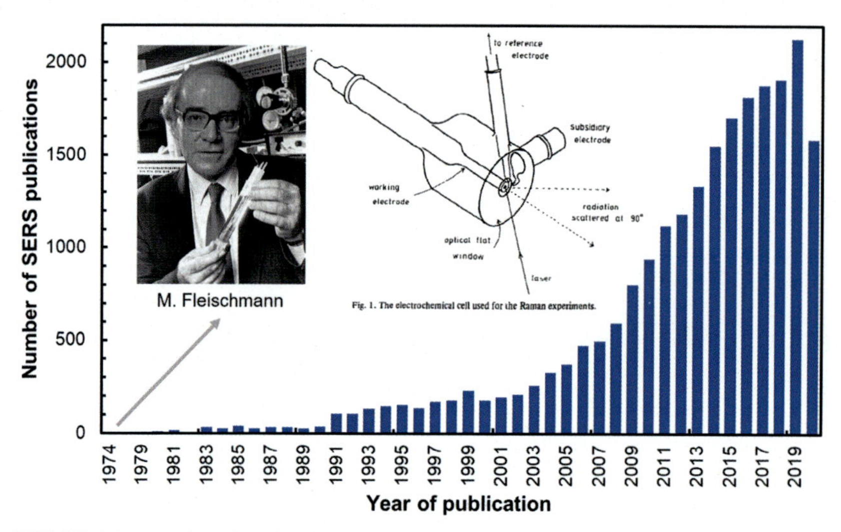

FIGURE 6.3 Number of publications per year since the first report of Raman enhancement. Insets of Martin Fleischmann photograph and the figure relative to the electrochemical cell used for Raman experiments, in 1974. [9] Web of Science database search in October, 2020.

nanofabrication techniques in the 1980s. The use of metallic nanoparticles for Raman signals enhancement, particularly Ag and Au has enabled increasing the intensities by several orders of magnitude. [12–15] Fig. 6.4 shows a timeline of the evolution of Raman spectroscopy and SERS in the early years of development, which contributed to the exponential growth of the use of this technique on a multitude of applications.

There are two accepted independent mechanisms for the phenomenon of Raman enhancement, that occur often at the same time: chemical and electromagnetic enhancement. [7,8] Chemical enhancement is a charge transfer mechanism where the orbital's overlapping of the adsorbed molecules and the metallic surface induces electronic interactions. This type of enhancement leads to the amplification of Raman signals, in approximately $10-10^3$ orders of magnitude. [7,16] Electromagnetic enhancement—the main contributor for Raman signal amplification—on the other hand, is caused by plasmon resonances in the metal, which induces the scattering of photons by the delocalized electrons present on the metallic nanoparticle surface, providing the largest enhancements, for up to 10^4-10^{12}. [7,8] Among all metallic nanostructures, Ag, Au, and Cu nanostructures are responsible for inducing the highest Raman enhancements due to their localized surface plasmon resonance (LSPR) in the visible and near infrared regions, which is strongly dependent on their size and morphology. [7] Moreover, when two metallic nanoparticles are at a certain distance, it generates a so-called hot spot that enhances even further Raman signals, particularly helpful for single-molecule detection. [6–8]

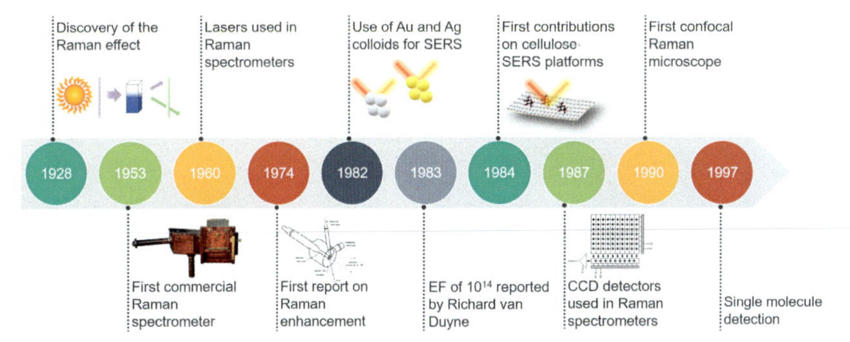

FIGURE 6.4 Timeline of important landmarks that contributed to the evolution of Raman and Surface-enhanced Raman spectroscopy.

The magnitude of the Raman signal amplification can be calculated and it is named enhancement factor (EF), which is one of the most important figures of merit to which SERS is concerned. This factor corresponds to the enhancement in Raman intensities of the molecules adsorbed on SERS surfaces, when compared to the same number of molecules on non-SERS surfaces. [6] Several factors can influence SERS enhancement, such as: characteristics of the excitation source, detection setup, type of SERS substrate, intrinsic properties of the analyte, among others. [6]

6.2 Raman enhancement nanostructures onto paper substrates: characteristics and fabrication methods

The recent advances on SERS research have shown an increasing focus on the development of novel SERS platforms that can offer high sensitivities, uniformity, reproducibility, easiness of fabrication, and an inherent low-cost. Cellulose-based materials in the form of paper substrates and other native or modified forms of cellulose have shown their versatility as a vehicle or support of SERS devices due to their fundamental characteristics such as flexibility, lightweight, biocompatibility, biodegradability, and availability. [17–20] Moreover, it can be easily tailored for different SERS purposes, offering advantageous properties, such as hydrophilicity or hydrophobicity, high surface area, and easy adaptation to nanofabrication purposes. [7,8] Ultimately, its low-cost and the aforementioned characteristics of these materials make it an important candidate for point-of-care applications based on SERS.

The first contributions using cellulose as SERS platform dates back to 1984. [21,22] A significant increase in the number of publications reporting cellulose and paper-based substrates on the fabrication of SERS devices has been observed (Fig. 6.5).

Concerning the last 10 years, a Web of Science search has shown that the type of enhancement structures that are commonly applied to paper and

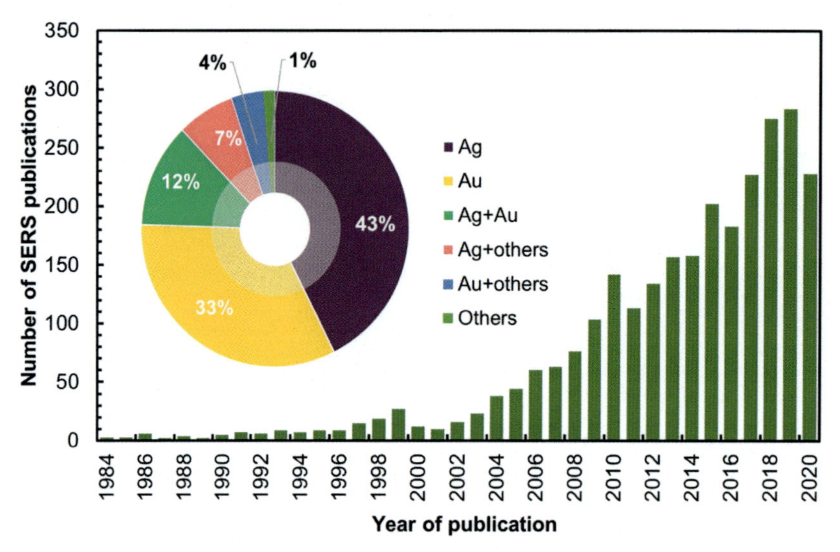

FIGURE 6.5 Number of publications per year of paper and cellulose-based SERS devices. Inset of percentage distribution of the type of enhancement nanostructure utilized on paper and cellulose-based SERS devices in the last 10 years. Web of Science database search in October, 2020.

cellulose platforms (Fig. 6.5, inset), from an universe of more than 150 publications, are Ag and Au nanostructures, due to their strong plasmon resonance. Among these, Ag accounts for the highest percentage, since it is the one that provides the larger signal enhancements. [6] Moreover, about 24% of paper and cellulose-based SERS contributions has reported the use of composites of Ag and Au nanostructures, [23–25] Ag and Au with other nanostructures such as metal oxides [26–29] or graphene-based materials [30–33] and even other type of nanostructures such as ZnO [34] or MoO_{3-x} nanosheets. [35]

Herein, contributions over the last 10 years reporting cellulose SERS platforms with a focus on the type of nanostructures used for Raman signal amplification are reviewed. Among these, different strategies for enhancement nanostructure's synthesis and their deposition methods onto cellulose and paper substrates achieving different analytical features are highlighted. Fig. 6.6 shows an infographic with the most used techniques, employed with cellulose substrates, to produce enhancement Raman nanostructures and their transference methods.

6.2.1 Gold-based nanostructures

Gold nanoparticles (AuNPs) are the most common gold-based nanostructures employed in SERS and paper-based SERS platforms. In their vast majority,

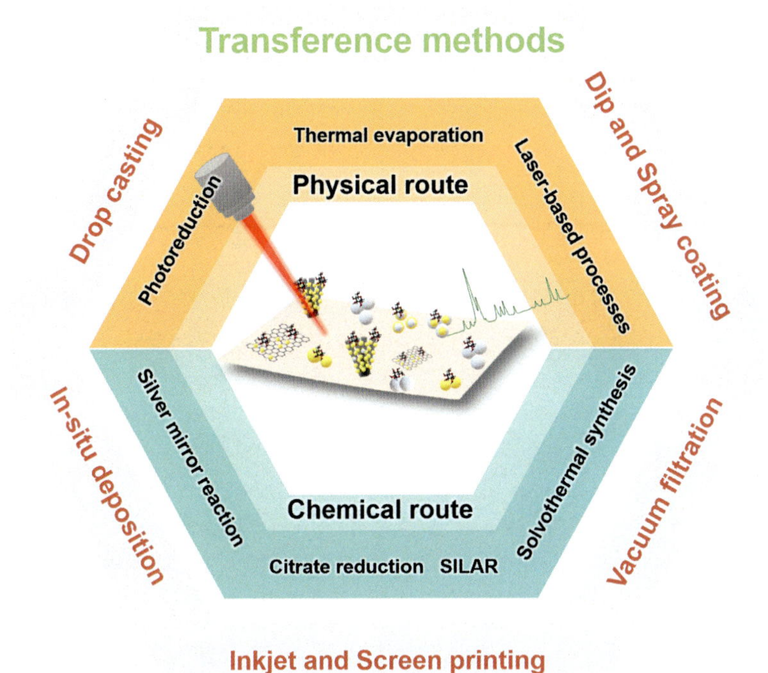

FIGURE 6.6 Infographic with the most common production and transference techniques employed in the production of cellulose-based SERS substrates.

these structures are mainly produced through citrate-based methods following modified Turkevich or Frens protocols. [15,36–56] Additionally, sodium borohydride ($NaBH_4$) has also been reported as reducing agent of AuNPs in paper-based SERS platforms [57–61] or a combination of both trisodium citrate (Na_3Ct) and $NaBH_4$. [62] These nanostructures were successfully applied to different forms of cellulose materials.

Table 6.1 resumes the main characteristics of representative cellulose-based SERS platforms with gold nanostructures.

In 2012, Ngo et al. [15] studied the effect of addition of AuNPs to paper substrates and their capability as amplifiers of Raman scattering. The authors followed the Turkevich method for AuNPs synthesis, with 1 mM $HAuCl_4 \cdot 3H_2O$ and 1% Na_3Ct, resulting in a surface plasmon resonance (SPR) centered at 530 nm, with an average diameter of 23.2 nm. The colloidal solution was diluted to different concentrations that were then transferred to a petri dish in which Whatman grade 1 paper disks were dipped for 24 h. Despite the fact that citrate-based AuNPs are negatively charged as well as paper, due to the presence of hydroxyl groups of cellulose, a high absorption of AuNPs on paper was achieved due to van der Waals interactions, which

TABLE 6. 1 Gold-based SERS devices fabricated onto cellulose substrates.

Reference	Nanomaterial	Cellulose format	Fabrication method	Analyte	LOD	EF
Teixeira and Poppi [50]	AuNPs	Office paper	Citrate reduction and drop casting	Thiabendazole	-	-
Chen et al. [60]	AuNPs	Whatman grade 1 paper	In situ seed-mediated growth with $NaBH_4$	4-MPBA Brassinoste-roids	10^{-11} M	1.13×10^8
Xiong et al. [63]	AuNPs	Nanofibrillated cellulose	Citrate reduction and direct deposition	Melamine aq. solution Melamine in milk	0.1 ppm1 ppm	-
Kwon et al. [64]	AuNRs	Cellulose nanofibers	Seed-mediated growth with $NaBH_4$ and deposition	R6G Thiram Tricyclazole Carbaryl	10 nM1 nM100 nM1 μM	1.4×10^7
He et al. [65]	AuNSs	Filter and office paper	Ascorbic acid reduction and immersion	CV	1 nM	1.2×10^7
Campu et al. [66]	AuBPs	Whatman grade 1 paper	Seed-mediated growth and plasmonic calligraphy	Biotin-streptavidin interaction	-	-

Godoy et al. [67]	AuSph	Hydrophobic chromatography paper	Seed-mediated growth with citrate reduction and inkjet printing	CV Thiram	-	$2.0 \times 10^6 2.0 \times 10^6$
Kang et al. [27]	$Au@Fe_3O_4$ NPs	Bacterial cellulose nanocrystals	In situ citrate reduction	MG isothiocyanate	5.8×10^{-10} M	-
Kim et al. [26]	ZnONRs decorated with AuNPs	Whatman paper	In situ hydrothermal method and SILAR synthesis	CV Human amniotic fluids	-	1.25×10^7
Ponlamuangdee et al. [30]	GO-AuNRs	Whatman grade 1 paper	Seed-mediated growth with $NaBH_4$ reduction and layer-by-layer immersion	R6G Mitoxantrone	0.1 nM 5 μM	4.56×10^7

4-MPBA, 4 mercaptophenylboronic acid; *AuBPs*, Gold nanobipyramids; *AuNPs*, Gold nanoparticles; *AuNRs*, Gold nanorods; *AuNSs*, Gold nanostars; *AuSph*, Gold nanospheres; *CV*, Crystal violet; *GO*, Graphene oxide; *MG*, Malachite green; *R6G*, Rhodamine 6G; *ZnONRs*, Zinc Oxide nanorods.

contradicted the need for cationically modified AuNPs to achieve efficient absorption onto paper, as previously reported. [68,69] As Raman probe molecule, 4-aminothiophenol (4-ATP) was used due to its strong affinity to AuNPs, and a linear increase of spectra intensities was observed to increasing concentrations of AuNPs loaded on paper. This was facilitated due to an increased number of hot spots resulting in a high electromagnetic enhancement. The determination of the limit of detection (LOD) was carried out with 0.20 mg/mL AuNPs-treated paper and calculated below 1 nM, due to the sill good signal-to-noise ratio of the obtained spectrum. Lastly, a direct comparison with AuNPs deposited on a silicon wafer showed a decrease on SERS efficiency with a limit of detection of 100 nM, due to the low amount of AuNPs that were able to be retained in the substrate and the higher spatial distance between the AuNPs—decrease in the number of hot spots.

An interesting approach was taken by Zhang, K. and coworkers [44] to obtain a closely packed monolayer of AuNPs on paper by the patterning self-assembled NPs on the tip of an arrow-shaped paper strip (Fig. 6.7A). The authors explored the oil/water interface for self-assembly of noble metals, which rendered closely packed NPs, and transferred it to an arrow-shaped Whatman grade 1 paper. The citrate-based AuNPs were obtained through a Frens' method with 0.01% (w/w) of $HAuCl_4 \cdot 3H_2O$ and 1% (w/w) Na_3Ct rendering 56 nm-sized NPs with an absorption plasmonic band centered at 534 nm. To evaluate the performance of the SERS substrate, the spectra of 2,2′-bipyridyl was collected showing superior intensities when compared with AuNPs deposited on silicon wafer and a paper-based SERS substrate obtained through a 24 h immersion method. The arrow shape of the paper-based SERS platform allowed for convenient sample loading on the wider side, whereas the sharp point was helpful for analyte concentration and also on diminishing possible coffee-ring effects avoiding uneven distribution of the analyte. Rhodamine 6G (R6G) was used to evaluate the limit of detection, due to its high Raman cross-section, and compared with 2,2,′-bipyridyl. Although the latter molecule has a lower Raman section, the SERS platform was able to detect it to as low as 1 fM and as low as 50 a for R6G. The capability of the fabricated SERS platform for swabbing and for chromatographic separation of complex samples was also successfully proven.

Jang et al. [53] prepared paper-based SERS platforms by spray-coating ~61 nm-sized AuNPs, synthesized with 1 wt.% $HAuCl_4 \cdot 3H_2O$ and 1 wt.% of Na_3Ct, onto hydrophilic Whatman grade 40 paper, and compared its performance with the conventional dip-coating method. Besides being less time-consuming (30 min instead of 24 h), the method allowed a deeper penetration of AuNPs throughout the three-dimensional porous structure of paper, inducing more local aggregations between the NPs, due to faster drying processes. The prepared plasmonic papers were tested with 4-nitrobenzenethiol (4-NBT) and a twofold increasement on Raman intensities was obtained for the spray-coated paper resulting in a higher EF. Additionally, the authors

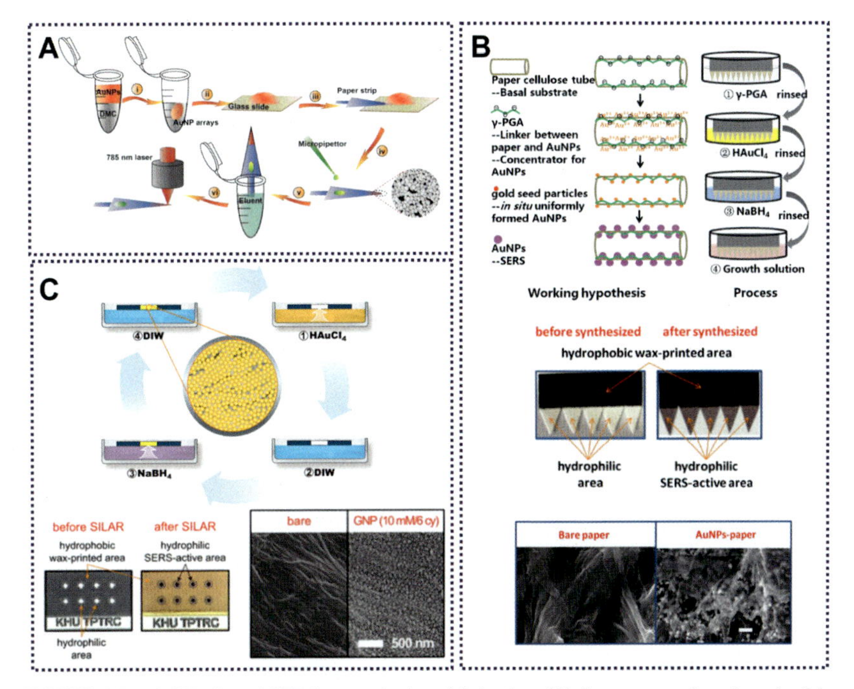

FIGURE 6.7 AuNPs-based SERS paper devices fabrication. (A) Strategy employed on the fabrication and sample detection: paper patterning with self-assembled interfacial AuNP array and on-card separation, preconcentration, and sample detection. (B) Representation of the AuNPs synthesis process through in situ seed-mediated growth, photographs of the paper SERS device before and after synthesis and SEM images of bare paper and AuNPs-paper. (C) Schematic representation of SILAR technique, photographs of the paper SERS device before and after synthesis, with wax patterning, and SEM images of bare paper and the optimized AuNPs-paper. *(Figure 6.7A) Reprinted with permission from Zhang, K et al., ACS Applied Materials and Interfaces, 7(30), 2015. Copyright 2015, American Chemical Society; (Figure 6.7B) Reprinted with permission from Chen, M. et al., Talanta, 165, 2017. Copyright 2017, Elsevier; (Figure 6.7C) Reprinted with permission from Kim, W et al., Analytical Chemistry, 88(10), 2016. Copyright 2016, American Chemical Society.*

tested the efficiency of the sample's loading onto the SERS membranes, by comparing its performance when the sample is applied dropwise and when the SERS membrane is dipped into the sample solution. In both dip-coated and spray-coated SERS membranes, when a low concentration of 4-NBT was added dropwise, relatively low SERS intensity were observed. On the contrary, for high concentrations the signal intensity increased with higher values for the spray-coated membranes. This observation was explained by the hydrophilic nature of filter paper where low concentrations of analyte penetrate easily throughout the paper's fibers whereas when higher concentrations are used the coverage of the plasmonic surface is maximized. When the substrates were soaked with the analyte solution, the SERS response increased, especially for low concentrations, with higher intensities for the spray-coated membranes. The required time for signal saturation was higher in the spray-coated matrix, due to the slower diffusion rate of the analyte through the fibers, explained by the deeper penetration of the AuNPs. These observations demonstrated the importance of analyte's loading onto SERS membranes. In addition, the authors evaluated the influence of adding a second layer of AuNPs, creating a sandwich of AuNPs-analyte-AuNPs, for electromagnetic enhancement. In fact, an increase in the signal amplification was observed especially for the sandwich structure fabricated by spray-coating, with a calculated EF higher than 4.5×10^6 for 1 μM 4 -NBT.

A seed-growth method was used for the synthesis of citrate-based AuNPs with an SPR peak at 526 nm and an average size of 37.91 ± 3.26 nm. [36] The AuNPs were loaded onto filter paper by an immersion method for different times and 4-mercaptobenzoic acid (4-MBA) was used as probe molecule. The characteristic distinct peak at 1068.18 cm^{-1} was used for the optimization and performance' evaluation of the prepared SERS substrate. An increase on SERS intensity at 1068.18 cm^{-1} was observed up to 30 min of immersion time followed by a decrease on SERS intensity for higher immersion times. Reproducibility of the 30 min-immersion SERS membrane was evaluated by recording spectra on 30 random points from three batches with an relative standard deviation (RSD) as low as 5.28%. Moreover, the optimized SERS membrane showed good stability over time and a limit of detection of 10^{-9} M for 4-MBA. As proof-of-concept, the SERS membrane was applied to the detection of methyl parathion both in solution and on apple surfaces, using the characteristic peak centered at 1336.79 cm^{-1}. A linear relationship was observed from 0.018 to 0.354 μg/cm^2 with a limit of detection of 0.011 μg/cm^2, and recovery rate on real-world surfaces ranging from 94.09% to 98.72%.

Exploiting the coffee-ring effect of AuNPs on cellulose nanofibers, Chen et al. [39] developed a SERS system for trace analysis and chemical identification. AuNPs were synthesized through a citrate-reduction method with 2.5 mM $HAuCl_4 \cdot 3H_2O$ and 0.25% (w/v) Na_3Ct and deposited onto bare glass substrates and onto cellulose nanofibers. In both cases, a coffee-ring was formed upon drying with a 10-fold increase in length when cellulose nanofibers were used. Recorded spectra of 4-ATP, at the coffee-ring zone, with concentrations from 1×10^{-4} to 1×10^{-12} M, showed its characteristic Raman peaks with high reproducibility. On the contrary, when the spectra were collected outside the coffee-ring it was not possible to obtain the Raman characteristic peaks at trace amounts (1×10^{-12} M). Additionally, the authors demonstrated the use of their SERS platform for multimolecular detection.

Chen and coworkers [60] prepared AuNPs, with $NaBH_4$ as reducing agent, with an in situ seed-mediated growth (ISSMG) approach on Whatman grade 1 paper (Fig. 6.7B). This approach consisted of four steps, briefly: (1) a paper strip was placed vertically in a poli(γ-glutamic acid) (γ-PGA) solution, used as linker and concentrator; (2) treatment with an $HAuCl_4 \cdot 3H_2O$ solution for 2 min; (3) treatment with $NaBH_4$ for 20 min; (4) treatment in the growth solution for 20 min for the in situ growth of larger AuNPs. Between each step, the paper strip was dipped in distilled water for washing. 4-Mercaptophenylboronic acid (4-MPBA) was used as Raman probe, following the characteristic peak 1071 cm^{-1}, to optimize the concentration of γ-PGA to 20 mg/mL. A direct comparison between bare paper, AuNPs-paper and γ-PGA-AuNPs-paper showed higher signal intensities for the latter system. The authors claimed that the presence of γ-PGA led to an increase in the number of hot spots and an evenly distribution of AuNPs on paper. Moreover, increasing concentrations of reactants led to a red shift on the SPR wavelength, translated in an increase of AuNPs size, which was accompanied by an increase in SERS performance up to 30 mM and a decrease for 40 mM. The sensitivity and performance of the SERS membrane was evaluated showing a linear correlation from 10^{-11} to 10^{-6} M and EF of 1.13×10^8. Finally, the SERS membrane was used for the rapid detection of brassinosteroids in plant tissues, with a linear correlation from 0.1 to 20 μM, RSD values between 3.0% and 13.8% and relative recoveries between 79.1% and 109.1%.

A successive ionic layer absorption and reaction (SILAR) technique (Fig. 6.7 C) was used for the early diagnosis of infectious keratoconjunctivitides [61] and subarachnoid hemorrhage-induced complications. [58] The SILAR technique uses $NaBH_4$ as reducing agent and is based on the successive dipping of a prewax printed paper substrates in a $HAuCl_4 \cdot 3H_2O$ solution followed by a $NaBH_4$ solution. After each step, the paper substrates are dipped in distilled water for elimination of unreacted reagents and nonattached particles. The particles produced in the first SILAR cycle act as nuclei, and their size gradually increased with additional SILAR cycles due to layer absorption and reaction. Therefore, the sizes of these particles can be controlled by the number of SILAR cycles used. For the early diagnosis of infectious keratoconjunctivitides [61] an optimization of the paper SERS membrane was performed by the study of the variation on the number of SILAR cycles and reactant's concentration. To evaluate SERS performance 2-naphthalenethiol (2-NAT) was used as Raman probe. Six SILAR cycles with 10 mM concentration of reactants showed the best SERS responses, in terms of intensity and reproducibility (RSD 7.53%). Different 2-NAT concentrations were used to evaluate the sensitivity of the optimized substrate, showing a linear correlation from 1 p to 1 μM, with 0.99 correlation coefficient and an EF calculated as 7.8×10^8. Lastly, a PCA analysis was used to distinguish between three types of infectious eyes disease achieving high sensitivity and selectivity. Later, for the diagnosis of subarachnoid hemorrhage-induced complications, [58] the authors used the same SILAR technique but with a pH-adjusted cellulose surface to 3.0, that corresponds to the pH of the most negatively charged cellulose fibers, leading to a strong absorption of Au^{+} and concomitant uniform and densely distribution of AuNPs throughout the cellulose fibers, which improved SERS performance—increase in the EF to 3.29×10^{9} with an RSD of 8.5% and LOD for CV of 0.74 p.

Other paper-based SERS platforms based in different AuNPs synthesis strategies have also been reported, such as for the detection of Mucin-1 in whole blood with AuNPs prepared by chemical plating [70]; deposited by a dip-coating method with polymerized dopamine used for pesticide's detection [71]; prepared by citrate-based reduction in regular office paper for thiabendazole screening [50]; in situ synthesized in cellulose nanofibrils for *Escherichia coli* screening in water samples, [72] among others. [52,59,71]

Gold nanorods (AuNRs) possess transversal and longitudinal SPR which can be tuned to match the Raman excitation sources and/or analyte absorption leading to high enhancement factors. [73] Generally, AuNRs are synthesized through a seed-mediated method, using $HAuCl_4 \cdot 3H_2O$ as precursor, $NaBH_4$ as reducing agent and cetyltrimethylammonium bromide (CTAB) as surfactant to obtain a seed solution. This solution is then transferred to a growth solution, in acidic conditions, that contains $HAuCl_4 \cdot 3H_2O$, CTAB, $AgNO_3$ and, usually a weaker reducing agent than $NaBH_4$ is used, such as ascorbic acid. [64,68,73–78]

Several reports have combined the use of AuNRs with different types of cellulose-based substrates with various transference methods for a multitude of applications.

Lee and coworkers [68] presented for the first time a SERS swab technique for effective sample collection on surfaces. The conformable nature of a common filter paper combined with AuNRs, with aspect ratio around 4 (80 nm long and 20 nm in diameter), was used for the detection of trace quantities of target analytes by easily swabbing the surface on the paper SERS membrane. The 1,4-benzenedithiol (1,4-BDT) residue was spread over a 4 cm^2 surface and analyzed by SERS, through the characteristic peak at 1058 cm^{-1}, achieving a limit of detection as low as 140 pg with an EF of approximately 5×10^{6}.

A plasmonic ink of AuNRs modified with poly(styrenesulfonate) (PSS) was loaded into a ballpoint pen and used to "write" in a Whatman grade 1 filter paper, through a pen-on-paper method. [74] An aspect ratio of around 3 (88 ± 7 nm in length and 33 ± 2 nm in width) was attained for the AuNRs with LSPR peaks at 513 and 683 nm. The prepared paper-based SERS substrates were inserted in a stainless-steel microneedles patch for the simultaneous collection and analysis of interstitial fluids. R6G was used as model analyte, which bounds to the negatively charged PSS-AuNRs through electrostatic interactions, achieving a linear detection in the μM range.

Wu et al. [73] soaked Whatman grade 1 filter paper with a AuNRs solution, for 24 h at 40°C, for the detection of Sudan compounds (Fig. 6.8A). The prepared AuNRs presented an aspect ratio around 4 ± 0.5 (54 ± 4 nm in length

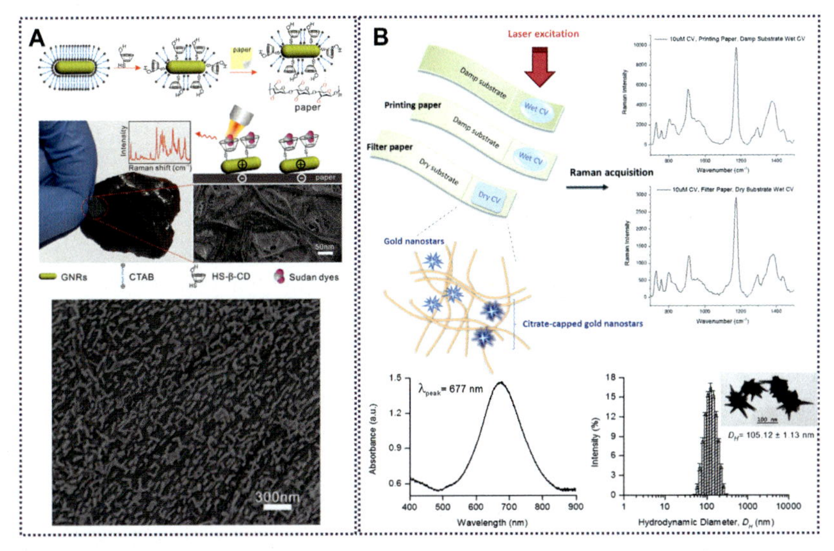

FIGURE 6.8 Au nanostructures-based SERS paper devices fabrication. (A) Workflow for the production of the AuNRs SERS device with photograph of adulterated Resina Draconin for detection of Sudan molecules and SEM image of AuNRs-paper. (B) Study of paper configurations for SERS measurement and characterization of the utilized AuNSs. *(Figure 6.8A) Reprinted with permission from Wu, M. et al., Spectrochimica Acta - Part A: Molecular and Biomolecular Spectroscopy,196, 2018. Copyright 2018, Elsevier. (Figure 6.8B) Reprinted, under a Creative Commons Attribution 3.0 Unported Licence, from Wu, M. et al., Spectrochimica Acta - Part A: Molecular and Biomolecular Spectroscopy,196, 2018. Published by the Royal Society of Chemistry.*

and 12 ± 2 nm in width) with plasmonic absorption peaks centered at 514 and 775 nm. Due to the hydrophobic nature of Sudan compounds, AuNRs were functionalized with mono-6-thio-cyclodextrin (HS-β-CD), through Au-S bonds. This compound served as an inclusion compound due to its cavity that features a hydrophilic exterior and hydrophobic interior, thus decreasing the distance between the analyte and AuNRs, yielding high enhancement factors. A quantification of Sudan III and Sudan IV was achieved using their characteristic peaks located at $1133\ cm^{-1}$ and $1097\ cm^{-1}$ with limits of detection of 0.1 and 0.5 μM, respectively. As a real-case scenario proof-of-concept, a small volume of different concentrations of Sudan III and IV was added to the surface of Resina Draconis samples. The paper SERS substrate was swabbed over the surface of the spiked Resina Draconis rendering a LOD of 17.62 and 38.04 ng/cm^2 for Sudan III and IV, respectively.

For the measurement of glucose levels in blood samples, Torul et al. [77] used a nitrocellulose membrane and wax printing to create a microfluidic channel to separate glucose form blood cells and proteins. The AuNRs were modified with 4-MBA and 1-decanethiol (1-DT) molecules, to facilitate the penetration of glucose molecules, and applied in the cellulose-based substrate by using a drop-dry technique. The B-OH band located at 1070 cm^{-1}, of the modified AuNRs, was found to decrease in intensity with increasing concentrations of glucose, allowing a limit of detection of 0.1 mM with a linear range from 0.5 to 10 mM.

Gold nanostars (AuNSs) can generate high Raman signal enhancements due to the increased number of hot spots created by their multiple branches and sharp tips. [65,79] The use of AuNSs for SERS enhancement combined with paper platforms has already been reported.

He et al. [65] used crystal violet (CV) as the Raman probe molecule for the full optimization of paper-based SERS platforms with AuNSs (Fig. 6.8B). The authors followed a one-pot protocol with the reduction of $HAuCl_4 \cdot 3H_2O$ by ascorbic acid in the presence of silver nitrate, to shape the nanostructures, followed by a capping process with Na_3Ct. The greenish blue AuNSs colloidal solution, with a plasmonic peak at 677 nm, was dipped onto filter and printing paper substrates, and several conditions were tested to evaluate their influence in the Raman EF. These included dry substrate-dry CV, dry substrate-wet CV, damp substrate-dry CV, and damp substrate-wet CV. In addition, to the influence of the capping agent, the quantity of AuNSs immobilized on the paper substrate and the acquisition conditions of Raman spectra were also evaluated. The authors concluded that with exception for the configuration damp substrate-wet CV, where the EF was comparable, filter paper substrates rendered higher EFs than printing paper. This behavior was justified by the high porosity of the filter paper, which facilitated the immobilization of AuNSs and the light penetration into the fibers to excite the AuNSs located deeper in the fibers. Higher EFs were obtained with wet CV, for both papers, and no variation was observed on filter paper when damp or dry, due to its hydrophilic nature that facilitated the interaction of CV with AuNSs even when the substrate was dry. On the contrary, a higher EF was observed for damp printing paper also due to the easier interaction between enhancement nanostructure and analyte. The optimized filter paper dry substrate-wet CV condition was directly compared with commercial SERS substrates, showing superior performance, with a calculated EF of 1.2×10^7 for 1 nM CV.

Gao and coworkers [79] applied an immunoassay to a lateral flow strip for detection of neuron-specific enolase in blood plasma, with AuNSs. The synthesis of AuNSs was carried out with a seed-growth method with $HAuCl_4 \cdot 3H_2O$ as precursor, Na_3Ct and $NaBH_4$ as stabilizing and reducing agents and PVP as shaping agent. AuNS@Raman Reporter@silica sandwich nanoparticles were used as SERS probe achieving a highly sensitive device with a limit of detection of 0.86 ng/mL and linear detection range from 1 to 75 ng/mL.

Other gold-based nanostructures have also been reported such as gold nanobipyramids, deposited in Whatman paper through plasmonic calligraphy, for streptavidin detection [66]; black phosphorus-Au deposited on filter paper

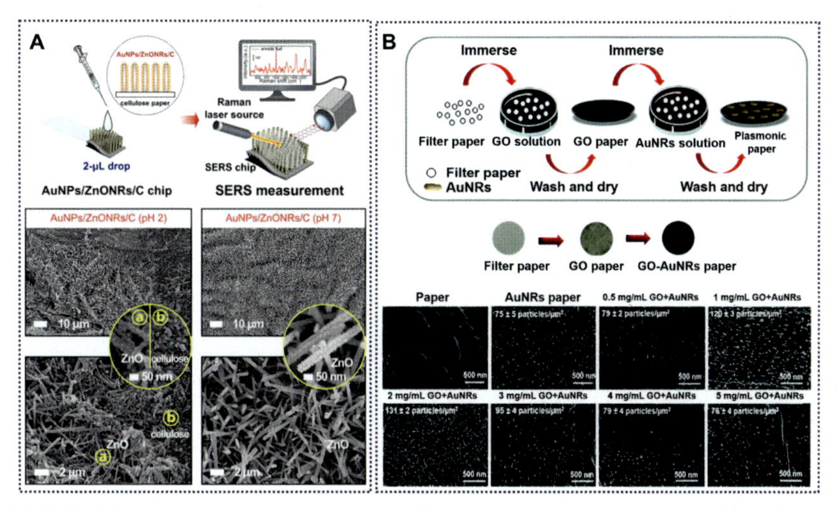

FIGURE 6.9 Au nanostructures with metal oxides and graphene-based materials. (**A**) ZnONRs decorated with AuNPs synthesis and measurement strategy and SEM images of the synthetized nanostructures at different pH values. (B) Workflow schematic for the preparation of GO-AuNRs paper and SEM images of bare and AuNRs paper with different quantities of GO. *(Figure 6.9A) Reprinted with permission from Kim, W. et al., ACS Nano,12(7), 2018. Copyright 2018, American Chemical Society; (Figure 6.9B) Reprinted, under a Creative Commons Attribution 3.0 Unported Licence, from Ponlamuangdee, K. et al., New Journal of Chemistry,44 (33), 2020. Published by the Royal Society of Chemistry (RSC) on behalf of the Centre National de la Recherche Scientifique (CNRS) and the RSC.*

for rapid detection of foodborne bacteria [80] and Au sputtered on paper for sensitive detection of heavy metal ions. [81] Moreover, Au nanostructures combined with other materials, on cellulose-based SERS platforms, have also been reported: Au@Fe_3O_4 NPs with bacterial cellulose nanocrystals for magnetic separation and detection of malachite green and isothiocyanate [27]; ZnONRs decorated with AuNPs (Fig. 6.9A) for human amniotic fluids analysis [26]; graphene oxide and AuNRs prepared by layer-by-layer immersion on filter paper (Fig. 6.9B) for mitoxantrone anticancer drug screening [30]; graphene oxide-AuNSs on filter paper [31] or graphene-isolated-Au-nanocrystals for free bilirubin measurement [82] and AuNRs with reduced graphene oxide for sulfur dioxide sensing in wine samples. [83]

6.2.2 Silver-based nanostructures

Silver nanoparticles (AgNPs) are reported in the literature as the contributors for the highest Raman amplifications. Several publications have shown their synthesis procedures similar to AuNPs synthesis, or even through physical processes, and different transference methods for cellulose and paper-based substrates.

Table 6.2 resumes the main characteristics of representative cellulose-based SERS platforms with silver nanostructures.

TABLE 6.2 Silver-based SERS devices fabricated onto cellulose substrates.

Reference	Nanomaterial	Cellulose format	Fabrication method	Analyte	LOD	EF
Chen et al. [84]	AgNPs	Nanofibrillated cellulose paper	In situ SILAR synthesis	RB2-NT	1.3×10^{-12} M 172×10^{-15} M	1.46×10^{9}
Das et al. [85]	AgNPs	Filter paper	In situ silver mirror reaction	R6GRB	10^{-12} M10^{-9} M	1.42×10^{10} 0.659×10^{6}
Ferreira et al. [86]	AgNPs	Nanocellulose	In situ hydrothermal synthesis	R6GExosomes	10^{-11} M10^{9} p/ml	10^{5}
Araújo et al. [87]	AgNPs	Cardboard packaging	In situ thermal evaporation	R6G	-	1×10^{6}
Oliveira et al. [88]	AgNSs	Office paper	Chemical synthesis and drop casting	R6G	11.4 ± 0.2 pg	3×10^{7}
Xu et al. [89]	AgNF/AgNP arrays	Whatman grade 1 paper	In situ chemical synthesis	CVR6G	10^{-21} M10^{-14} M	-
Xiao et al. [90]	S-rGO/AgNPs	Weighing paper	Hummers method and commercially obtained AgNPs with gravure and inkjet printing	MGR6G	10^{-7} M	10^{9}
Araújo et al. [29]	ZnONRs decorated with AuNPs	Cellulose paper	In situ hydrothermal method and thermal evaporation	R6G	$8.9 \pm 0.9 \times 10^{-9}$ ng	10^{7}
Ogundare and van Zyl [91]	AgNPs@SiO_2	Nanocrystalline cellulose	In situ chemical synthesis	MG	0.9 nM	-

2-NT, 2-Naph-thalenethiol; *AgNF*, silver nanoflower; *AgNPs*, silver nanoparticles; *AgNSs*, silver nanostars; *CV*, crystal violet; *MG*, malachite green; *R6G*, rhodamine 6G; *RhB*, rhodamine B; *SILAR*, successive ionic layer absorption and reaction; *S-rGO*, sulfonated reduced graphene oxide; *ZnONRs*, zinc oxide nanorods.

Rajapandiyan et al. reported a photochemical method for the fabrication of AgNPs-filter paper for SERS applications. [92] In this work, 3 mm circular pieces of filter paper were immersed in a solution containing $AgNO_3$ and Na_3Ct and illuminated with a 365 nm, 3 W light emitting diode. The additional presence of sodium hydroxide (NaOH) in the mixing solution was evaluated showing a higher coverage of the filter paper with AgNPs. *Para*-nitrothiophenol (pNTP) was used as model molecule to optimize the parameters involved in the preparation of this SERS platform. Different illumination times were tested (1, 3, and 5 h) rendering an increase in the AgNPs diameter (from 27.4 ± 9.5 to 75.0 ± 17.5 nm), resulting in an increase on AgNPs density, which led to an increase in SERS intensity. Optimal concentrations of all reactants were also carefully studied. The fabricated SERS substrate was applied to adenine sensing where a linear range was observed for the characteristic band located at 735 cm^{-1}, with a correlation coefficient higher than 0.99, a limit of detection of 160 nM, an RSD of 7.6% and an EF of $2.1 \pm 0.1 \times 10^7$.

AgNPs-decorated filter paper, with a hydrophobic treatment, demonstrated an subnanomolar sensitivity for pesticide detection (Fig. 6.10A). [93]

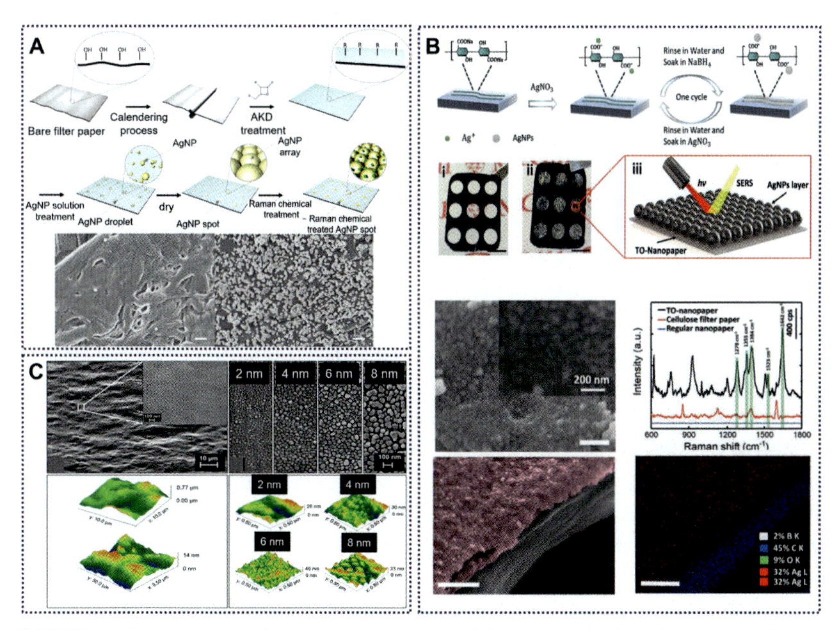

FIGURE 6.10 AgNPs-based SERS paper devices fabrication. (A) Schematic workflow for the production of hydrophobic SERS paper and SEM image of bare paper and AgNPs-treated paper. (B) TEMPO-oxidized nanopaper for AgNPs synthesis schematic strategies and respective SEM and chemical analysis and Raman spectra. (C) Cardboard packing as substrate for thermally evaporated AgNPs, SEM, and AFM analysis. *(Figure 6.10A) Reprinted with permission from Lee, M. et al., ACS Sensors, 3(1), 2018. Copyright 2018, American Chemical Society. (Figure 6.10B) Reprinted with permission from Chen, L. et al., Advanced Materials Interfaces, 6 (24), 2019. Copyright 2019, John Wiley and Sons. (Figure 6.10C) Reprinted with permission from Araújo, A. et al., Nanotechnology, 25(41), 2014. Copyright 2014, IOP Publishing.*

AgNPs were obtained by a citrate-based reduction of $AgNO_3$, showing a plasmonic band centered at 422 nm, with an average diameter of 92 ± 21 nm. The fabrication of the paper SERS platform consisted on a calendaring process, to reduce its roughness (9.4 to 4.0 μm), followed by treatment with alkyl ketene dimer, to create a hydrophobic surface (contact angle from 15 to 114 degrees), and deposition of 2 μl AgNPs droplets, dried at room temperature. The hydrophobic treatment allowed for a higher density of AgNPs on the paper surface and higher retaining of the analyte. As a test analyte, 4-ATP was used to evaluate the effect of the hydrophobic treatment compared to untreated filter paper, showing higher SERS intensities due to the high number of hot-spots formed by AgNPs clusters. The influence of AgNPs concentration on SERS response was evaluated through simulation, correlated with E-field enhancement, showing an increase with increased concentrations of AgNPs up to an optimal value. Then a decrease is observed due to delocalization of the E-field throughout large structures and hindrance of irradiation and the scattering of light. The reproducibility and sensitivity of the optimized substrate was studied, showing an RSD of 6.19% and a LOD of 0.60 nM. Lastly, the pesticides thiram and ferbam were analyzed attaining calculated limits of detection of 0.46 and 0.49 nM, respectively.

Cellulose fibers recycled from waste paper were used as support for SERS detection. [94] The cellulose fibers were treated with glycidyltrimethylammonium chloride to add positive charges on their surface (cationization) to be decorated with negatively charged AgNPs, by electrostatic interactions. One percent Na_3Ct was added to a $AgNO_3$ aqueous solution to obtain AgNPs with a 414 nm maximum absorption band, and diameters between 40 and 60 nm. The Raman reporter molecule *p*-aminothiophenol was used to evaluate SERS performance achieving a LOD of 10 ppb. Additionally, the characteristic peaks of bisphenol A (BPA) were used for its detection on aqueous solutions, obtaining a linear range from 100 to 0.01 ppm, contrary to AgNPs assembled on a glass substrate, where at 0.1 ppm practically no SERS signal was obtained. The higher sensitivity of the cellulose fibers was explained by the higher number of hot spots created by their shrinking. Interference studies were performed with thiram, showing distinct peaks from BPA, and soda water was spiked with different concentrations of BPA, showing obvious SERS signal at 0.05 ppm, which is the limit defined by the European Union.

Xian et al. [95] reported a composite paper-based SERS device employing citrate-AgNPs obtained by in situ reduction, with cellulose nanocrystals (CNC) for phenylethanolamine A and metronidazole detection. The CNC-Ag composites, obtained by the thermal reduction of silver nitrate ($AgNO_3$) with Na_3Ct were coated onto filter paper through vacuum filtration. The composites filled the filter paper porous, thus enhancing reproducibility (RSD of 4.29%), and the additional immersion on hydrophobic dodecyl mercaptan improved sample collection, preventing analyte random diffusion trough the paper-SERS substrate. An optimization process on CNC-Ag composite volume, to obtain different thicknesses, and on dodecyl mercaptan concentration was carried out. SERS spectra of R6G was recorded at different concentrations, showing a linear range from 10^{-7} to 10^{-9} M with a $R^2 = 0.978$. Lastly, the SERS substrate was tested for practical use achieving limits of detection of 5×10^{-9} and 2×10^{-7} M for phenylethanolamine A and metronidazole, respectively.

Based on a SILAR technique with $NaBH_4$ as reducing agent, Chen and coworkers developed a nanofibrillated cellulose (NFC)-paper-based SERS multiwell plate (Fig. 6.10B). [84] The highly mechanically stable NFC paper, due to its densely packed nanofibers with nanometer-sized porous, was patterned by wax printing to create multiwells for the in situ reduction of $AgNO_3$ to AgNPs. Prior to the in situ reduction, the NFC paper was treated with 2,2,6,6-(tetramethylpi-peridin-1-yl)oxyl (TEMPO) to promote the partial oxidation of cellulose, rendering a high amount of carboxyl groups. These carboxyl groups facilitate the formation of a uniform continuous layer of AgNPs with high density per surface area, due to the electrostatic interaction between Ag^+ and negatively charged carboxyl groups. Two Raman model analytes were used to evaluate the performance of the prepared NFC SERS paper—rhodamine B (RhB) and 2-naph-thalenethiol (2-NT), with an EF of 1.46×10^9, an RSD $\leq$ 11% and limits of detection of 1×10^{-12} and 172×10^{-12} M, respectively.

A report on the optimization of a spray deposition method to fabricate SERS substrates was described for the first time, in 2013, by Li, B. and coworkers. [96] A wax-patterned Whatman grade 1 paper was used for the spraying of colloidal solutions of AgNPs, prepared by the chemical reduction of $AgNO_3$ with two different reductants: hydroxylamine hydrochloride (HH) and $NaBH_4$, with Na_3Ct as stabilizing agent, both at room temperature. A difference on the NPs size was obtained with 69 and 38 nm, for HH and $NaBH_4$, respectively, which resulted in higher SERS signals for the bigger particles, when tested with R6G. A direct comparison between the spray and droplet method showed a significant drop on RSD (64.4% to 14.5%, $n = 20$), due to a more even distribution of AgNPs throughout the paper surface, with the spray method, avoiding coffee-ring effects. The number of spray cycles was optimized, showing an increase in SERS intensity up to 25 cycles, followed by a plateau from 25 to 35 cycles. The 25 cycles were selected as the optimal number of cycles. An evaluation on SERS performance was carried out showing an EF of 2×10^7. Additionally, the spray method was applied to several types of paper substrates, displaying the versatility and adaptability of the method.

Silver mirror reaction was reported in several papers as a technique to prepare Ag/paper-based SERS platforms. [85,97–100] Cheng et al. tested several types of filter papers with a silver mirror reaction for SERS sensing of tyrosine in aqueous solutions. [100] The silver mirror reaction was based on the vertical placement of filter paper strips in a plating solution, called Tollen's reagent, composed by 50 mM $AgNO_3$, 300 mM ammonia and 500 mM glucose, as the reducing agent—which is placed in a 55°C water bath, for a certain time. The as-prepared substrates were immersed in 1 mM probe solutions of either *p*NTP, *p*ATP, or *p*MBA. The recorded spectra showed an increase in SERS signal of 5, 40, and 200 times, respectively. While the fivefold increase for *p*NTP can be explained by the multilayer structure of the AgNPs on fibrous filter paper, the higher SERS signals increased for *p*ATP and *p*MBA is also explained by the formation of dynamic hot spots, induced by their dual functional groups, increasing SERS signals. Filter papers with different pores sizes were pretreated with 5% ammonia for 1 h, prior to the silver mirror reaction, to study the influence on SERS performance. No significant changes were observed on SERS intensities showing the independence of SERS performance from pores sizes, as long as the filter papers are rough enough to allow the synthesis of AgNPs in between the paper fibers. The EF values for *p*NTP, *p*ATP, and *p*MBA were calculated as 1.9×10^7, 1.4×10^7, and 3.2×10^7, respectively, an order of magnitude higher than the counterpart glass slides with silver mirror reaction. Finally, the optimized paper SERS platform was used for tyrosine detection following the characteristic band located at 1594 cm^{-1}, from υ (C = C). A 40-fold increase was observed, explained by the formation of dynamic hot spots, by the three functional groups of tyrosine, achieving a limit of detection of 625 nM.

Saarinen and coworkers established cost-effective routes for roll-to-roll fabrication of paper-based SERS devices with AgNPs. [101] Liquid flame spray technique was used for the direct synthesis and deposition of AgNPs onto graphite carbon-coated paper. Chemical reduction of $AgNO_3$ was achieved with gas flow ratios of 55 l/min (H_2) and 5 l/min (O_2) for the production of 90 nm-sized AgNPs and 40 l/min (H_2) and 20 l/min (O_2) for 50 nm-sized AgNPs. To evaluate the SERS performance of the as-prepared substrates, CV was measured at different concentrations showing its characteristic peaks with 100 nM.

Cellulose self-reducing properties to obtain AgNPs has been reported for paper-based SERS sensing. [102–104] Cellulose nanofibril (CNF) was used for the hydrothermal reduction of $AgNO_3$ to ~22 nm-sized AgNPs attached to the CNF network. [102] The reduction capabilities of CNF was explained by the high number of active functional groups (hydroxyl and carboxyl groups) and a high specific surface area, which maximizes the interaction of Ag^+ ions with the functional groups. The CNF-AgNPs suspension was subject to a low temperature freezing process followed by freeze drying to obtain a sponge-like structure, with a porous structure. In addition to using this CNF-AgNPs sponge for catalytic degradation, its SERS properties was evaluated with RhB. The characteristic peaks of RhB were observed from 5×10^{-3} to 5×10^{-7} M, with a logarithmic linear relationship. For the detection of riboflavin, Ogundare et al. isolated nanocrystalline cellulose (NCC) from discarded cigarette filter, and used it as reducing and stabilizing agent of AgNPs. [103] To a 0.1 wt.% NCC dispersion, adjusted to several pH values, different concentrations of $AgNO_3$ were added, and the mixing solution was maintained at 80°C, for a certain period of time. A study on the effect of pH, precursor concentration, and reaction time on the AgNPs synthesis was carried out. It was demonstrated that alkaline-induced synthesis renders monodispersed and highly concentrated AgNPs, due to an increase of hydroxyl groups available to reduce Ag^+ ions. Moreover, a red shift on the absorption band was observed for higher reaction times and precursor concentration. Finally, the as-prepared SERS membranes were used for riboflavin quantification, with increasing concentration from 0.00001 to 10 μM, showing a LOD of 0.3 μM, with a linear range up to 5 μM ($R^2 = 0.9221$).

Besides chemically obtained AgNPs, some works have reported physical techniques to prepare NPs onto paper-based substrates. Using a pulsed laser deposition (PLD) technique, Khan et al. [105] prepared paper SERS substrates, with a 20 W fiber laser (1064.5 nm) using a high repetition rate, 33 kHz, 250 ns pulse duration, and 30 J cm^{-2} laser fluence. This substrate was used to enhance R6G signal, and directly compared with a commercial SERS substrate, showing significant similarity concerning EF and signal-to-noise ratio. From cardboard packaging substrates, Araújo et al. [87] fabricated a AgNPs-based substrate for Raman enhancement (Fig. 6.10C). The fabrication method employed—thermal evaporation assisted by an electron beam—induced a direct arrangement of the AgNPs with good control on shape and size, in a one-step approach. The deposition of different thicknesses of Ag film (2 to 8 nm), at 150°C, rendered densely distributed AgNPs with an elongated shape, with increasing diameters. An electromagnetic model based on the single-particle Mie theory formalism was used to study the localized near-field light enhancement originated in the vicinity of plasmonic NPs. This has shown that for this system the optimal NP diameter is ~60 nm so, the 6 nm-Ag film thickness was selected for further SERS measurements since it renders, on average, 60 nm elongated AgNPs. Additionally, the LSPR of the as-prepared cardboard SERS platform, located at 660 nm, closed to the incident laser of 633 nm, allowed achieving a Raman EF up to 10^6 and RSD lower than 5%, with R6G as model molecule.

Silver nanoflowers (AgNFs) have also been reported associated with paper-based SERS platforms, due to their high surface area and large size and capability to fill the paper' porous, thus avoiding losses of incident lasers. [89,106] Xu et al. [89] published a method for paper-based SERS sensing with synthesized three-dimensional AgNFs/AgNPs arrays, to enhance electromagnetic coupling (Fig. 6.11A). For the fabrication of nanoarrays a piece of paper was soaked three times in acidic aqueous solutions of 0.2 M $SnCl_2$ followed by 0.2 M $AgNO_3$, for two minutes, and washed with acetone and water. Afterward, the piece of paper was soaked in a mixed solution of 0.1 M

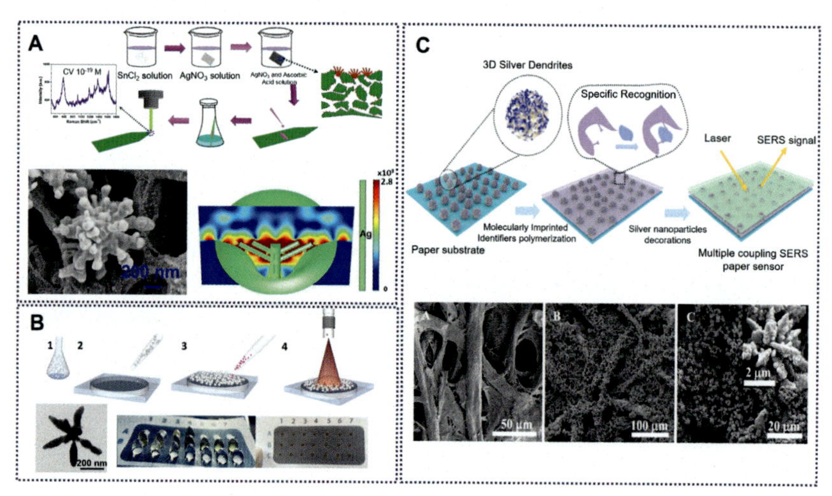

FIGURE 6.11 Ag nanostructures-based SERS paper devices fabrication. (A) Workflow for the production of 3D AgNF/AgNP arrays onto paper substrate, respective SEM image and simulation image of the electric field distribution. (B) Fabrication process and use of AuNSs-paper (1—synthesis, 2— wax patterning and AuNSs deposition, 3—addition of R6G, and 4 – SERS measurement) with TEM image of AuNs and photographs of AuNSs-paper before and after drying. (C) Workflow of the construction of multiple coupling enhancement SERS paper substrates and SEM images of bare paper and paper decorated with SDs. *(Figure 6.11A) Reprinted with permission from Xu, Y. et al., Sensors and Actuators, B: Chemical, 265, 2018. Copyright 2018, Elsevier. (Figure 6.11B) Reprinted with permission from Oliveira, M. et al., Scientific Reports, 7(1), 2017. Copyright 2017, Springer Nature. (Figure 6.11C) Reprinted with permission from Zhao, P. et al., ACS Applied Materials and Interfaces, 12(7), 2020. Copyright 2020, American Chemical Society.*

$AgNO_3$ and 0.1 M ascorbic acid, for a certain time. R6G and CV were used as Raman probe molecules to evaluate the performance of the fabricated substrate. Their characteristic peaks were still visible at 10^{-14} M, with an RSD of 13.23% and 11.68% for R6G and CV, respectively. Additionally, the authors created a preconcentration device, with AgNFs/AgNPs arrays, by cutting a rectangular paper strip with a 30 degree sharp tip in one side, to facilitate sample loading and concentration of the analyte in the sharp tip. Using CV and R6G as model analyte to test the developed architecture, it was possible to detect concentrations as low as 10^{-19} M and 10^{-20} M, respectively.

Silver nanostars (AgNSs), that are structurally similar to AgNFs, have been reported associated with a regular office paper substrate for SERS applications (Fig. 6.11B). [107] Citrate-capped AgNSs were obtained by the reaction of an hydroxylamine alkaline solution with $AgNO_3$, followed by addition of Na_3Ct, rendering a nanostructure with approximately 8 tips per star and a tip-to-tip length of ~200 nm. A 10 μl volume of AgNSs were drop-casted onto prewax printed regular office paper, at different concentrations, ranging from 0.22 to 7.04 nM. The as-prepared SERS platform was directly compared with Whatman grade 1 paper with AgNSs and 23 nm-sized AgNPs and with office paper with AgNPs, prepared with the same conditions. By evaluation of the EF, using R6G as model molecule, the office paper prepared with AgNSs rendered the highest value [$(3.0 \pm 0.7) \times 10^7$], showing a superior SERS performance when compared with the other architectures. This was explained by the anisotropy of the AgNSs and by the hydrophobic nature of paper, which facilitates a higher concentration of nanostructures on its surface.

Silver dendrites (AgDs) represents a multiple hot spot nanostructure that can significantly enhance Raman signals with its sharp tips and edges and narrow gaps between each trunks and branches. [108] In 2019, Wang et al. [108] fabricated paper-based superhydrophobic SERS substrates for the detection of nitenpyram. The fern-like silver dendrites (F-AgNDs) structures were fabricated by a replacement reaction method using zinc and silver nitrate. Precleaned Zn plates were immersed in an $AgNO_3$ solution, for a certain time, and the as-produced silver was collected from solution and purified. The prepared F-AgNDs were deposited onto polyvinylpyrrolidone (PVP)-treated filter paper, which was then immersed in an octyltrimethoxysilane (OTMOS) solution, for a certain time, to achieve a rough superhydrophobic paper substrate. The optimal reaction time for the formation of F-AgNDs was studied, obtaining with 40 s, $\sim 12\ \mu m$ trunks with $\sim 6\ \mu m$ branches. The measurement of different concentrations of nitenpyram' SERS spectra in the as-prepared SERS substrate revealed an RSD of 9.9%, an EF of 3.4×10^7 and a limit of detection of $\sim 10^{-9}$ M, obtained with PCA. Additionally, the authors used their paper SERS substrate on the surface of apples, spiked with different nitenpyram concentrations, obtaining a LOD of 4.6×10^{-8} M.

More recently, in 2020, Zhao and coworkers [109] developed a paper sensing platform with three-dimensional AgDs, a molecularly imprinted identifier and AgNPs sandwich for the quantification of the insecticide neonicotinoids (Fig. 6.11C). Whatman grade 2 paper was used as platform for the in situ fabrication of AgDs through chemical reduction of 0.4 M $AgNO_3$ by 1.6 M hydroxylamine. On the other side of the paper substrate, conductive graphite ink was screen-printed on its surface, where the molecular identifiers were electropolymerized with pyrrole as monomer and imidacloprid as target analytes. Finally, after adsorption of target analytes, a layer of AgNPs was fabricated on top of the electropolymerized identifiers, through chemical reduction of $AgNO_3$ with ascorbic acid, to amplify the SERS signal intensities. A detection limit as low as 0.02811 ng/mL was obtained for imidacloprid with high sensitivity and selectivity when tested against other interferents.

Other silver-based nanostructures have also been reported, such as Ag triangular nanoplates dropped on filter paper [110]; Ag nanocubes electrosprayed onto cellulose acetate microspheres [111]; and Ag nanotriangles loaded on filter paper for the detection of ultratrace levels of the explosive

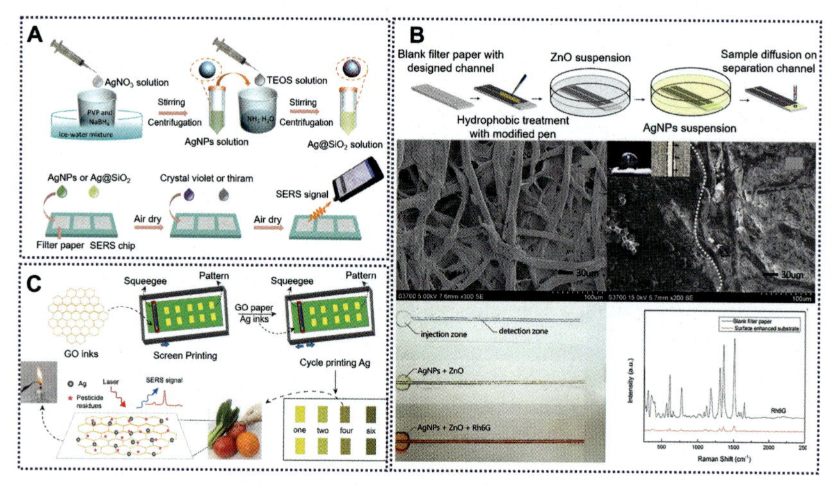

FIGURE 6.12 Ag nanostructures with metal oxides and graphene-based materials. (A) Preparation workflow of Ag@SiO_2 NPs SERS chips with a smartphone Raman analyzer. (B) Schematic representation of paper separation channel SERS device workflow with SEM images of bare filter paper and separation channel, photograph of the separation channel, and comparison of SERS spectra for R6G in bare filter paper and SERS paper. (C) Schematic representation of the procedure for the preparation of AgNPs/GO paper SERS devices. *(Figure 6.12A) Reprinted with permission from Sun, M. et al., Journal of Materials Science and Technology, 35 (10), 2019. Copyright 2019, Elsevier. (Figure 6.12B) Reprinted with permission from Jin, X. et al., Spectrochimica Acta - Part A: Molecular and Biomolecular Spectroscopy,240, 2020. Copyright 2020, Elsevier. (Figure 6.12C) Reprinted, under a Creative Commons Attribution 3.0 Unported Licence, from Ma, Y. et al., Analytical Methods, 10(38), 2018. Published by the Royal Society of Chemistry.*

piric acid. [112] Additionally, the combination of Ag nanostructures with other materials, such as metal oxides and graphene-based materials has also been linked with paper-based SERS devices. As example, Ag@SiO_2 has been used extensively as nanocubes deposited on filter paper, by vacuum filtration, for melamine screening in milk samples [28] and with a Fe-TiO_2 nanosheets modified paper for thiabendazole analysis, [113] or combined with nanocrystalline cellulose for malachite green detection, [91] or with a strawberry-like morphology deposited onto filter paper for acrylamide detection [114] or as core-shell nanoparticles, on filter paper, for thiram residues quantification (Fig. 6.12A). [115] Metal oxide ZnO has also been reported for Raman enhancement decorated with AgNPs and prepared on a cardboard packaging substrate, [29] applied together with an AgNPs suspension on a paper microfluidic device for the separation and detection of pesticides (Fig. 6.12B) [116] or combined with a graphene-nano-mesh for environmental pollutants screening, achieving an EF of 10^{11}. [32] Finally, an AgNPs ink with graphene oxide was screen-printed on a paper substrate for thiram, thiabendazole, and methyl parathion detection (Fig. 6.12C), [117] other work used arrays of graphene oxide isolated AgNPs, prepared by wet chemical synthesis, and deposited on hydrophobic paper, achieving an limit of detection of 10^{-19} M for R6G [33] and weighing paper coated with reduced graphene oxide/AgNPs for malachite green sensing [90] have also been reported.

6.2.3 Gold and silver composites nanostructures

The fabrication of paper-based SERS substrates based on plasmonic nanostructures has been extensively reported in the past years. The combination of Au and Ag nanostructures for Raman enhancement can benefit from the characteristics of both metals individually, such as from the high ability of Ag nanostructures for Raman enhancement and from the chemically stability and easiness of bio-functionalization of Au nanostructures. [13,118,119]

Table 6.3 resumes the main characteristics of representative cellulose-based SERS platforms with gold and silver composites nanostructures.

Different strategies and architectures of these nanostructures have been reported. For instance, the combination of paper substrates and Au@AgNPs has been reported with different morphologies, such as nanorods, [121] nanocubes, [122] and nanoparticles. [25,120,123] Zhu and coworkers developed a paper-based microfluidic device for the screening of thiram residues with core-shell Au@AgNPs (Fig. 6.13A). [123] To citrate-based 30 nm-sized AuNPs, used as core NPs, the reducing agent ascorbic acid and the precursor $AgNO_3$ were added, for the synthesis of AgNPs on the surface of AuNPs, reaching 7 nm thickness. A plasmon resonance from 350 to 560 nm was observed for the as-prepared NPs, with a zeta potential value of -21.7 mV. A patterned microfluidic device was fabricated, using Whatman grade 1 paper, with three independent regions: region 1 for drop-casted Au@AgNPs, region 2 for analyte adding, and region 3 for the reaction of the analyte with the plasmonic structures and subsequent SERS measurement. The adsorption of thiram onto the synthesized Au@AgNPs, in the reaction chamber, augmented the chemical enhancement, with a 2×10^6 factor and a LOD down to 1×10^{-9} M. Additionally, the authors successfully measured thiram residues in adulterated tea samples, with recoveries from 95% to 110%.

In addition to Au@AgNPs core-shell nanostructures, the reverse was also reported in the literature—Ag@AuNPs—associated with cellulose-based SERS devices. [13,20,23] Based on the SILAR technique, Whatman grade 1 paper was used as the substrate for the preparation of Ag@AuNPs (Fig. 6.13B). [13] Briefly, AgNPs were first in situ synthesized onto paper, with six cycles of 20/20 mM $AgNO_3$/$NaNB_4$ (conditions previously optimized by the authors) [14], followed by the AuNPs synthesis. Despite the previously optimized conditions for the SILAR synthesis of AuNPs onto paper substrates (six cycles 10/10 mM $HAuCl_4 \cdot 3H_2O$/$NaBH_4$) [61], herein the number of cycles was decreased to two and the concentration of reactants to 1 mM. This was explained by the fact that with a higher number of cycles and higher reactant's concentration the AuNPs formed had a superior

TABLE 6.3 Gold and silver composites SERS devices fabricated onto cellulose substrates.

Reference	Nanomaterial	Cellulose format	Fabrication method	Analyte	LOD	EF
Sha et al. [25]	Au@Ag core-shell NPs	Filter paper	Au Seed-mediated growth and Ag chemical synthesis and self-assembly	Estazolam	5 mg/l	-
Asgari et al. [120]	Au@Ag core-shell NPs	Nanofibrillar cellulose	Chemical synthesis and mixing	ThiramParaquat	71 μg/l46 μg/l	$\sim 10^4$
Park et al. [24]	Au/Ag alloy	Chromatography paper	In situ concurrent thermal evaporation	Folic acid	1 pM	-
Kim et al. [13]	Ag@AuNPs	Whatman grade 1 paper	In situ SILAR synthesis	R6GAnilineSodium azideMGBCa progression	1 fM1uM10 nM100 pM-	3.98×10^8

BCa, breast cancer; *MG*, malachite green; *R6G*, rhodamine 6G; *SILAR*, successive ionic layer absorption and reaction.

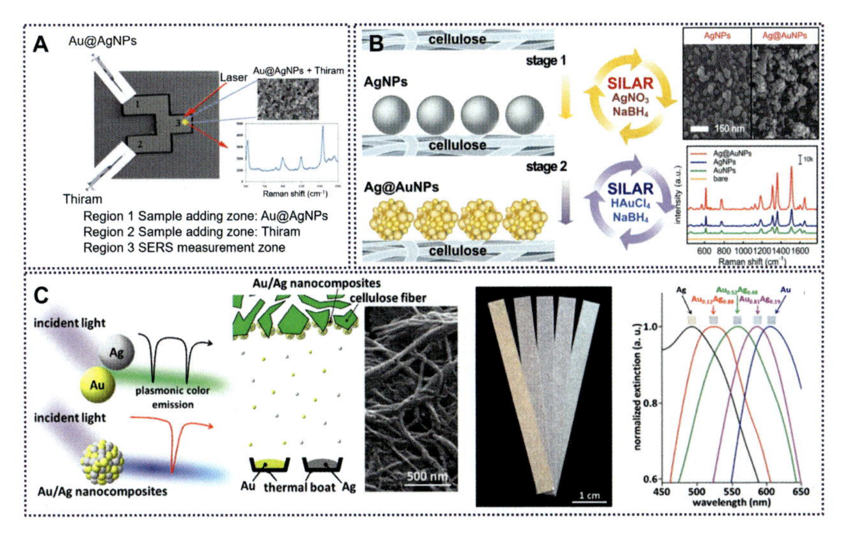

FIGURE 6.13 Au and Ag composites-paper SERS devices. (A) Paper-based microfluidic device with Au@AgNPs for thiram detection. (B) Schematic representation of the Au@AgNPs production, onto paper substrates, by a SILAR technique and SEM images and SERS comparison between AgNPs and Ag@AuNPs. (C) Paper-based Au/Ag nanocomposites with plasmonic tunable by Ag/Au ratio composition with respective photograph and extinction spectra. *(Figure 6.13A) Reprinted, under a Creative Commons Attribution 3.0 Unported Licence, from Zhu, J. et al., Analytical Methods, 9(43), 2017. Published by the Royal Society of Chemistry; (Figure 6.13B) Reprinted with permission from Kim, W. et al., Analytical Chemistry, 89(12), 2017. Copyright 2017, American Chemical Society; (Figure 6.13C) Reprinted with permission from Park, M. et al., ACS Applied Materials and Interfaces, 10(1), 2018. Copyright 2018, American Chemical Society.*

diameter than AgNPs, making AuNPs the main contributors for SERS activity. In fact, finite element method simulations that where corroborated experimentally, showed a decrease on electric field from two cycles to four cycles, which was translated in a decrease of Raman enhancement. With two SILAR cycles, clusters of 4 nm-sized AuNPs were formed on the AgNPs surface, decreasing the interparticle distance and increasing the number of hot spots. However, four SILAR cycles rendered a thick layer of 6-nm gold, disappearing the nanoparticles gap and surface roughness, which ultimately decreased SERS signals. For the evaluation of SERS activity of the bimetallic paper, R6G was measured at different concentrations (1 fM to 1 μM), showing a linear behavior with an EF, for 1 fM, of 3.98×10^8, an RSD of 6.1%. Moreover, it was observed a small decrease of 10% of Raman intensities after 30 days, whereas for the AgNPs paper platform a 75% signal decrease occurred. The platform was used on real-life applications for the detection of aniline, sodium azide, and malachite green, through their characteristic prominent peaks, achieving linear ranges within the established maximum allowed concentrations. Additionally, the platform was used for breast cancer screening in urine samples. Similarly, Rijo et al. used the same SILAR technique applied to regular office paper for breast cancer screening through exosomes profiling. [20,124]

Through thermal evaporation Au/Ag nanocomposites were deposited on chromatography paper and employed in the detection of folic acid for cancer screening (Fig. 6.13C). [24] The existence of a single SPR peak proved the formation of a plasmonic alloy with tunable plasmon resonance wavelength (from 499 to 606 nm), by varying the Au/Ag ratio, showing a red shift as the Au content increased. By matching the SPR with the Raman extinction wavelength the highest SERS intensities were obtained. In this, $Au_{0.12}Ag_{0.88}$ was used, with a 514 nm excitation laser, to detect of folic acid, using the characteristic peak at 1604 cm^{-1} and rendering a limit of detection down to 1 pM.

6.3 Applications of paper and cellulose-based SERS devices

The high selectivity provided by Raman spectroscopy coupled to the high sensitivity enabled by SERS substrates has prompted its use on a multitude of applications in different fields, such as in healthcare—for diagnostic and therapeutic studies and follow-up—and in environmental settings—for pesticides detection, antibiotics control, pathogen agents, and heavy metals screening, among others. Herein, a revision of the state-of-art on applications with cellulose and paper-based SERS substrates is presented.

6.3.1 Health applications: diagnostic and therapeutics

Cancer constitutes the second leading cause of mortality worldwide, only surpassed by cardiovascular disease. [125] Based on GLOBOCAN estimates, about 18 million new cancer cases and 9.5 million deaths occurred in 2018 worldwide. [126] The diagnosis of cancer in the current scenario is made by detection techniques that are usually expensive, invasive and time-consuming. [127] Thus, the development of simple, reliable, cost-effective and noninvasive detection methods for early cancer screening/diagnosis and posterior follow-up is particularly important, due to the disease's prevalence and potential lethality.

For the quantitative detection of Mucin-1 in whole blood, a cancer biomarker, Hu et al. [70] developed a platform based on the in situ synthesis of AuNPs onto filter paper. The performance of the SERS strip was evaluated with 4-MBA showing an RSD lower than 8% thus demonstrating a high uniformity of the platform. Their detection strategy relied on the conjugation of AuNPs with aptamer of Mucin-1, which was then hybridized with a complementary DNA labeled with a Raman reporter. In the case of target's presence, a replacement of complementary DNA occurs causing a decrease on the intensity of the SERS signal. This method rendered a log linear correlation from 50 ng/mL to 50 μg/mL and was successfully tested against other common proteins, with no significant signal loss.

AuNPs where also used as enhancement layer on a filter paper substrate for the label-free detection of adenosine in urine. [38] The presence of this nucleoside in urine samples has been correlated to several types of cancer [38,128] and its screening might be used for the noninvasive early diagnosis and follow-up of cancer. The preparation of the AuNPs followed a chemical route through the well-known citrate-reduction method [129] resulting in 45 nm-average diameter spherical particles with an SPR centered at 532 nm. The deposition of the plasmonic nanostructures was achieved with a ring-oven pre-concentration system, thus improving sampling efficiency. Due to the complex nature of urine matrix samples, the authors proposed a SERS-chemometric method based on multivariate curve resolution-alternating least squares, combined with the standard addition method which allowed to achieve a sensitive, reproducible and selective method for adenosine detection in urine samples.

As it is known, the anisotropy of the plasmonic structures can enhance even further the SERS signal. [65,75,78,89,106] In particular, AuNFs deposited on hydrophobic paper were used for cervical cancer diagnosis [130] through an immunoassay combined with Au-Ag nanoshuttles as SERS tags. This sandwich was able to quantitatively detect two common cervical cancers biomarkers, squamous cell carcinoma antigen and osteopontin in serum samples, with a LOD of 8.6 pg/mL and 4.4 pg/mL, respectively.

AuNRs have also been successfully employed on cancer diagnosis through immersion onto filter papers, rendering EFs in the range of 10^6-10^8. [75,78] The intrinsic SERS spectra of HT-29 cell line from colorectal cancer was used in combination with principal component analysis and k-nearest-neighbor algorithm, to differentiate it from the negative controls, achieving sensitivities, specificities and accuracies above 80% (Fig. 6.14A). [75]

Liu and coworkers [78] established a method also based on AuNRs for the distinction between normal and cancerous cells through a peak ratio algorithm, achieving a sensitivity and selectivity of 100%.

Although Au nanostructures are the most commonly used amongst biosensing applications, due to its inherent biocompatibility and easy conjugation with bioreceptors, Ag-based nanostructures or Au-Ag alloys have also been used. Ahn et al. [131] developed a dual optical system (SERS and colorimetric) based on Ag nanoplates (AgNPls) for the detection of hydrogen sulfide (H_2S) gas, released from live prostate cancer cells. The AgNPls that were chemically synthesized through a seed-mediated growth method were dropped in PDMS-patterned filter paper detection zones. The SERS detection allowed monitoring quantitatively H_2S with a linear range of 82–330 nM. Similarly, Ferreira and coworkers [86] fabricated a SERS platform combining bacterial nanocellulose, from a *nata de coco* source, with in situ grown anisotropic Ag nanostructures, through a microwave assisted hydrothermal method, achieving an EF of 10^5. This platform was successfully applied for breast cancer diagnosis through exosomes profiling. Exosomes are small

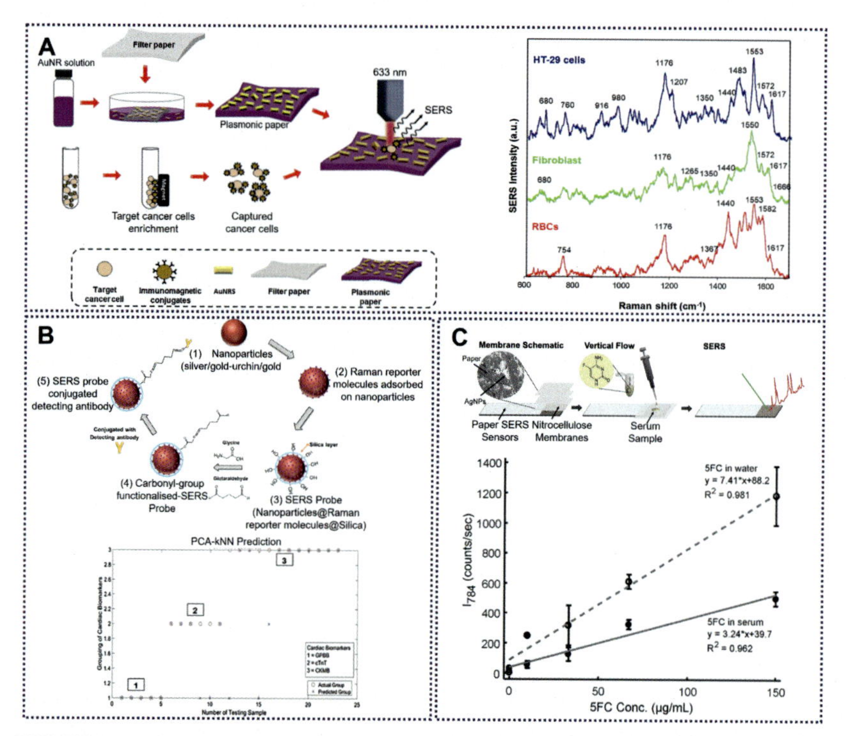

FIGURE 6.14 Health applications—diagnostic and therapeutic. (A) AuNR-paper for the detection of cancer cell trough an immunomagnetic assay and resultant SERS spectra of HT-29 cells, fibroblast, and RBCs. (B) Schematic workflow for the preparation of SERS probe for cardiac biomarkers detection and PCA-kNN prediction. (C) Schematic procedure of a vertical-flow paper SERS device for screening of an anticancer drug and calibrations curve in water and serum. *(Figure 6.14A) Reprinted with permission from Reokrungruang, P. et al., Sensors and Actuators, B: Chemical, 285, 2019. Copyright 2020, Elsevier. (Figure 6.14B) Reprinted with permission from Lim, W. et al., Biosensors and Bioelectronics, 147, 2020. Copyright 2020, Elsevier. (Figure 6.14C) Reprinted with permission from Berger, A. et al., Analytical Chimica Acta, 949, 2017. Copyright 2017, Elsevier.*

(50−150 nm in diameter) vesicles secreted from cells into body fluids such as blood, saliva, and urine. These extracellular vesicles show potential for cancer diagnostics, as they transport several molecular contents of the cells from which they originate. [125,132−134] The SERS spectra of exosomes retrieved from tumorous and nontumorous cell lines were subjected to a statistical principal component analysis allowing for data grouping with 95% confidence.

The implementation of plasmonic bimetallic nanoparticles for SERS sensing was proven advantageous, due to the combination of features which allowed tailoring LSPR. [118,120] Reports on these nanostructures showed their use on cancer diagnosis either for folic acid detection with picomolar sensitivity [24] or also for exosomes profiling. [20,124] Through a statistical principal components analysis method, a platform based on Ag@Au core-shell nanostructures was fabricated [20,124], following a two-stage SILAR method published by Kim et al. [13], for breast cancer screening. Contrary to these authors that implemented the method on filter paper, other work used regular office paper as substrate. The probable presence of partially oxidized cellulose monomers moieties, on office paper, allowed to suppress the first reduction process: Ag(I) to Ag(0), to form the core of the nanostructures, which decreased the total cost of the platforms and preparation time.

Cardiovascular diseases are the leading cause of death worldwide [135] and efforts on the development of paper-based SERS platform for its screening and early detection have been recently reported. Mabbott and coworkers [51] reported the detection of miR-29a, a microRNA associated with myocardial infarction, achieving a LOD down to 47 pg/μL. AuNPs, obtained through a citrate-reduction method, were used as the Raman enhancement structures, and conjugated with complementary DNA to miR-29a and with a Raman reporter (malachite green isothiocyanate) to facilitate SERS read out. This system was applied in a three-dimensional paper-based microfluidic device comprising layers for sample loading, conjugation, incubation, capture and washing steps. The developed device also allowed the colorimetric discrimination between positive and negative results, offering a simple and easy-to-interpret solution to untrained personnel. Also, for myocardial infarction early diagnosis, Lim et al. [136] established a method based on the simultaneous detection of multiple cardiac biomarkers—GPBB, CK-MB, and CTnT (Fig. 6.14B). Antibodies against each of these biomarkers where conjugated with Ag, Au urchin, and AuNPs and three different Raman reporters—4-nitroaniline, *tert*-butylhydroquinone, and methyl red, respectively. The characteristic Raman spectral signatures of each reporter were further processed with partial least squares models allowing the quantification of each biomarker. The obtained LOD of the developed paper-based SERS device was calculated as 8, 10, and 1 pg/mL for GPBB, CK-MB, and cTnT, respectively.

Besides the aforementioned methods, several other types of paper-based SERS substrates combining different types of papers and plasmonic structures have been reported in the biosensing field. For instance, the use of AuNPs onto hydrophilic or hydrophobic papers has been reported for the detection of L-citrulline, [137] uric acid, [41] and virus-infected tears. [43] For the detection of free bilirubin, graphene-isolated-Au-nanocrystals [82] and graphene oxide-AuNSs [31] were used. ZnO decorated with AuNPs has been reported for the diagnostic of prenatal diseases in women through the analysis of human amniotic fluids. [26] Ag was used in the form of nanoparticles deposited on regular office paper for glucose sensing [138], and for procalcitonin detection, in the form of nanowires combined with Au hollow nanocages. [139] Additionally, nanoshells combining Au and Ag where reported as Raman enhancement structures, for the detection of Myxovirus protein A through an immunoassay. [140]

Therapeutic drug monitoring (TDM) allows optimizing prescriptions and monitoring the safety and effectiveness use of life-saving drugs. SERS detection coupled with paper-based devices can contribute to efficient, low-cost, noninvasive and on-site TDM. A few strategies for the development of these devices has been reported in the literature. Berger et al. [141] built a system with a vertical-flow architecture to monitor the antifungal flucytosine levels in undiluted serum, avoiding the need for complex and time-consuming sample preparation (Fig. 6.14C). The developed solution comprises a filtering membrane, made of nitrocellulose, for sample pretreatment, placed on top of an inkjet printed paper-based SERS sensor. This device combined with a portable Raman spectrometer allowed on-site detection. For the Raman signal enhancement, chemically synthesized AgNPs were printed in an array pattern allowing the identification of flucytosine characteristic Raman peaks. The most intense peak, at 784 cm^{-1}, was used for the quantification of the target levels achieving and LOD of 10 μg/mL. For even less invasive TDM applications, Yokoyama et al. [142,143] has shown preliminary results of the use of tear samples, using the intrinsic capillary forces of cellulose. AuNRs absorbed on filter paper were used for screening antiepileptic and anticonvulsant agents, sodium valproic acid [142] and sodium phenobarbital, [143] respectively.

Ponlamuangdee and coworkers combined graphene oxide and AuNRs with a layer-by-layer immersion method on filter paper, for the fabrication of paper-based SERS devices. [30] Using R6G as the probe molecule, the authors were able to achieve an EF of 4.56×10^7, due to the electromagnetic and chemical enhancement effects arising from the AuNRs and graphene oxide, respectively. As proof-of-concept, the developed SERS substrate was tested in the detection of an anticancer drug, mitoxantrone, attaining a LOD of 5 μM. For the determination of ketoprofen in aqueous solutions and human saliva, Díaz-Liñán et al. immobilized AgNFs on nylon-coated paper [106] achieving a dual detection system based on SERS and ambient pressure mass spectrometry analysis, with a LOD and limit of quantification (LOQ) of 0.023 and 0.076 mg/L, respectively.

6.3.2 Environmental applications: food, dairy products, beverages and water safety

Inexpensive, accurate, and simple-use sensing devices, such as paper-based SERS platforms, are of paramount relevance in the food safety scenario. Although there is regulation of governmental and nongovernmental organizations around the world, reports on the presence of pesticides, dyes, antibiotics feed to livestock and/or in the aquaculture industry and other environmental pollutants are still recurrent. [144–147]

Thiram is a very well-known pesticide that is use as fungicide for the preservation of fruits, that can cause several health issues, such as lethargy and limited motor activity. Several groups have reported the use of paper-based

SERS devices for detecting and monitoring this compound. Based on Au nanostructures, different strategies can be found in the literature. For instance, Xiong and coworkers [56] optimized a composite based on cellulose nanofibers and chemically synthesized AuNPs, via electrostatic interactions. The direct comparison between this composite and a filter paper/AuNPs substrate, using 4-ATP as probe molecule, showed an enhancement of 20 times on the SERS spectra intensities. The application of the developed three-dimensional SERS platform on apple juice allowed detecting the presence of thiram with a limit of detection of 52 ppb. For the simultaneous detection of thiram and other pesticides, [64] such as tricyclazole and carbaryl, nanoporous cellulose was used as a matrix for AuNRs, rendering a uniform and well-distribution of the nanostructures throughout the substrate (Fig. 6.15A).

An EF of 1.4×10^7 was obtained for R6G as probe molecule and LODs of 1 nM, 100 nM and 1 μM were calculated for thiram, tricyclazole, and carbaryl, respectively. Additionally, the authors performed a real-scenario swab test to trace residues on apple peels, accomplishing limits of detection for thiram, tricyclazole and carbaryl of 6, 60, and 600 ng/cm^2, respectively. These values were well below the permitted maximum residual level on apple peels, of

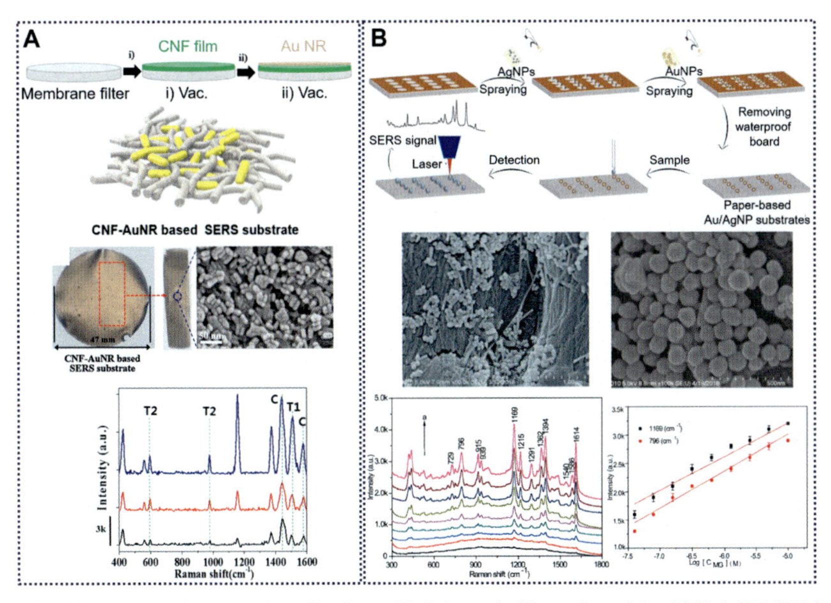

FIGURE 6.15 Environmental applications. (A) Schematic illustration of the CNF-AuNR SERS platform, its photograph and respective SEM image and SERS spectra for the simultaneous detection of pesticides. (B) Schematic workflow to produce the paper-based Au/AgNPs, SEM image and representative concentration-dependent SERS spectra with correspondent calibration curve. *(Figure 6.15A) Reprinted with permission from Kwon, G. et al., Cellulose, 26(8), 2019. Copyright 2019, Springer Nature. (Figure 6.15B) Reprinted with permission from Yang, G. et al., Microchimica Acta, 187(5), 2020. Copyright 2020, Springer Nature.*

2000 ng/cm^2. Luo and coworkers [62] reported a method for the simultaneous detection of thiram, parathion methyl and malachite green on apple peels, based on in situ synthesized AuNPs on pseudo-paper films composed of polyethylenimine and microcrystalline cellulose. The platform achieved an EF of 3.02×10^6 and limits of detection of 1.1 ng/cm^2 for thiram, well below the maximum residue limit. More recently, Godoy et al. [67] used inkjet printing to deposit 77 nm-diameter sized Au nanospheres (AuSph) onto hydrophobic paper to enhance the device's sensing performance, due to the confinement of the analyte in the AuSph platform. A full optimization of the printing process tuning the amount of nanoparticles in the sensing areas, allowed obtaining an EF up to 2.0×10^6, using 4-nitrophenol as model analyte, and a quantification region threshold of 10^{-11} M for thiram and for crystal violet. Hydrophobic paper was also used by Lee et al. [148] as substrate for the deposition of UV reduced AuNPs@GO flakes. A LOD of 10 nM and 1 μM was calculated for R6G and thiram, respectively.

Based on Ag nanostructures, Zhu et al. [99] reported an AgNPs-decorated filter paper combined with dynamic SERS by collecting Raman spectra during the drying up process after sample loading onto the SERS substrate, for increased sensitivity. CV was used as probe molecule, demonstrating stronger signals with dynamic SERS than with after absolute evaporation of analyte. The produced SERS membranes, based on a simple silver mirror reaction, were swabbed on different fruit's peels—apples, bananas and tomatoes – detecting thiram with LODs of 24, 7.2, and 36 ng/cm^2, respectively. A AgNPs/GO paper was prepared by a screen printing method [117] for the detection of thiram and also thiabendazole and methyl parathion with LODs of 0.26, 28, and 7.4 ng/cm^2, respectively, in aqueous solutions and 1.2, 56, and 30 ng/cm^2, respectively, on the surface of fruits and vegetables. Sun and coworkers developed SERS substrates with filter paper and Ag@SiO_2 core-shell nanoparticles, [115] for shell-isolated NP-enhanced Raman spectroscopy (SHINERS), to avoid potential disturbing interactions between analytes and the metal core. CV was used as probe molecule to optimize the size of the Ag core and SiO_2 thickness, reaching a LOD of 10^{-9} M with 3 nm shell thickness, for both crystal violet and thiram.

The combination of Ag and Au has also been reported in paper-based SERS platforms for thiram detection. As an example, Lin et al. [122] assembled an Au@Ag nanocubes monolayer on a patterned paper through a liquid-liquid interface self-assembly technique, for the detection of pesticides in soil, without the need for sample preparation. The SERS performance was evaluated with 4-MBA as probe molecule with an EF of 0.56×10^5 and an RSD of 12.8%. For the quantitative detection of thiram in soil, the paper untreated areas allowed sample injection and filtration of impurities, achieving a LOD of 0.148 mg/kg using thiram' characteristic peak centered at 1386 cm^{-1}. Asgari et al. [120] fabricated a nanofibrillar cellulose/Au@Ag core-shell nanoparticles composite for detection of thiram and also paraquat in lettuce. EF of

$\sim 10^4$ was obtained for 4-MBA and an increase on Raman intensities of about 8 times was observed comparing the bimetallic nanoparticles with the AuNPs on the nanofibrillar cellulose. The nanocomposite achieved limits of detection of 71 and 46 μg/L for thiram and paraquat, respectively.

Synthetic dyes such as malachite green, methylene blue, and metalin yellow, among others, used in the textile, paper, aquaculture and food industries can have harmful effects on human health. [91] The need for regular on-site monitoring of these dyes can be aided by paper-based SERS devices. Regenerated cellulose fibers, from wastepaper, were decorated with citrate-stabilized AuNPs and used on the detection of malachite green in water samples. [37] The composite achieved a limit of detection lower than 10 ppb, using malachite green' characteristic peak located at $1173\ cm^{-1}$. For the simultaneous detection of malachite green, methylene blue, and crystal violet in fish, Yang et al. fabricated a filter paper-based SERS platform with sprayed AgNPs and AuNPs with an calculated EF of 9.0×10^7 (Fig. 6.15B). [149] The combination of the high enhancement from AgNPs with the chemical stability of AuNPs rendered an RSD, among peak intensities, lower than 3% after 4 weeks of storage under ambient conditions with limits of detection of 4.3×10^{-9}, 2.0×10^{-8}, and 8.1×10^{-8} M for malachite green, methylene blue, and crystal violet, respectively. Zhu et al. built a poly(4-styrenesulfonate) (PSS)-functionalized filter paper SERS platform for the simultaneous separation, preconcentration and detection of samples, achieved due to the capillary forces of cellulose and the surface chemical gradient and electrostatic interaction generated by PSS. [150] As enhancement layer, AgNPs were prepared through the silver mirror reaction. As proof-of-concept, the authors applied the developed SERS substrate for screening the colorants sunset yellow and lemon yellow in drinks, with limits of detection of 10^{-5} M. Nanocomposites of nanocrystalline cellulose, AgNPs and SiO_2 were prepared by Ogundare and coworkers for the detection of malachite green. [91] The addition of SiO_2 resulted in the clustering of AgNPs, increasing the hot spots, which decreased the LOD from 5.2 to 0.9 nM. The integration of Ag on graphene nano-mesh-Ag-ZnO hybrid papers was carried out by Bharadwaj et al., for the detection of methyl orange, R6G, and also paraquat. [32] The developed nanohybrid was deposited on filter paper by suction filtration and rendered a high amount of hot spots, attaining an high EF of 1.97×10^{11} for paraquat molecules and, allowed detecting low levels of pollutants, down to 10^{-11}, 10^{-13}, and 10^{-14} M, for methyl orange, R6G, and paraquat, respectively. For the detection of CV and malachite green on fish, a hydrophobic paper-based semiconducting substrate was fabricated with two-dimensional MoO_{3-x} nanosheets through a regular printing process, achieving limits of detection as low as 10^{-6} M for both molecules. [35]

Other examples of paper-based SERS platforms for dye's monitoring includes drop casted AgNPs on high-pressure homogenized nanocellulose from *nata de coco* for methylene blue in water, [151] in situ grown AgNPs onto polydopamine templated filter paper for malachite green, [152] faceted AgNPs with cellulose nanofibrils for common dyes in water, [153] AgNPs grown in filter paper through silver mirror reaction, [85] inkjet Ag nanostructures on filter paper for metalin yellow and malachite green, [154] AgNPs in situ synthesized on glass fiber paper for malachite green in fish, [155] AgNPs produced with the SILAR method for malachite green detection in aquaculture settings [14] and brush painted AgNPs on paper for malachite green. [156]

Additional reports of paper-based SERS platforms for environmental applications shows the detection of melamine or antibiotics residues, such as tetracycline in milk samples. Several systems have been described for melamine screening in milk. These include nanofibrillated cellulose with AuNPs, which combined with partial least squares statistical analysis rendered a limit of detection of 1 ppm. [63] Other works include Ag@SiO_2 nanocubes on filter paper with 0.17 mg/L as limit of detection [28] and AgNPs with commercially available filter paper, made hydrophobic and achieving 1 ppm as limit of detection. [157] For the monitoring of tetracycline in milk a cardboard substrate covered with an aluminum thin layer served as substrate for the thermal evaporation of AgNPs. Through a peak ratio method, a LOD of 0.1 ppm was obtained. [158] For the quality monitoring of wine, AuNRs-rGO deposited on paper through vacuum filtration were used for sulfur dioxide detection [83] and semiconductor ZnO was used for sulfites quantification. [34] Black phosphorus Au deposited on filter paper was reported for foodborne bacteria *Staphylococcus aureus*, *Listeria monocytogenes*, and *E. coli* screening [80] and Au thin films deposited onto filter paper by sputtering were applied on heavy metals detection. [81]

6.3.3 Other applications

Paper-based SERS devices have also demonstrated their many advantages in several other applications, due to its simplicity, low-cost, easy tailoring, high selectivity and sensitivity and on-site detection.

For instance, Au@Ag core-shell NPs self-assembled on filter paper was used for screening the drug estazolam in beverages. [25] AuNPs deposited on office paper, with an EF of 10^6, was applied for cocaine detection to fight drugs trafficking, with a dual-detection device combining SERS with electrochemical measurements. [54] For explosives monitoring, AuNPs and AuNSs, deposited on filter paper, was used for piric acid detection. [159] In other works, Ag/AuNPs produced through laser ablation and soaked on filter paper were used in the detection of piric acid, 2,4-dinitrotoluene and 3-nitro-1,2,4-triazol-5-one [160], whereas Ag nanotriangles, also deposited on filter paper, were used for piric acid detection. [112] For cosmetics applications, filter paper with AgNPs was used for nicotinamide detection [161] and AgNPs, produced by a green synthesis method with glucose and combined with methylcellulose, were used for artwork analysis. [162]

6.4 Concluding remarks and perspectives

The outstanding properties of SERS, including high sensitivity and selectivity has been applied in a vast area of scenarios. Analytical approaches based on this spectroscopy can be label-free, by the intrinsic Raman spectra of analytes or coupled with more complex assays, such as immunoassays, allowing single or simultaneous detection of compounds in several types of samples, from simple to more complex liquid or solid matrix.

Within the current pandemic scenario, there is a high need for SARS-CoV-2 virus detection solutions, either in clinical or environmental settings. Devices able to offer rapid and sensitive results are of high value. Paper-based SERS devices can be of paramount importance due to their relevant analytical features that can meet the present needs.

Relevant figures of merit for SERS platforms include high enhancement factors with a low relative standard deviation revealing a high uniformity and reproducibility of the SERS substrate. The literature revision presented in this chapter shows a multitude of strategies for cellulose and paper-based SERS devices fabrication. Efforts have been made to increase the uniformity and reproducibility of paper-based SERS substrates. Starting from the different synthesis approaches for enhancement layers production—based on chemical or physical processes—to several transference methods applied to a wide variety of cellulose formats. In most cases, Ag- and Au-based nanostructures are used for Raman amplification due to their strong LSPR, which results in high enhancement factors. The different approaches for their production rendered nanostructures with different shapes and sizes, which had a direct influence in Raman signals enhancement. Fabrication methods based on one-step approaches have been reported, such as in situ synthesis, thermal evaporation, or laser-based techniques, allowing for a mask less, fast and simple production of paper-based SERS platforms. Additionally, tactics for the induction of hot spots on SERS devices has also been reported since it can enhance even further Raman intensities. The use of different forms and formats of cellulose as substrate has shown to be very versatile, since their characteristics can be easily tuned to meet several criteria such as, hydrophilicity or hydrophobicity, different porosities, thicknesses or roughness, which influenced the performance of SERS devices. A strong influence of the transference methods of these enhancement structures to the substrate has also been shown. Paper facilitates transference methods based on printing (inkjet printing or screen printing), coating processes (dip-coating or spray-coating), immersion methods, and also in situ physical and chemical depositions.

Furthermore, the use of paper as substrate allowed the production of low-cost SERS devices when compared with other standard materials such as glass or silicon wafers.

Multiple applications have been reported in the last years based on paper SERS platforms, especially in the areas of diagnostic and environmental analysis. The lightweight nature of paper combined with its flexibility is compatible with point-of-care analysis, which can be combined with portable Raman spectrometers for on-site detections.

6.5 Acknowledgements

The authors acknowledge financial support from National Funds through the FCT - Fundação para a Ciência e a Tecnologia, I.P., within the scope of the project refª UIDB/50025/2020–2023. We also acknowledge the project SYNERGY H2020-WIDESPREAD-2020–5, CSA, proposal nº 952169 and funding from the European Research Council (ERC) under the European Union's Horizon 2020 research and innovation program (DIGISMART grant agreement No. 787410). A. C. Marques acknowledges the funding from the Portuguese Science Foundation through the PhD Grant SFRH/BD/115173/2016.

References

[1] C.V. Raman, A new radiation, Indian. J. Phys. 2 (1928) 387–398.
[2] The Nobel Prize in Physics 1930 https://www.nobelprize.org/prizes/physics/1930/summary/.
[3] Indian Academy of Sciences. *C V Raman – A Pictorial Biography*; 1988.
[4] C.V. Raman, K.S. Krishnan, A new type of secondary irradiation, Nature (1928) 501–502.
[5] American Chemical Society. American Chemical Society International Historic Chemical Landmarks. The Raman Effect http://portal.acs.org/portal/PublicWebSite/education/%0Awhatischemistry/landmarks/ramaneffect/index.htm.
[6] E. Le Ru, P. Etchegoin, Principles of Surface Enhanced Raman Spectroscopy and Related Plasmonic Effects, Elsevier, 2009.
[7] S.A. Ogundare, W.E. van Zyl, A review of cellulose-based substrates for SERS: fundamentals, design principles, applications, Cellulose 26 (11) (2019) 6489–6528. Available from: https://doi.org/10.1007/s10570-019-02580-0.
[8] C.C. Huang, C.Y. Cheng, Y.S. Lai, Paper-based flexible surface enhanced Raman Scattering platforms and their applications to food safety, Trends Food Sci. Technol. 100 (2020) 349–358. Available from: https://doi.org/10.1016/j.tifs.2020.04.019.
[9] M. Fleischmann, P.J. Hendra, A.J. McQuillan, Raman spectra of pyridine adsorbed at a silver electrode, Chem. Phys. Lett. 26 (2) (1974) 163–166. Available from: https://doi.org/10.1016/0009-2614(74)85388-1.
[10] D.L. Jeanmaire, R.P. Van Duyne, Surface Raman Spectroelectrochemistry. Part I. heterocyclic, aromatic, and aliphatic amines adsorbed on the anodized silver electrode, J. Electroanal. Chem. 84 (1) (1977) 1–20. Available from: https://doi.org/10.1016/S0022-0728(77)80224-6.
[11] A.J. Creighton, G.M. Albrecht, Anomalously intense Raman spectra of pyridine at a silver electrode, J. Am. Chem. Soc. 99 (15) (1977) 5215.
[12] Y. Sanguansap, K. Karn-orachai, R. Laocharoensuk, Tailor-made porous striped gold-silver nanowires for surface enhanced Raman scattering based trace detection of β-hydroxybutyric acid, Appl. Surf. Sci. (2020) 500. Available from: https://doi.org/10.1016/j.apsusc.2019.144049.
[13] W. Kim, J.C. Lee, G.J. Lee, H.K. Park, A. Lee, S. Choi, Low-cost label-free biosensing bimetallic cellulose strip with SILAR-synthesized silver core-gold shell nanoparticle structures, Anal. Chem. 89 (12) (2017) 6448–6454. Available from: https://doi.org/10.1021/acs.analchem.7b00300.

[14] W. Kim, Y.H. Kim, H.K. Park, S. Choi, Facile fabrication of a silver nanoparticle immersed, surface-enhanced Raman scattering imposed paper platform through successive ionic layer absorption and reaction for on-site bioassays, ACS Appl. Mater. Interfaces 7 (50) (2015) 27910–27917. Available from: https://doi.org/10.1021/acsami.5b09982.

[15] Y.H. Ngo, D. Li, G.P. Simon, G. Garnier, Gold nanoparticle-paper as a three-dimensional surface enhanced Raman scattering substrate, Langmuir 28 (23) (2012) 8782–8790. Available from: https://doi.org/10.1021/la3012734.

[16] P.L. Stiles, J.A. Dieringer, N.C. Shah, R.P. Van Duyne, Surface-enhanced Raman spectroscopy, Annu. Rev. Anal. Chem. 1 (1) (2008) 601–626. Available from: https://doi.org/10.1146/annurev.anchem.1.031207.112814.

[17] W. Suntornsuk, L. Suntornsuk, Recent applications of paper-based point-of-care devices for biomarker detection, Electrophoresis 41 (5–6) (2020) 287–305. Available from: https://doi.org/10.1002/elps.201900258.

[18] W.K. Tomazelli Coltro, C.M. Cheng, E. Carrilho, D.P. de Jesus, Recent advances in low-cost microfluidic platforms for diagnostic applications, Electrophoresis 35 (16) (2014) 2309–2324. Available from: https://doi.org/10.1002/elps.201400006.

[19] T. Ozer, C. McMahon, C.S. Henry, Advances in paper-based analytical devices, Annu. Rev. Anal. Chem. 13 (1) (2020) 061318–114845. Available from: https://doi.org/10.1146/annurev-anchem-061318-114845.

[20] A.C. Marques, T. Pinheiro, G.V. Martins, A.R. Cardoso, R. Martins, M.G. Sales, et al., Non-enzymatic lab-on-paper devices for biosensing applications, Comprehensive Analytical Chemistry, 89, Elsevier B.V, 2020, pp. 189–237. Available from: https://doi.org/10.1016/bs.coac.2020.05.001.

[21] T. VoDinh, M.Y.K. Hiromoto, G.M. Begun, R.L. Moody, Surface-enhanced Raman spectrometry for trace organic analysis, Anal. Chem. 56 (9) (1984) 1667–1670. Available from: https://doi.org/10.1021/ac00273a029.

[22] C.D. Tran, Subnanogram detection of dyes on filter paper by surface-enhanced Raman scattering spectrometry, Anal. Chem. 56 (4) (1984) 824–826. Available from: https://doi.org/10.1021/ac00268a057.

[23] J.A. Lartey, J.P. Harms, R. Frimpong, C.C. Mulligan, J.D. Driskell, J.H. Kim, Sandwiching analytes with structurally diverse plasmonic nanoparticles on paper substrates for surface enhanced Raman spectroscopy, RSC Adv. 9 (56) (2019) 32535–32543. Available from: https://doi.org/10.1039/c9ra05399a.

[24] M. Park, C.S.H. Hwang, K.H. Jeong, Nanoplasmonic alloy of Au/Ag nanocomposites on paper substrate for biosensing applications, ACS Appl. Mater. Interfaces 10 (1) (2018) 290–295. Available from: https://doi.org/10.1021/acsami.7b16182.

[25] X. Sha, S. Han, H. Zhao, N. Li, C. Zhang, W.L.J. Hasi, A rapid detection method for on-site screening of estazolam in beverages with Au@Ag core-shell nanoparticles paper-based SERS substrate, Anal. Sci. 36 (6) (2020) 667–674. Available from: https://doi.org/10.2116/ANALSCI.19P361.

[26] W. Kim, S.H. Lee, J.H. Kim, Y.J. Ahn, Y.-H. Kim, J.S. Yu, et al., Paper-based surface-enhanced Raman spectroscopy for diagnosing prenatal diseases in women, ACS Nano 12 (7) (2018) 7100–7108. Available from: https://doi.org/10.1021/acsnano.8b02917.

[27] S. Kang, A. Rahman, E. Boeding, P.J. Vikesland, Synthesis and SERS application of gold and iron oxide functionalized bacterial cellulose nanocrystals (Au@Fe3O4@BCNCs), Analyst 145 (12) (2020) 4358–4368. Available from: https://doi.org/10.1039/d0an00711k.

[28] M.L. Mekonnen, W.N. Su, C.H. Chen, B.J. Hwang, Ag@SiO2 nanocube loaded miniaturized filter paper as a hybrid flexible plasmonic SERS substrate for trace melamine

detection, Anal. Methods 9 (48) (2017) 6823–6829. Available from: https://doi.org/10.1039/c7ay02192e.

[29] A. Araújo, A. Pimentel, M.J. Oliveira, M.J. Mendes, R. Franco, E. Fortunato, et al., Direct growth of plasmonic nanorod forests on paper substrates for low-cost flexible 3D SERS platforms, Flex. Print. Electron. 2 (1) (2017). Available from: https://doi.org/10.1088/2058-8585/2/1/014001.

[30] K. Ponlamuangdee, G.L. Hornyak, T. Bora, S. Bamrungsap, Graphene oxide/gold nanorod plasmonic paper – a simple and cost-effective SERS substrate for anticancer drug analysis, N. J. Chem. 44 (33) (2020) 14087–14094. Available from: https://doi.org/10.1039/d0nj02448a.

[31] X. Pan, L. Li, H. Lin, J. Tan, H. Wang, M. Liao, et al., A graphene oxide-gold nanostar hybrid based-paper biosensor for label-free SERS detection of serum bilirubin for diagnosis of jaundice, Biosens. Bioelectron. (2019) 145. Available from: https://doi.org/10.1016/j.bios.2019.111713.

[32] S. Bharadwaj, A. Pandey, B. Yagci, V. Ozguz, A. Qureshi, Graphene nano-Mesh-Ag-ZnO hybrid paper for sensitive SERS sensing and self-cleaning of organic pollutants, Chem. Eng. J. 336 (2018) 445–455. Available from: https://doi.org/10.1016/j.cej.2017.12.040.

[33] C. Yang, Y. Xu, M. Wang, T. Li, Y. Huo, C. Yang, et al., Multifunctional paper strip based on GO-veiled Ag nanoparticles with highly SERS sensitive and deliverable properties for high-performance molecular detection, Opt. Express 26 (8) (2018) 10023. Available from: https://doi.org/10.1364/oe.26.010023.

[34] M. Chen, H. Yang, L. Rong, X. Chen, A gas-diffusion microfluidic paper-based analytical device (MPAD) coupled with portable surface-enhanced Raman scattering (SERS): Facile Determination of Sulphite in Wines, Analyst 141 (19) (2016) 5511–5519. Available from: https://doi.org/10.1039/c6an00788k.

[35] L. Lan, X. Hou, Y. Gao, X. Fan, T. Qiu, Inkjet-printed paper-based semiconducting substrates for surface-enhanced Raman spectroscopy, Nanotechnology 31 (5) (2020). Available from: https://doi.org/10.1088/1361-6528/ab4f11.

[36] J. Xie, L. Li, I.M. Khan, Z. Wang, X. Ma, Flexible paper-based SERS substrate strategy for rapid detection of methyl parathion on the surface of fruit, Spectrochim. Acta - Part. A Mol. Biomol. Spectrosc. (2020) 231. Available from: https://doi.org/10.1016/j.saa.2020.118104.

[37] Q. Yu, X. Kong, Y. Ma, R. Wang, Q. Liu, J.P. Hinestroza, et al., Multi-functional regenerated cellulose fibers decorated with plasmonic Au nanoparticles for colorimetry and SERS assays, Cellulose 25 (10) (2018) 6041–6053. Available from: https://doi.org/10.1007/s10570-018-1987-9.

[38] J.E.L. Villa, C. Pasquini, R.J. Poppi, Surface-enhanced Raman spectroscopy and MCR-ALS for the selective sensing of urinary adenosine on filter paper, Talanta 187 (2018) 99–105. Available from: https://doi.org/10.1016/j.talanta.2018.05.022.

[39] R. Chen, L. Zhang, X. Li, L. Ong, Y.G. Soe, N. Sinsua, et al., Trace analysis and chemical identification on cellulose nanofibers-textured SERS substrates using the "Coffee Ring" effect, ACS Sens. 2 (7) (2017) 1060–1067. Available from: https://doi.org/10.1021/acssensors.7b00403.

[40] Z. Xiong, X. Chen, P. Liou, M. Lin, Development of nanofibrillated cellulose coated with gold nanoparticles for measurement of melamine by SERS, Cellulose 24 (7) (2017) 2801–2811. Available from: https://doi.org/10.1007/s10570-017-1297-7.

[41] J.E.L. Villa, R.J. Poppi, A portable SERS method for the determination of uric acid using a paper-based substrate and multivariate curve resolution, Analyst 141 (6) (2016) 1966–1972. Available from: https://doi.org/10.1039/c5an02398j.

[42] J.E.L. Villa, D.P. Santos, R.J. do Poppi, Fabrication of gold nanoparticle-coated paper and its use as a sensitive substrate for quantitative SERS analysis, Microchim. Acta 183 (10) (2016) 2745–2752. Available from: https://doi.org/10.1007/s00604-016-1918-0.
[43] W.S. Kim, J.H. Shin, H.K. Park, S. Choi, A low-cost, monometallic, surface-enhanced Raman scattering-functionalized paper platform for spot-on bioassays, Sens. Actuators, B Chem. 222 (2016) 1112–1118. Available from: https://doi.org/10.1016/j.snb.2015.08.030.
[44] K. Zhang, J. Zhao, H. Xu, Y. Li, J. Ji, B. Liu, Multifunctional paper strip based on self-assembled interfacial plasmonic nanoparticle arrays for sensitive SERS detection, ACS Appl. Mater. Interfaces 7 (30) (2015) 16767–16774. Available from: https://doi.org/10.1021/acsami.5b04534.
[45] Y.H. Ngo, D. Li, G.P. Simon, G. Garnier, Effect of cationic polyacrylamide dissolution on the adsorption state of gold nanoparticles on paper and their surface enhanced Raman scattering properties, Colloids Surf. A Physicochem. Eng. Asp. 420 (2013) 46–52. Available from: https://doi.org/10.1016/j.colsurfa.2012.12.018.
[46] Y.H. Ngo, D. Li, G.P. Simon, G. Garnier, Effect of cationic polyacrylamides on the aggregation and SERS performance of gold nanoparticles-treated paper, J. Colloid Interface Sci. 392 (1) (2013) 237–246. Available from: https://doi.org/10.1016/j.jcis.2012.09.080.
[47] C. Li, Y. Liu, X. Zhou, Y. Wang, A paper-based SERS assay for sensitive duplex cytokine detection towards the atherosclerosis-associated disease diagnosis, J. Mater. Chem. B 8 (16) (2020) 3582–3589. Available from: https://doi.org/10.1039/c9tb02469g.
[48] Y.H. Ngo, W.L. Then, W. Shen, G. Garnier, Gold nanoparticles paper as a SERS bio-diagnostic platform, J. Colloid Interface Sci. 409 (2013) 59–65. Available from: https://doi.org/10.1016/j.jcis.2013.07.051.
[49] L. Wang, S. Feng, N. Liu, J. Lei, H. Lin, L. Sun, et al., Microwave synthesis of Au nanoparticles as promising SERS substrates, High-Power Lasers Appl. 8551 (2012) 85511H. Available from: https://doi.org/10.1117/12.2001032. *VI.*
[50] C.A. Teixeira, R.J. Poppi, Paper-based SERS substrate and one-class classifier to monitor thiabendazole residual levels in extracts of mango peels, Spectrochim. Acta - Part. A Mol. Biomol. Spectrosc. (2020) 229. Available from: https://doi.org/10.1016/j.saa.2019.117913.
[51] S. Mabbott, S.C. Fernandes, M. Schechinger, G.L. Cote, K. Faulds, C.R. Mace, et al., Detection of cardiovascular disease associated MiR-29a using paper-based microfluidics and surface enhanced Raman scattering, Analyst 145 (3) (2020) 983–991. Available from: https://doi.org/10.1039/c9an01748h.
[52] D.S. Burr, W.L. Fatigante, J.A. Lartey, W. Jang, A.R. Stelmack, N.W. McClurg, et al., Integrating SERS and PSI-MS with dual purpose plasmonic paper substrates for on-site illicit drug confirmation, Anal. Chem. 92 (9) (2020) 6676–6683. Available from: https://doi.org/10.1021/acs.analchem.0c00562.
[53] W. Jang, H. Byun, J.H. Kim, Rapid preparation of paper-based plasmonic platforms for SERS applications, Mater. Chem. Phys. (2020) 240. Available from: https://doi.org/10.1016/j.matchemphys.2019.122124.
[54] W.A. Ameku, W.R. De Araujo, C.J. Rangel, R.A. Ando, T.R.L.C. Paixão, Gold nanoparticle paper-based dual-detection device for forensics applications, ACS Appl. Nano Mater. 2 (9) (2019) 5460–5468. Available from: https://doi.org/10.1021/acsanm.9b01057.
[55] F.M. De Melo, A.S. Fante, V.D.M. Zamarion, H.E. Toma, SERS-active carboxymethyl cellulose-based gold nanoparticles: high-stability in hypersaline solution and selective response in the Hofmeister Series, N. J. Chem. 43 (21) (2019) 8093–8100. Available from: https://doi.org/10.1039/c9nj01552c.

[56] Z. Xiong, M. Lin, H. Lin, M. Huang, Facile synthesis of cellulose nanofiber nanocomposite as a SERS substrate for detection of thiram in juice, Carbohydr. Polym. 189 (2018) 79–86. Available from: https://doi.org/10.1016/j.carbpol.2018.02.014.

[57] Y. Mu, X. Zhang, A paper-fiber-supported 3D SERS substrate, Plasmonics 15 (3) (2020) 889–896. Available from: https://doi.org/10.1007/s11468-019-01097-3.

[58] W. Kim, S.H. Lee, Y.J. Ahn, S.H. Lee, J. Ryu, S.K. Choi, et al., A label-free cellulose SERS biosensor chip with improvement of nanoparticle-enhanced LSPR effects for early diagnosis of subarachnoid hemorrhage-induced complications, Biosens. Bioelectron. 111 (2018) 59–65. Available from: https://doi.org/10.1016/j.bios.2018.04.003.

[59] J.C. Lee, W. Kim, S. Choi, Fabrication of a SERS-encoded microfluidic paper-based analytical chip for the point-of-assay of wastewater, Int. J. Precis. Eng. Manuf. - Green. Technol. 4 (2) (2017) 221–226. Available from: https://doi.org/10.1007/s40684-017-0027-9.

[60] M. Chen, Z. Zhang, M. Liu, C. Qiu, H. Yang, X. Chen, In situ fabrication of label-free optical sensing paper strips for the rapid surface-enhanced Raman scattering (SERS) detection of brassinosteroids in plant tissues, Talanta 165 (2017) 313–320. Available from: https://doi.org/10.1016/j.talanta.2016.12.072.

[61] W. Kim, J.C. Lee, J.H. Shin, K.H. Jin, H.K. Park, S. Choi, Instrument-free synthesizable fabrication of label-free optical biosensing paper strips for the early detection of infectious keratoconjunctivitides, Anal. Chem. 88 (10) (2016) 5531–5537. Available from: https://doi.org/10.1021/acs.analchem.6b01123.

[62] W. Luo, M. Chen, N. Hao, X. Huang, X. Zhao, Y. Zhu, et al., In situ synthesis of gold nanoparticles on pseudo-paper films as flexible SERS substrate for sensitive detection of surface organic residues, Talanta 197 (2019) 225–233. Available from: https://doi.org/10.1016/j.talanta.2018.12.099.

[63] Z. Xiong, X. Chen, P. Liou, M. Lin, M.Á. Sers, Á. Gold, Development of nanofibrillated cellulose coated with gold nanoparticles for measurement of melamine by SERS, Cellulose 24 (7) (2017) 2801–2811. Available from: https://doi.org/10.1007/s10570-017-1297-7.

[64] G. Kwon, J. Kim, D. Kim, Y. Ko, Y. Yamauchi, J. You, Nanoporous cellulose paper-based SERS platform for multiplex detection of hazardous pesticides, Cellulose 26 (8) (2019) 4935–4944. Available from: https://doi.org/10.1007/s10570-019-02427-8.

[65] S. He, J. Chua, E.K.M. Tan, J.C.Y. Kah, Optimizing the SERS enhancement of a facile gold nanostar immobilized paper-based SERS substrate, RSC Adv. 7 (27) (2017) 16264–16272. Available from: https://doi.org/10.1039/c6ra28450g.

[66] A. Campu, L. Susu, F. Orzan, D. Maniu, A.M. Craciun, A. Vulpoi, et al., Multimodal biosensing on paper-based platform fabricated by plasmonic calligraphy using gold nanobypiramids ink, Front. Chem. FEB (2019) 7. Available from: https://doi.org/10.3389/fchem.2019.00055.

[67] N.V. Godoy, D. García-Lojo, F.A. Sigoli, J. Pérez-Juste, I. Pastoriza-Santos, I.O. Mazali, Ultrasensitive inkjet-printed based SERS sensor combining a high-performance gold nanosphere ink and hydrophobic paper, Sens. Actuators, B Chem. (2020) 320. Available from: https://doi.org/10.1016/j.snb.2020.128412.

[68] C.H. Lee, L. Tian, S. Singamaneni, Paper-based SERS swab for rapid trace detection on real-world surfaces, ACS Appl. Mater. Interfaces 2 (12) (2010) 3429–3435. Available from: https://doi.org/10.1021/am1009875.

[69] R.J.B. Pinto, P.A.A.P. Marques, A.M. Barros-Timmons, T. Trindade, C.P. Neto, Novel SiO2/Cellulose nanocomposites obtained by in situ synthesis and via polyelectrolytes

assembly, Compos. Sci. Technol. 68 (3–4) (2008) 1088–1093. Available from: https://doi.org/10.1016/j.compscitech.2007.03.001.

[70] S.W. Hu, S. Qiao, J.B. Pan, B. Kang, J.J. Xu, H.Y. Chen, A paper-based SERS test strip for quantitative detection of Mucin-1 in whole blood, Talanta 179 (2018) 9–14. Available from: https://doi.org/10.1016/j.talanta.2017.10.038.

[71] V.T.N. Linh, J. Moon, C.W. Mun, V. Devaraj, J.W. Oh, S.G. Park, et al., A facile low-cost paper-based SERS substrate for label-free molecular detection, Sens. Actuators, B Chem. 291 (2019) 369–377. Available from: https://doi.org/10.1016/j.snb.2019.04.077.

[72] S.N. Tanis, H. Ilhan, B. Guven, E.K. Tayyarcan, H. Ciftci, N. Saglam, et al., A disposable gold-cellulose nanofibril platform for SERS mapping, Anal. Methods 12 (24) (2020) 3164–3172. Available from: https://doi.org/10.1039/d0ay00662a.

[73] M. Wu, P. Li, Q. Zhu, M. Wu, H. Li, F. Lu, Functional paper-based SERS substrate for rapid and sensitive detection of Sudan dyes in herbal medicine, Spectrochim. Acta - Part. A Mol. Biomol. Spectrosc. 196 (2018) 110–116. Available from: https://doi.org/10.1016/j.saa.2018.02.014.

[74] C. Kolluru, R. Gupta, Q. Jiang, M. Williams, H. Gholami Derami, S. Cao, et al., Plasmonic paper microneedle patch for on-patch detection of molecules in dermal interstitial fluid, ACS Sens. 4 (6) (2019) 1569–1576. Available from: https://doi.org/10.1021/acssensors.9b00258.

[75] P. Reokrungruang, I. Chatnuntawech, T. Dharakul, S. Bamrungsap, A simple paper-based surface enhanced Raman scattering (SERS) platform and magnetic separation for cancer screening, Sens. Actuators, B Chem. 285 (2019) 462–469. Available from: https://doi.org/10.1016/j.snb.2019.01.090.

[76] L. Susu, A. Campu, A.M. Craciun, A. Vulpoi, S. Astilean, M. Focsan, Designing efficient low-cost paper-based sensing plasmonic nanoplatforms, Sens. (Switz.) 18 (9) (2018). Available from: https://doi.org/10.3390/s18093035.

[77] H. Torul, H. Çiftçi, D. Çetin, Z. Suludere, I.H. Boyacı, U. Tamer, Paper Membrane-based SERS platform for the determination of glucose in blood samples, Anal. Bioanal. Chem. 407 (27) (2015) 8243–8251. Available from: https://doi.org/10.1007/s00216-015-8966-x.

[78] Q. Liu, J. Wang, B. Wang, Z. Li, H. Huang, C. Li, et al., Paper-based plasmonic platform for sensitive, noninvasive, and rapid cancer screening, Biosens. Bioelectron. 54 (2014) 128–134. Available from: https://doi.org/10.1016/j.bios.2013.10.067.

[79] X. Gao, P. Zheng, S. Kasani, S. Wu, F. Yang, S. Lewis, et al., Paper-based surface-enhanced Raman scattering lateral flow strip for detection of neuron-specific enolase in blood plasma, Anal. Chem. 89 (18) (2017) 10104–10110. Available from: https://doi.org/10.1021/acs.analchem.7b03015.

[80] D. Huang, Z. Zhuang, Z. Wang, S. Li, H. Zhong, Z. Liu, et al., Black phosphorus-Au filter paper-based three-dimensional SERS substrate for rapid detection of foodborne bacteria, Appl. Surf. Sci. (2019) 497. Available from: https://doi.org/10.1016/j.apsusc.2019.143825.

[81] Y. Song, Z. Ma, H. Fang, Q. Zhang, Q. Zhou, Z. Chen, et al., Au sputtered paper chromatography tandem Raman platform for sensitive detection of heavy metal ions, ACS Sens. 5 (5) (2020) 1455–1464. Available from: https://doi.org/10.1021/acssensors.0c00395.

[82] Y. Zou, Y. Zhang, Y. Xu, Y. Chen, S. Huang, Y. Lyu, et al., Portable and label-free detection of blood bilirubin with graphene-isolated-Au-nanocrystals paper strip, Anal. Chem. 90 (22) (2018) 13687–13694. Available from: https://doi.org/10.1021/acs.analchem.8b04058.

[83] D. Li, H. Duan, Y. Ma, W. Deng, Headspace-sampling paper-based analytical device for colorimetric/surface-enhanced Raman scattering dual sensing of sulfur dioxide in wine,

Anal. Chem. 90 (9) (2018) 5719–5727. Available from: https://doi.org/10.1021/acs.analchem.8b00016.

[84] L. Chen, B. Ying, P. Song, X. Liu, A nanocellulose-paper-based SERS multiwell plate with high sensitivity and high signal homogeneity, Adv. Mater. Interfaces 6 (24) (2019). Available from: https://doi.org/10.1002/admi.201901346.

[85] D. Das, S. Senapati, K.K. Nanda, "Rinse, repeat": an efficient and reusable SERS and catalytic platform fabricated by controlled deposition of silver nanoparticles on cellulose paper, ACS Sustain. Chem. Eng. 7 (16) (2019) 14089–14101. Available from: https://doi.org/10.1021/acssuschemeng.9b02651.

[86] N. Ferreira, A. Marques, H. Águas, H. Bandarenka, R. Martins, C. Bodo, et al., Label-free nanosensing platform for breast cancer exosome profiling, ACS Sens. 4 (8) (2019) 2073–2083. Available from: https://doi.org/10.1021/acssensors.9b00760.

[87] A. Araújo, C. Caro, M.J. Mendes, D. Nunes, E. Fortunato, R. Franco, et al., Highly efficient nanoplasmonic SERS on cardboard packaging substrates, Nanotechnology 25 (41) (2014) 415202. Available from: https://doi.org/10.1088/0957-4484/25/41/415202.

[88] M.J. Oliveira, P. Quaresma, M.P. De Almeida, A. Araújo, E. Pereira, E. Fortunato, et al., Office paper decorated with silver nanostars-an alternative cost effective platform for trace analyte detection by SERS, Sci. Rep. 7 (1) (2017). Available from: https://doi.org/10.1038/s41598-017-02484-8.

[89] Y. Xu, P. Man, Y. Huo, T. Ning, C. Li, B. Man, et al., Synthesis of the 3D AgNF/AgNP arrays for the paper-based surface enhancement Raman scattering application, Sens. Actuators, B Chem. 265 (2018) 302–309. Available from: https://doi.org/10.1016/j.snb.2018.03.035.

[90] G. Xiao, Y. Li, W. Shi, L. Shen, Q. Chen, L. Huang, Highly sensitive, reproducible and stable SERS substrate based on reduced graphene oxide/silver nanoparticles coated weighing paper, Appl. Surf. Sci. 404 (2017) 334–341. Available from: https://doi.org/10.1016/j.apsusc.2017.01.231.

[91] S.A. Ogundare, W.E. van Zyl, Amplification of SERS "hot spots" by silica clustering in a silver-nanoparticle/nanocrystalline-cellulose sensor applied in malachite green detection, Colloids Surf. A Physicochem. Eng. Asp. 570 (2019) 156–164. Available from: https://doi.org/10.1016/j.colsurfa.2019.03.019.

[92] P. Rajapandiyan, J. Yang, Photochemical method for decoration of silver nanoparticles on filter paper substrate for SERS application, J. Raman Spectrosc. 45 (7) (2014) 574–580. Available from: https://doi.org/10.1002/jrs.4502.

[93] M. Lee, K. Oh, H.K. Choi, S.G. Lee, H.J. Youn, H.L. Lee, et al., Subnanomolar sensitivity of filter paper-based SERS sensor for pesticide detection by hydrophobicity change of paper surface, ACS Sens. 3 (1) (2018) 151–159. Available from: https://doi.org/10.1021/acssensors.7b00782.

[94] S. Liu, R. Cui, Y. Ma, Q. Yu, A. Kannegulla, B. Wu, et al., Plasmonic cellulose textile fiber from waste paper for BPA sensing by SERS, Spectrochim. Acta - Part. A Mol. Biomol. Spectrosc. (2020) 227. Available from: https://doi.org/10.1016/j.saa.2019.117664.

[95] L. Xian, R. You, D. Lu, C. Wu, S. Feng, Y. Lu, Surface-modified paper-based SERS substrates for direct-droplet quantitative determination of trace substances, Cellulose 27 (3) (2020) 1483–1495. Available from: https://doi.org/10.1007/s10570-019-02855-6.

[96] B. Li, W. Zhang, L. Chen, B. Lin, A fast and low-cost spray method for prototyping and depositing surface-enhanced Raman scattering arrays on microfluidic paper based device, Electrophoresis 34 (15) (2013) 2162–2168. Available from: https://doi.org/10.1002/elps.201300138.

[97] Y. Lu, Y. Luo, Z. Lin, J. Huang, A silver-nanoparticle/cellulose-nanofiber composite as a highly effective substrate for surface-enhanced Raman spectroscopy, Beilstein J. Nanotechnol. 10 (2019) 1270–1279. Available from: https://doi.org/10.3762/BJNANO.10.126.

[98] J.D. Weatherston, R.K.O. Seguban, D. Hunt, H.J. Wu, Low-cost and simple fabrication of nanoplasmonic paper for coupled chromatography separation and surface enhanced Raman detection, ACS Sens. 3 (4) (2018) 852–857. Available from: https://doi.org/10.1021/acssensors.8b00098.

[99] Y. Zhu, M. Li, D. Yu, L. Yang, A novel paper rag as "D-SERS" substrate for detection of pesticide residues at various peels, Talanta 128 (2014) 117–124. Available from: https://doi.org/10.1016/j.talanta.2014.04.066.

[100] M.L. Cheng, B.C.; Tsai, J. Yang, Silver nanoparticle-treated filter paper as a highly sensitive surface-enhanced Raman scattering (SERS) substrate for detection of tyrosine in aqueous solution, Anal. Chim. Acta 708 (1–2) (2011) 89–96. Available from: https://doi.org/10.1016/j.ac.2011.10.013.

[101] J.J. Saarinen, D. Valtakari, S. Sandén, J. Haapanen, T. Salminen, J.M. Mäkelä, et al., Roll-to-roll manufacturing of disposable surface enhanced Raman scattering (SERS) sensors on paper based substrates, Nord. Pulp Pap. Res. J. 32 (2) (2017) 222–228. Available from: https://doi.org/10.3183/npprj-2017-32-02-p222-228.

[102] S.W. Chook, C.H. Chia, C.H. Chan, S.X. Chin, S. Zakaria, M.S. Sajab, et al., A porous aerogel nanocomposite of silver nanoparticles-functionalized cellulose nanofibrils for SERS detection and catalytic degradation of rhodamine B, RSC Adv. 5 (108) (2015) 88915–88920. Available from: https://doi.org/10.1039/c5ra18806g.

[103] S.A. Ogundare, W.E. van Zyl, Nanocrystalline cellulose as reducing- and stabilizing agent in the synthesis of silver nanoparticles: application as a surface-enhanced Raman scattering (SERS) substrate, Surf. Interfaces 13 (2018) 1–10. Available from: https://doi.org/10.1016/j.surfin.2018.06.004.

[104] Y. Meng, Y. Lai, X. Jiang, Q. Zhao, J. Zhan, Silver nanoparticles decorated filter paper via self-sacrificing reduction for membrane extraction surface-enhanced Raman spectroscopy detection, Analyst 138 (7) (2013) 2090–2095. Available from: https://doi.org/10.1039/c3an36485b.

[105] T.M. Khan, J.G. Lunney, D. O'Rourke, M.C. Meyer, J.R. Creel, K.E. Siewierska, Various pulsed laser deposition methods for preparation of silver-sensitised glass and paper substrates for surface-enhanced Raman spectroscopy, Appl. Phys. A Mater. Sci. Process. 125 (9) (2019). Available from: https://doi.org/10.1007/s00339-019-2968-z.

[106] M.C. Díaz-Liñán, M.T. García-Valverde, A.I. López-Lorente, S. Cárdenas, R. Lucena, Silver nanoflower-coated paper as dual substrate for surface-enhanced Raman spectroscopy and ambient pressure mass spectrometry analysis, Anal. Bioanal. Chem. 412 (15) (2020) 3547–3557. Available from: https://doi.org/10.1007/s00216-020-02603-x.

[107] M.J. Oliveira, P. Quaresma, M. Peixoto de Almeida, A. Araújo, E. Pereira, E. Fortunato, et al., Office paper decorated with silver nanostars - an alternative cost effective platform for trace analyte detection by SERS, Sci. Rep. 7 (1) (2017) 2480. Available from: https://doi.org/10.1038/s41598-017-02484-8.

[108] Q. Wang, Y. Liu, Y. Bai, S. Yao, Z. Wei, M. Zhang, et al., Superhydrophobic SERS substrates based on silver dendrite-decorated filter paper for trace detection of Nitenpyram, Anal. Chim. Acta 1049 (2019) 170–178. Available from: https://doi.org/10.1016/j.ac.2018.10.039.

[109] P. Zhao, H. Liu, L. Zhang, P. Zhu, S. Ge, J. Yu, Paper-based SERS sensing platform based on 3D silver dendrites and molecularly imprinted identifier sandwich hybrid for neonicotinoid quantification, ACS Appl. Mater. Interfaces 12 (7) (2020) 8845–8854. Available from: https://doi.org/10.1021/acsami.9b20341.

[110] G. Weng, Y. Feng, J. Zhao, J. Li, J. Zhu, J. Zhao, Size dependent SERS activity of Ag triangular nanoplates on different substrates: glass vs paper, Appl. Surf. Sci. 478 (2019) 275–283. Available from: https://doi.org/10.1016/j.apsusc.2019.01.142.

[111] Y. Ke, G. Meng, Z. Huang, N. Zhou, Electrosprayed large-area membranes of Ag-nanocubes embedded in cellulose acetate microspheres as homogeneous SERS substrates, J. Mater. Chem. C. 5 (6) (2017) 1402–1408. Available from: https://doi.org/10.1039/c6tc04579k.

[112] C. Wang, B. Liu, X. Dou, Silver nanotriangles-loaded filter paper for ultrasensitive SERS detection application benefited by interspacing of sharp edges, Sens. Actuators, B Chem. 231 (2016) 357–364. Available from: https://doi.org/10.1016/j.snb.2016.03.030.

[113] M.L. Mekonnen, C.H. Chen, M. Osada, W.N. Su, B.J. Hwang, Dielectric nanosheet modified plasmonic-paper as highly sensitive and stable SERS substrate and its application for pesticides detection, Spectrochim. Acta - Part. A Mol. Biomol. Spectrosc. (2020) 225. Available from: https://doi.org/10.1016/j.saa.2019.117484.

[114] L. Wu, W. Zhang, C. Liu, M.F. Foda, Y. Zhu, Strawberry-like SiO2/Ag nanocomposites immersed filter paper as SERS substrate for acrylamide detection, Food Chem. (2020) 328. Available from: https://doi.org/10.1016/j.foodchem.2020.127106.

[115] M. Sun, B. Li, X. Liu, J. Chen, T. Mu, L. Zhu, et al., Performance enhancement of paper-based SERS chips by shell-isolated nanoparticle-enhanced Raman spectroscopy, J. Mater. Sci. Technol. 35 (10) (2019) 2207–2212. Available from: https://doi.org/10.1016/j.jmst.2019.05.055.

[116] X. Jin, P. Guo, P. Guan, S. Wang, Y. Lei, G. Wang, The fabrication of paper separation channel based SERS substrate and its recyclable separation and detection of pesticides, Spectrochim. Acta - Part. A Mol. Biomol. Spectrosc. (2020) 240. Available from: https://doi.org/10.1016/j.saa.2020.118561.

[117] Y. Ma, Y. Wang, Y. Luo, H. Duan, D. Li, H. Xu, et al., Rapid and sensitive on-site detection of pesticide residues in fruits and vegetables using screen-printed paper-based SERS swabs, Anal. Methods 10 (38) (2018) 4655–4664. Available from: https://doi.org/10.1039/c8ay01698d.

[118] K. Sivashanmugan, W.L. Huang, C.H. Lin, J.D. Liao, C.C. Lin, W.C. Su, et al., Bimetallic nanoplasmonic gap-mode SERS substrate for lung normal and cancer-derived exosomes detection, J. Taiwan. Inst. Chem. Eng. 80 (2017) 149–155. Available from: https://doi.org/10.1016/j.jtice.2017.09.026.

[119] S.Y. Chen, M. Yang, X.Y. Liu, L.S. Zha, Study on Au@Ag core-shell composite bimetallic nanorods laoding filter paper as SERS substrate, Guang Pu Xue Yu Guang Pu Fen. Xi/Spectroscopy Spectr. Anal. 38 (6) (2018) 1747–1752. Available from: https://doi.org/10.3964/j.issn.1000-0593(2018)06-1747-06.

[120] S. Asgari, L. Sun, J. Lin, Z. Weng, G. Wu, Y. Zhang, et al., Nanofibrillar cellulose/Au@Ag nanoparticle nanocomposite as a SERS substrate for detection of paraquat and thiram in lettuce, Microchim. Acta 187 (7) (2020). Available from: https://doi.org/10.1007/s00604-020-04358-9.

[121] S. Lin, W. Hasi, S. Han, X. Lin, L. Wang, A dual-functional PDMS-assisted paper-based SERS platform for the reliable detection of thiram residue both on fruit surfaces and in

juice, Anal. Methods 12 (20) (2020) 2571–2579. Available from: https://doi.org/10.1039/d0ay00483a.
[122] X. Lin, S. Lin, Y. Liu, H. Zhao, B. Liu, L. Wang, Lab-on-paper surface-enhanced Raman spectroscopy platform based on self-assembled Au@Ag nanocube monolayer for on-site detection of thiram in soil, J. Raman Spectrosc. 50 (7) (2019) 916–925. Available from: https://doi.org/10.1002/jrs.5595.
[123] J. Zhu, Q. Chen, F.Y.H. Kutsanedzie, M. Yang, Q. Ouyang, H. Jiang, Highly sensitive and label-free determination of thiram residue using surface-enhanced Raman Spectroscopy (SERS) coupled with paper-based microfluidics, Anal. Methods 9 (43) (2017) 6186–6193. Available from: https://doi.org/10.1039/c7ay01637a.
[124] Rijo, C. A Paper-Based Low-Cost Label-Free Biosensing Silver Core-Gold Shell Nanostructure for SERS to Be Applied to Breast Cancer Diagnostics, NOVA University of Lisbon, 2018.
[125] H. Peinado, S. Lavotshkin, D. Lyden, The secreted factors responsible for pre-metastatic niche formation: old sayings and new thoughts, Semin. Cancer Biol. 21 (2) (2011) 139–146. Available from: https://doi.org/10.1016/j.semcancer.2011.01.002.
[126] The Global Cancer Observatory; 2019.
[127] Selwyna, P.G.C.; Loganathan, P.R.; Begam, K.H. Development of electrochemical biosensor for breast cancer detection using gold nanoparticle doped CA 15–3 antibody and antigen interaction. *Int. Conf. Signal Process. Image Process. Pattern Recognit. 2013, ICSIPR 2013* 2013, *1*, 1–7. https://doi.org/10.1109/ICSIPR.2013.6497963.
[128] Urinary Nucleosides as Biomarkers of Breast, Colon, Lung, and Gastric Cancer in Taiwanese. *PLoS One* **2013**.
[129] J. Turkevich, P.C. Stevenson, J. Hillier, A study of the nucleation and growth processes in the synthesis of colloidal gold, Discuss. Faraday Soc. 11 (c) (1951) 55–75. Available from: https://doi.org/10.1039/DF9511100055.
[130] D. Lu, M. Ran, Y. Liu, J. Xia, L. Bi, X. Cao, SERS spectroscopy using Au-Ag nanoshuttles and hydrophobic paper-based Au nanoflower substrate for simultaneous detection of dual cervical cancer–associated serum biomarkers, Anal. Bioanal. Chem. (2020). Available from: https://doi.org/10.1007/s00216-020-02843-x.
[131] Y.J. Ahn, Y.G. Gil, Y.J. Lee, H. Jang, G.J. Lee, A dual-mode colorimetric and SERS detection of hydrogen sulfide in live prostate cancer cells using a silver nanoplate-coated paper assay, Microchem. J. 155 (2020) 104724. Available from: https://doi.org/10.1016/j.microc.2020.104724.
[132] B. Costa-Silva, N.M. Aiello, A.J. Ocean, S. Singh, H. Zhang, B.K. Thakur, et al., Pancreatic cancer exosomes initiate pre-metastatic niche formation in the liver, Nat. Cell Biol. 17 (6) (2015) 1–7. Available from: https://doi.org/10.1038/ncb3169.
[133] B.K. Thakur, H. Zhang, A. Becker, I. Matei, Y. Huang, B. Costa-Silva, et al., Double-stranded DNA in exosomes: a novel biomarker in cancer detection, Cell Res. 24 (6) (2014) 766–769. Available from: https://doi.org/10.1038/cr.2014.44.
[134] S. Sharma, H.I. Rasool, V. Palanisamy, C. Mathisen, ḰM. Schmidt, D.T. Wong, et al., Structural-mechanical characterization of nanoparticle exosomes in human saliva, using correlative AFM, FESEM, and force spectroscopy, ACS Nano 4 (4) (2010) 1921–1926.
[135] K. Mc Namara, H. Alzubaidi, J.K. Jackson, Cardiovascular disease as a leading cause of death: how are pharmacists getting involved? Integr. Pharm. Res. Pract. 8 (2019) 1–11. Available from: https://doi.org/10.2147/IPRP.S133088.
[136] W.Y. Lim, C.H. Goh, T.M. Thevarajah, B.T. Goh, S.M. Khor, Using SERS-based microfluidic paper-based device (MPAD) for calibration-free quantitative measurement of

AMI cardiac biomarkers, Biosens. Bioelectron. (2020) 147. Available from: https://doi.org/10.1016/j.bios.2019.111792.

[137] Locke, A.K.; Coté, G.L.; Deutz, N.E.P. Development of a paper-based vertical flow SERS assay for citrulline detection using aptamer-conjugated gold nanoparticles. In *Optical Diagnostics and Sensing XVIII: Toward Point-of-Care Diagnostics*; Cote, GL, (Ed.); Proceedings of SPIE; 2018; Vol. 10501, p 4. https://doi.org/10.1117/12.2290604.

[138] N. Chamuah, A. Hazarika, D. Hatiboruah, P. Nath, SERS on paper: an extremely low cost technique to measure Raman signal, J. Phys. D. Appl. Phys 50 (48) (2017). Available from: https://doi.org/10.1088/1361-6463/aa8fef.

[139] D. Lu, Z. Deng, J. Xia, X. Cao, SERS immunoassay for detection of procalcitonion in serum from patients with premature rupture of membranes using hollow Au nanocages and Ag nanowires-decorated filter paper, Vib. Spectrosc. (2019) 104. Available from: https://doi.org/10.1016/j.vibspec.2019.102963.

[140] L. Russo, M. Sánchez-Purrà, C. Rodriguez-Quijada, B.M. Leonardo, V. Puntes, K. Hamad-Schifferli, Detection of resistance protein A (MxA) in paper-based immunoassays with surface enhanced Raman spectroscopy with AuAg nanoshells, Nanoscale 11 (22) (2019) 10819–10827. Available from: https://doi.org/10.1039/c9nr02397f.

[141] A.G. Berger, S.M. Restaino, I.M. White, Vertical-flow paper SERS system for therapeutic drug monitoring of flucytosine in serum, Anal. Chim. Acta 949 (2017) 59–66. Available from: https://doi.org/10.1016/j.ac.2016.10.035.

[142] Yokoyama, M.; Nishimura, T.; Yamada, K.; Kido, M.; Ohno, Y.; Sakurai, Y.; et al. Feasibility study of paper-based surface enhanced Raman spectroscopy of tear fluids for onsite therapeutic drug monitoring. In *World Automation Congress Proceedings*; World Automation Congress; 2014; pp 474–477. https://doi.org/10.1109/WAC.2014.6936005.

[143] Yokoyama, M.; Yamada, K.; Nishimura, T.; Kido, M.; Jeong, H.; Ohno, Y. Paper-based surface enhanced Raman spectroscopy of pnenobarbital sodium for point-of-care therapeutic drug monitoring. In *Optical Diagnostics and Sensing XV: Toward Point-of-Care Diagnostics*; Cote, GL, (Ed.); Proceedings of SPIE; 2015; Vol. 9332, p 933210. https://doi.org/10.1117/12.2080619.

[144] S. Mebdoua, Pesticide residues in fruits and vegetables, in: J. Mérillon, K. Ramawat (Eds.), Bioactive Molecules in Food, Springer, Cham, 2018, pp. 1–39. Available from: https://doi.org/10.1007/978-3-319-54528-8_76-1.

[145] Monitoring of pesticide residues in commonly used fruits and vegetables in Kuwait. *Int. J. Environ. Res. Public Health* 2017, *14* (8), 833. https://doi.org/10.3390/ijerph14080833.

[146] R.C. Okocha, I.O. Olatoye, O.B. Adedeji, Food safety impacts of antimicrobial use and their residues in aquaculture, Public. Health Rev. 39 (1) (2018) 21. Available from: https://doi.org/10.1186/s40985-018-0099-2.

[147] P.K. Rai, S.S. Lee, M. Zhang, Y.F. Tsang, K.-H. Kim, Heavy metals in food crops: health risks, fate, mechanisms, and management, Environ. Int. 125 (2019) 365–385. Available from: https://doi.org/10.1016/j.envint.2019.01.067.

[148] D.J. Lee, D.Y. Kim, Hydrophobic paper-based SERS sensor using gold nanoparticles arranged on graphene oxide flakes, Sens. (Switz.) 19 (24) (2019). Available from: https://doi.org/10.3390/s19245471.

[149] G. Yang, X. Fang, Q. Jia, H. Gu, Y. Li, C. Han, et al., Fabrication of paper-based SERS substrates by spraying silver and gold nanoparticles for SERS determination of malachite green, methylene blue, and crystal violet in fish, Microchim. Acta 187 (5) (2020). Available from: https://doi.org/10.1007/s00604-020-04262-2.

[150] Y. Zhu, L. Zhang, L. Yang, Designing of the functional paper-based surface-enhanced Raman spectroscopy substrates for colorants detection, Mater. Res. Bull. 63 (2015) 199–204. Available from: https://doi.org/10.1016/j.materresbull.2014.12.004.
[151] A. Purwidyantri, M. Karina, C.-H. Hsu, Y. Srikandace, B.A. Prabowo, C.-S. Lai, Facile bacterial cellulose nanofibrillation for the development of a plasmonic paper sensor, ACS Biomater. Sci. Eng. 6 (5) (2020) 3122–3131. Available from: https://doi.org/10.1021/acsbiomaterials.9b01890.
[152] L. Zhang, J. Liu, G. Zhou, Z. Zhang, Controllable in-situ growth of silver nanoparticles on filter paper for flexible and highly sensitive sers sensors for malachite green residue detection, Nanomaterials 10 (5) (2020). Available from: https://doi.org/10.3390/nano10050826.
[153] J. Gu, A. Dichiara, hybridization between cellulose nanofibrils and faceted silver nanoparticles used with surface enhanced Raman scattering for trace dye detection, Int. J. Biol. Macromol. 143 (2020) 85–92. Available from: https://doi.org/10.1016/j.ijbiomac.2019.12.018.
[154] A. Kumar, V. Santhanam, Paper swab based SERS detection of non-permitted colourants from dals and vegetables using a portable spectrometer, Anal. Chim. Acta 1090 (2019) 106–113. Available from: https://doi.org/10.1016/j.ac.2019.08.073.
[155] D. Deng, Q. Lin, H. Li, Z. Huang, Y. Kuang, H. Chen, et al., Rapid detection of malachite green residues in fish using a surface-enhanced Raman scattering-active glass fiber paper prepared by in situ reduction method, Talanta 200 (2019) 272–278. Available from: https://doi.org/10.1016/j.talanta.2019.03.021.
[156] W. Zhang, B. Li, L. Chen, Y. Wang, D. Gao, X. Ma, et al., Brushing, a simple way to fabricate SERS active paper substrates, Anal. Methods 6 (7) (2014) 2066–2071. Available from: https://doi.org/10.1039/c4ay00046c.
[157] C. Zhang, T. You, N. Yang, Y. Gao, L. Jiang, P. Yin, Hydrophobic paper-based SERS platform for direct-droplet quantitative determination of melamine, Food Chem. 287 (2019) 363–368. Available from: https://doi.org/10.1016/j.foodchem.2019.02.094.
[158] A. Marques, B. Veigas, A. Araújo, B. Pagará, P.V. Baptista, H. Águas, et al., Paper-based SERS platform for one-step screening of tetracycline in milk, Sci. Rep. 9 (1) (2019). Available from: https://doi.org/10.1038/s41598-019-54380-y.
[159] S.S.B. Moram, C. Byram, V.R. Soma, Gold-nanoparticle- and nanostar-loaded paper-based SERS substrates for sensing nanogram-level picric acid with a portable Raman spectrometer, Bull. Mater. Sci. 43 (1) (2020). Available from: https://doi.org/10.1007/s12034-019-2017-8.
[160] S.S.B. Moram, C. Byram, S.N. Shibu, B.M. Chilukamarri, V.R. Soma, Ag/Au nanoparticle-loaded paper-based versatile surface-enhanced Raman spectroscopy substrates for multiple explosives detection, ACS Omega 3 (7) (2018) 8190–8201. Available from: https://doi.org/10.1021/acsomega.8b01318.
[161] L.F. Sallum, F.L.F. Soares, J.A. Ardila, R.L. Carneiro, Optimization of SERS scattering by Ag-NPs-coated filter paper for quantification of nicotinamide in a cosmetic formulation, Talanta 118 (2014) 353–358. Available from: https://doi.org/10.1016/j.talanta.2013.10.039.
[162] B. Doherty, B.G. Brunetti, A. Sgamellotti, C. Miliani, A detachable SERS active cellulose film: a minimally invasive approach to the study of painting lakes, J. Raman Spectrosc. 42 (11) (2011) 1932–1938. Available from: https://doi.org/10.1002/jrs.2942.

Chapter 7

Chemiluminescence paper-based analytical devices

Waleed Alahmad[1,2], Pakorn Varanusupakul[1,2] and Takashi Kaneta[3]
[1]Chemical Approaches for Food Applications Research Group, Faculty of Science, Chulalongkorn University, Bangkok, Thailand, [2]Department of Chemistry, Faculty of Science, Chulalongkorn University, Bangkok, Thailand, [3]Department of Chemistry, Graduate School of Natural Science and Technology, Okayama University, Okayama, Japan

7.1 Introduction

Since Whitesides and coworkers proposed the utility of paper as a substrate for fabricating microfluidics [1], microfluidic paper-based analytical devices (μPADs) have gained much attention because of their feasibility in point-of-care testing and on-site analyses. Among several analytical methods using μPADs, chemiluminescence (CL) detection is one of the most sensitive and selective techniques. The CL measurement is intrinsically more sensitive than fluorescence detection because there is no background signal originating from an external light source. The further advantages of CL detection include low reagent costs and a wide dynamic range, so CL is promising in the enhancement of sensitivity for μPADs [2]. In general, CL-μPADs need measurements of light generated by chemical reactions, whereas both fabrication and operation are similar to μPADs with different detection schemes. Like the standard μPADs, CL-μPADs generally employ paper substrates that form microfluidic channels and/or spots limited by hydrophobic barriers. A hydrophilic cellulose support acts as a network of capillaries, where liquid is transported without the need for an external driving force. Therefore, a suitable design of the microfluidic channels can control the mixing of the reagents and sample that causes the CL reactions.

7.2 Design and fabrication techniques for chemiluminescence-microfluidic paper-based analytical devices

The main purpose of the fabrication step is the formation of hydrophilic channels surrounded by hydrophobic barriers on cellulose paper. Numerous

Paper-Based Analytical Devices for Chemical Analysis and Diagnostics.
DOI: https://doi.org/10.1016/B978-0-12-820534-1.00005-0
© 2022 Elsevier Inc. All rights reserved.

techniques have been reported for fabricating μPADs, including wax printing, inkjet printing, photolithography, flexographic printing, plasma treatment, laser treatment, wet etching, screen printing, and wax screen printing [3]. Both two-dimensional (2D) [4–8] and three-dimensional (3D) μPADs [9–12] can be fabricated using these techniques to transport fluids in both horizontal and vertical directions, depending on the complexity of the application. For the fabrication of a 2D channel, two parallel hydrophobic lines were drawn on the paper substrate, as to surround the hydrophilic channels. Because the hydrophobic barriers prevent the penetration of the hydrophilic solution out of the channels, the solution, consequently, flows in the channels under the effects of capillary action [10]. In these types of devices, sample zones, flow channels, and detection zones are located in horizontal directions to each other. Conversely, 3D-μPADs are fabricated by stacking alternating layers of paper pieces that bind with water-impermeable, double-sided adhesive tape. Therefore, an aqueous solution flows horizontally in a single layer and vertically between the layers [13]. A low-cost "origami" method has been proposed for the production of a 3D CL-μPAD via folding the paper substrate [14]. For a more detailed overview of the principles and comparisons of the main fabrication techniques, please refer to the review articles [15,16].

7.3 Detection of chemiluminescence Signals on microfluidic paper-based analytical devices

CL detectors measure light originating from the conversion of chemical energy into the emission of visible light (luminescence), resulting from an oxidation or hydrolysis reaction. The detectors measure CL in the dark to exclude any background signal from ambient light.

There have been several reported approaches in the literature for capturing and detecting the CL signal on paper-based microfluidic devices, including (1) a commercial luminescence analyzer, (2) a miniaturized photomultiplier tube, (3) a charge-coupled device (CCD) camera, (4) a digital camera, and (5) a smartphone camera.

7.3.1 Commercial luminescence analyzers

Most researchers use commercial analyzers due to their acceptable sensitivity and easy availability. Yu et al. [17] used a computerized, ultra-weak luminescence analyzer to measure the CL emission on a μPAD to establish a biosensor for uric acid determination. As shown in Fig. 7.1A, the biosensor generates CL signals via an enzyme reaction of urate oxidase that produces hydrogen peroxide (H_2O_2) in the presence of the substrate, uric acid, followed by the CL reaction between a rhodanine derivative and generated H_2O_2 in acid medium. The CL-μPAD biosensor consisted of one layer of

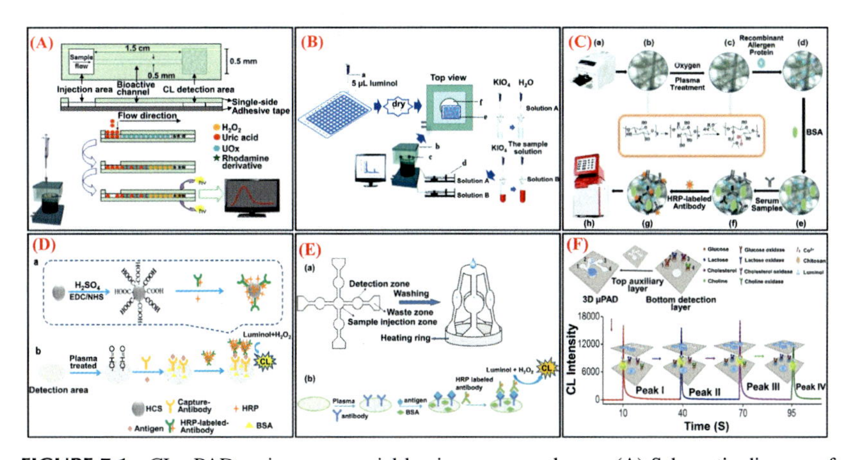

FIGURE 7.1 CL-μPADs using commercial luminescence analyzers. (A) Schematic diagram of uric acid determination, based on an enzymatic reaction. (B) Schematic of a typical CL diminishment assay procedure for the determination of β-agonists: (a) pipette, (b) CL analyzer, (c) injection hole, (d) syringe pump, (e) PAD, and (f) photomultiplier. (C) Schematic illustration of a paper-based immunoassay for the detection of allergen-specific IgE: (a–h) fabrication steps and operation of device. (D) Schematic diagram of the preparation procedure of a CL-μPAD immunodevice for the carcinoembryonic antigen consisting of two steps (a and b). (E) Schematic diagram of the 3D washing strategy with a ring-oven (a), with the immunoassay procedure for the detection area of the paper-based chip (b). (F) Fabrication steps of a 3D CL-μPAD for multiplexed measurements and the analysis procedure. *(Figure 7.1A) Reproduced from J. Yu, S. Wang, L. Ge, S. Ge, A novel chemiluminescence paper microfluidic biosensor based on enzymatic reaction for uric acid determination, Biosens. Bioelectron. 26 (2011) 3284–3289 with permission from Elsevier. (B) Reproduced from X. Chen, Y. Luo, B. Shi, X. Liu, Z. Gao, Y. Du, et al., Chemiluminescence diminishment on a paper-based analytical device: high throughput determination of β-agonists in swine hair, Anal. Methods 6 (2014) 9684–9690 with permission from The Royal Society of Chemistry. (C) Reproduced from X. Han, M. Cao, M. Wu, Y.J. Wang, C. Yu, C. Zhang, et al., A paper-based chemiluminescence immunoassay device for rapid and high-throughput detection of allergen-specific IgE, Analyst 144 (2019) 2584–2593 with permission from The Royal Society of Chemistry. (D) Reproduced from Y. Chen, W. Chu, W. Liu, X. Guo, Y. Jin, B. Li, Paper-based chemiluminescence immunodevice for the carcinoembryonic antigen by employing multienzyme carbon nanosphere signal enhancement, Microchim. Acta 185 (2018) 187 with permission from Springer. (E) Reproduced from W. Chu, Y. Chen, W. Liu, L. Zhang, X. Guo, Three-dimensional ring-oven washing technique for a paper based immunodevice, Luminescence (2019) 1–9. doi:10.1002/bio.3756 with permission from Wiley. (F) Reproduced from F. Li, J. Liu, L. Guo, J. Wang, K. Zhang, J. He, et al., High-resolution temporally resolved chemiluminescence based on double-layered 3D microfluidic paper-based device for multiplexed analysis, Biosens. Bioelectron. 141 (2019) 111472 with permission from Elsevier.*

paper, with a channel to flow a sample, stacked between two layers of water-impermeable, single-sided adhesive tapes. The top layer of water-impermeable, single-sided tape had a patterned square hole that served as the sample injection zone. A rectangular knife cutter was used for fabricating the sample injection zone on the top layer. The middle paper layer was composed of three independent components: a sample injection zone, a bioactive channel, and a CL detection zone. The CL measurement took place in a homemade, dark device-holder in which the μPAD was aligned to locate the position of the sample injection zone, the CL detection zone at the injection hole, and the sensing area of the photomultiplier in the analyzer, respectively. Moreover, Chen et al. [18] presented a high-throughput CL-μPAD sensor for the determination of β-agonists in swine hair. In this work, β-agonists diminished CL emission generated from the reaction between luminol and KIO_4, where KIO_4 was reduced to KIO_3 through the oxidation of β-agonists. The CL-μPAD sensor, which was fabricated by wax printing, was designed to be compatible with a 96-well plate. On the CL-μPAD, two detection zones, one for the sample and the other for the blank, were employed for a single sample measurement, so 48 samples were measured with a single CL-μPAD. The blank and sample solutions were delivered to the CL-μPAD by syringe pumps, as shown in Fig. 7.1B.

Several groups have reported CL-μPADs using CL analyzers [19,20,22–24]. Han et al. reported a CL-μPAD immunoassay device for rapid and high-throughput detection of allergen-specific lgE [19]. Herein, a paper substrate was treated with oxygen plasma to produce aldehyde groups on paper prior to the fabrication of a paper microzone plate via wax printing, followed by baking at 100°C. Incubation with recombinant Can f 1 (one of the main dog allergens) resulted in binding to the aldehyde groups via Schiff base formation. Thereafter, subsequent incubation with bovine serum albumin, for blocking unspecific bonding, sIgEs, and horseradish peroxidase (HRP)-labeled anti-IgE yielded a traditional enzyme-linked immunosorbent assay (ELISA). The HRP-labeled anti-IgE catalyzed the reaction between luminol and H_2O_2, resulting in CL, as shown in Fig. 7.1C. Also, Wang et al. [22] and Zhao et al. [23] used chitosan and plasma treatment, respectively, to modify the paper substrates to covalently immobilize antibodies on μPADs.

Even though the luminescence analyzers provide high sensitivity, Chen et al. [20] enhanced the sensitivity of a CL-μPAD by employing multienzyme carbon nanospheres for the determination of carcinoembryonic antigens. Simply, the paper substrate was treated with plasma to immobilize the capture antibody (Ab1) on the surface. The antigen reacted with Ab1, followed by binding with an HRP-labeled antibody in a standard sandwich immunoassay, where carbon nanospheres functionalized with both the HRP-labeled antibody and HRP were bound to the antibody on the paper, enhancing the CL signal. The luminol/H_2O_2 CL system, catalyzed by the functionalized carbon nanospheres, increased the sensitivity 10-fold compared to a commercial kit. A schematic illustration of the reactions is described in Fig. 7.1D.

ELISA, in general, requires a washing process that removes nonspecific-binding antibodies and antigens completely to reduce the blank signal. For this washing process in μPADs, a washing solution must flow continuously in the reaction zone. To realize this washing step in μPADs, Chu et al. [21] reported

CL-based ELISA in μPADs using a novel 3D ring-oven washing technique after first demonstrating the washing strategy in a 2D format. The 3D-μPAD was placed on a plastic holder, contacting the waste area to a heating ring, as shown in Fig. 7.1E. Continuous flow, generated by gravity and capillary action, flushed nonspecific-binding proteins to the waste area. The developed 3D washing strategy permitted carcinoembryonic antigen detection based on the luminol/H_2O_2 CL system catalyzed by HRP, as shown in Fig. 7.1E.

Different types of CL-μPADs have also been used for multiplexed analysis [13,25–27], although they required 3D fabrication due to the limited size of the luminol analyzer's photomultiplier. Li et al. [13] presented a double-layered 3D-μPAD that consisted of four detection zones with high temporal resolution. The 3D-μPAD resolved CL emissions for multiplexed (sequential) CL measurements in glucose, lactate, cholesterol, and choline assays. The temporally resolved CL emissions were obtained by virtue of the 3D-branched microfluidic channel design that created time delays for luminol transport from one detection zone to another, as shown in Fig. 7.1F. The peak intensity and peak shape of the temporally resolved CL emissions were quite stable with baseline separation. The chemistry used in the 3D-μPAD was based on the cobalt ion-catalyzed CL reactions between luminol and H_2O_2, generated from specific enzymatic reactions between oxidase and the analytes. On the other hand, a 3D origami μPAD, combined with a CL detection system, was also proposed by Ge et al. [14]. The developed 3D microfluidic paper-based immunodevice integrated channels for blood plasma separation from whole blood samples, automation of rinse steps, and multiplexed CL detections. The device was combined with a typical luminol-H_2O_2 CL system with a catalytic reaction by Ag nanoparticles, and it successfully demonstrated simultaneous detection of four tumor markers.

7.3.2 Miniaturized photomultiplier

The miniaturized photomultiplier is the best choice for miniaturized sensors and detection systems with high sensitivity in CL chemistry. Alahmad et al. [28] developed a miniaturized CL detection system for μPAD applications, consisting of a μPAD (2D), a small photomultiplier tube (22 × 22 × 60 mm), a power supply with a 15 V output, and a holder made of dark black acrylic for the μPAD. Each μPAD consisted of six channels that allowed for the determination of six different samples successively. Each channel was composed of injection, reaction, and waste zones, as shown in Fig. 7.2A. The main advantage of adding a waste zone is to generate continuous flow at the reaction zone, resulting in peak-shape CL signals because the reagent drained to the waste zone instead of staying at the reaction zone. Moreover, the holder was equipped with six optical fibers that were bundled and connected to the photomultiplier tube to achieve six successive measurements of the samples and standards. In their work, six sequential injections were

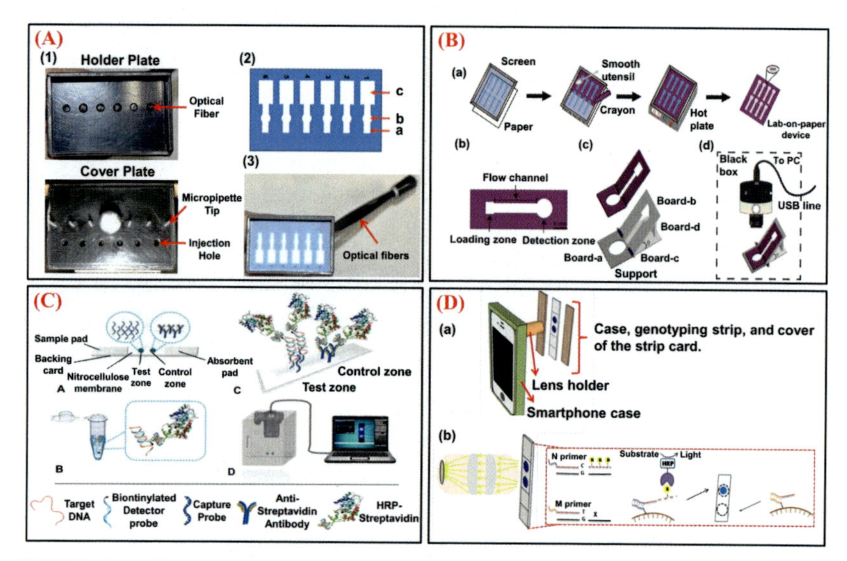

FIGURE 7.2 (A) CL-μPADs used miniaturized detection systems. (1) Photographic image of the μPAD holder. (2) The design of the μPAD consisted of (a) an injection zone, (b) reaction zone, and (c) waste zone. (3) The holder plate after connection with optical fibers and with the μPAD placed on it. (B) Schematics of fabrication and assembly of a paper-based GCF-CL. (a) Fabrication of paper-based microfluidic devices using a wax screen printing fabrication technique. (b) The layout of the unfolded paper-based GCF-CL. (c) Diagram of the paper-based device folded into the desired shape to produce gravity flow. (d) Schematic of the assembled paper device placed under the CCD in the black box for the GCF-CL assay. (C) Schematic illustration of the synthetic DNA identification based on a nucleic acid lateral flow biosensor (NALFB) and CL analysis. (a) Schematic illustration of the CL-NALFB configuration for quantitative assessment of target DNA. (b) The detector probe, target DNA, and HRP-Streptavidin conjugate were mixed to generate the HRP conjugate-detector probe-target DNA complexes. (c) The HRP label-detector probe-target DNA complexes were captured by the oligonucleotide capture probes on the test zone, and the excess HRP-Streptavidin conjugates were captured by an antistreptavidin antibody on the control zone. (d) CL detection using a bench-top dark box, equipped with a digital camera. (D) Scheme for CL imaging with a smartphone camera. (a) Layout of the CL portable device. (b) Close-up of the lens holder with the paper-based genotyping strip. *(A) Reproduced from W. Alahmad, K. Uraisin, D. Nacapricha, T. Kaneta, A miniaturized chemiluminescence detection system for a microfluidic paper-based analytical device and its application to the determination of chromium (III), Anal. Methods 8 (2016) 5414–5420 with permission from The Royal Society of Chemistry. (B) Reproduced from Q. Shang, P. Zhang, H. Li, R. Liu, C. Zhang, A flow chemiluminescence paper based microfluidic device for detection of chromium (III) in water, J. Innov. Opt. Health Sci. 6 (2019) 1950016. (C) Represented from F. Jahanpeyma, M. Forouzandeh, M.J. Rasaee, N. Shoaie, An enzymatic paper-based biosensor for ultrasensitive detection of DNA, Front. Biosci. 11 (2019) 122–135. (D) Reproduced from E.M. Spyrou, D.P. Kalogianni, S.S. Tragoulias, P.C. Ioannou, T.K. Christopoulos, Digital camera and smartphone as detectors in paper-based chemiluminometric genotyping of single nucleotide polymorphisms, Anal. Bioanal. Chem. 408 (2016) 7393–7402 with permission from Springer.*

completed within 1 min for each μPAD with short analysis time and high throughput. This system was successfully applied for the determination of Cr (III) in water samples, based on the Cr(III)-catalyzed oxidation of luminol by H_2O_2 in alkaline medium. Moreover, Wang et al. [32] used a homemade portable photomultiplier for CL detection, generated on a paper-based microdisk plate. They reported a novel paper-based molecular imprinted polymer (MIP)-grafted multidisk microdisk plate (P-MIP MMP) for sensitive and specific CL detection of pesticides through an indirect competitive assay using 2,4-dichlorophenoxyacetic acid.

7.3.3 Charge-coupled devices and digital cameras

CCDs for imaging, as well as digital cameras, were used as inexpensive detectors for CL detection. Herein, the CL signals are recorded as videos first. After that, video-cutting software is used to cut the videos into pictures. These pictures are analyzed by an automated image-processing program to find a picture with the maximum luminescence intensity. However, instead of recording videos, continuous photos of the detection zone can be used by taking them with an optimized shutter speed (exposure time). Finally, the mean gray or RGB (red, green, blue) values are obtained by an image-processing program to quantify the CL assay. Shang et al. [29] presented a gravity and capillary force-driven flow chemiluminescence (GCF-CL) paper-based microfluidic device for the detection of Cr(III) in water. The paper was folded and placed on a plastic holder to achieve gravity flow. In their work, a low-cost and luggable CCD camera was used for recording the CL signal generated from the Cr(III)-catalyzed oxidation of luminol by H_2O_2. Data acquisition and analysis methods were the same as explained previously, and the mean gray value of the whole luminous zone was represented as the CL intensity of each selected picture. Fig. 7.2B shows the fabrication, assembly, and detection procedure of the GCF-CL paper-based microfluidic device.

Digital cameras can also be used as an alternative detector to measure the CL signals. Jahanpeyma et al. [30] used a digital camera for detection in the development of an enzymatic, paper-based biosensor for the ultrasensitive detection of DNA. In this work, the CL paper-based lateral flow biosensor consisted of two-step hybridization: a primary hybridization of a 5′-biotinylated detection probe to the target DNA and a secondary hybridization of the resulting complex with the immobilized capture probe. Quantitative analysis of DNA was provided via an HRP-catalyzed reaction with the CL substrate, followed by shooting images with a digital camera. Herein, the authors took photos of the detection zone continuously using the camera, which was adjusted with a 30 s integration time to take long exposure images. This yielded pictures with the highest feasible level of CL signal, as shown in Fig. 7.2C. In the quantification step, CL signals were generated using image-processing software.

However, it is worth mentioning that CCDs and digital cameras are not as sensitive as photomultiplier tubes. To overcome this drawback, Liu and Zhang [33] enhanced the CL signals on a paper-based biosensor to detect 198-bp DNA fragments obtained by PCR amplification of the Listeria monocytogenes hlyA gene. A wax screen printing technique was used to fabricate the paper-based microfluidic biosensor. This work was based on hybridization reactions with a covalently-immobilized DNA probe, biotin-labeled signal DNA strands, and CL reactions catalyzed by a HRP-streptavidin conjugate. The p-iodine phenol was added to the HRP-luminol-H_2O_2 system to enhance the generated CL signal.

7.3.4 Smartphones

Because smartphones are ubiquitous and cost-effective, many researchers have developed smartphone-based μPADs. All smartphones are provided with a camera, which typically uses a complementary metal oxide semiconductor sensor with an image size up to 108 megapixels. The portability and remote operability of the camera, which has the ability to transfer photographic data either by image sharing or Bluetooth, make smartphones potentially powerful tools for remote monitoring in biochemical analyses [34,35]. Spyrou et al. [31] used a smartphone as a detector for a CL lateral flow μPAD assay and developed a method for chemiluminometric genotyping of the C677T single nucleotide polymorphism (SNP). To detect the emerging CL from the sensing areas, they constructed a 3D-printed smartphone attachment that houses inexpensive lenses, which helped to convert the smartphone into a portable CL imager, as shown in Fig. 7.2D. Herein, the CL emission was captured immediately by the smartphone with an exposure time of 8 s. Their work highlighted the ability to discriminate two alleles of an SNP in a single shot by imaging the strip, and this avoided the necessity of dual-labeling.

7.3.5 Other tools and instruments

Zhang et al. [36] replaced a laboratory CL analyzer with an optical detection module to convert optical signals into electrical signals and achieve signal amplification. The authors highlighted many advantages for this module, such as low power consumption, portability, miniaturization, integration of signal conversion and amplification, robustness, high dynamic response, and cost-effectiveness (cheaper than a smartphone). In their work, they developed a sensitive PAD system with CL detection and electrokinetic preconcentration for the detection of hemin, based on the luminol–H_2O_2 system. Liu et al. [37] used a microplate reader for measuring the CL signal on the detection zone of a μPAD. The microplate reader was programmed to measure the signal at a predetermined location in the wells. They developed a laminated paper-based analytical device (LPAD) with an origami-enabled CL immunoassay for the detection of cotinine. Herein, a craft-cutter was used to cut and fabricate a paper piece with flow channels, followed by the lamination of the paper piece with polyester thermal bonding lamination film. To isolate the detection zone from the flow channels, the author introduced a protocol of localized incision and paper-folding, based on origami art. The simple origami step eliminated possible reagent diffusion and flow during antibody immobilization steps and numerous washings.

Spectrofluorometers can also measure CL emission when the radiation source is switched off. Hassanzadeh et al. successfully determined CL emission for the measurement of total phenolic content in food samples on their developed μPAD. In the CL system, the reaction implemented on the paper device consisted of a H_2O_2 − rhodamine B (RhoB) − cobalt metal-organic framework (CoMOF) system [38]. They proved that the reaction of H_2O_2 with RhoB molecules, loaded into the nanopores of CoMOFs (R@CoMOFs), could produce an intensive CL emission. Conversely, Li et al. [39] used a mini imager to capture the CL images on paper, developed by a research center as an alternative tool to smartphones, CCDs, and digital cameras for high-sensitivity C-reactive protein detection, based on an ELISA and enhanced by a biotin-streptavidin system. Similar to the reports using smartphones, CCDs, and digital cameras, the images were converted into an 8-bit grayscale before quantitative analysis, and the intensity was measured in a uniform circle at each detection zone.

Table 7.1 summarizes the performances of several CL-μPADs reported in this chapter by highlighting the precision and limits of detection. As seen in Table 7.1, the limits of detection are much better than standard colorimetry due to inherent low background while the reproducibilities are comparable.

7.4 Conclusion

We have reviewed μPADs with CL detection, where different types of optical sensing were employed for measuring the emission. The most popular reaction system is luminol-H_2O_2, due to its simplicity. The advent of new materials and technologies accelerate the development of CL μPADs for different types of applications. One of the most important parts in the CL system is the detector, for sensitive sensing of CL emission. As this chapter shows, commercial luminescence analyzers are the most frequently used detectors, due to their high sensitivity compared to others. However, a homemade miniaturized detection system based on a small photomultiplier tube gives similar sensitivity to the commercial analyzers. Smartphones, CCDs, and digital cameras also show good sensitivity to some substances; although, the detection systems require signal enhancement. We hope that insight from this chapter will inspire readers to develop more portable and miniaturized CL detection methods in all research fields.

TABLE 7.1 Summarizes of the CL-μPADs developed for the determination of different analytes in different matrices.

Analytes and (samples)	Materials and structures	Fabrication methods	Detection tool	Detection limit	RSD%	Recovery %	References
Uric acid (Artificial urine)	Chromatography paper, 2D	Knife cutting	Ultra-weak luminescence analyzer	1.9 mmol L^{-1}	0.29	97–104	[17]
Carcinoembryonic antigen (Human serum)	Chromatography paper, 2D	Craft cutting	CL analyzer	3 pg mL^{-1}	NA	NA	[20]
Glucose, lactate, cholesterol, and choline (Human serum)	Chromatography paper, 3D	Wax screen printing	Ultra-weak luminescence analyzer	8, 15, 6, and 0.07 μmol L^{-1}	3.29	94–109	[13]
AFP, CA 153, CA 199, and CEA (whole blood)	Chromatography paper, 3D origami	Wax printing	CL analyzer	1.0 ng mL^{-1}, 0.4, and 0.06 U mL^{-1} and 0.02 ng mL^{-1}, respectively	5.2	NA	[14]
Cr(III) (Water samples)	Chromatography paper, 2D	Wax printing	Miniaturized photomultiplier	0.02 mg L^{-1}	6.5	99–107	[28]
Cr(III) (Water samples)	Chromatography paper, 2D	wax screen printing	CCD camera	0.03 mg L^{-1}	5	88–107	[29]
rRNA gene (*E. coli*)	Nitrocellulose membrane, 2D	Cutting	Digital camera	0.4 ng mL^{-1}	6	NA	[30]

Total cholesterol (Human serum)	Filter paper, 2D	Cutting	Smartphone	20 mg dL^{-1}	5	NA	[34]
Hemin (Iron supplements and human serum)	Glass fiber filter, 2D	Cutting	Optical detection module	0.58 mmol L^{-1}	14.3	85–116	[36]
Cotinine (Mouse serum)	Chromatography paper, Origami	Digital craft cutting	Microplate reader	5 ng mL^{-1}	5.00	97–104	[37]
Gallic acid, quercetin, catechin, kaempferol, and caffeic acid (Food)	Filter paper, 2D	Wax printing	Spectrofluorometer	0.98, 1.36, 1.48, 1.81, and 2.55 ng mL^{-1}	3.74–3.99	96–104	[38]
C-reactive protein (Human serum)	Chromatography paper, 2D	Wax printing	Mini imager	0.49 ng mL^{-1}	7.00	95–108	[39]

NA, Not available.

Acknowledgments

This research is supported by the Rachadapisek Sompot Fund for Postdoctoral Fellowship, Chulalongkorn University. T.K. acknowledges the support by JSPS KAKENHI, Grant numbers JP19H04675 and JP20H02766.

References

[1] A.W. Martinez, S.T. Phillips, M.J. Butte, G.M. Whitesides, Patterned paper as a platform for inexpensive, low-volume, portable bioassays, Angew. Chem. Int. Ed. 46 (2007) 1318–1320.

[2] T. Kaneta, W. Alahmad, P. Varanusupakul, Microfluidic paper-based analytical devices with instrument-free detection and miniaturized portable detectors, Appl. Spectrosc. Rev. 54 (2019) 117–141.

[3] G. Sriram, M.P. Bhat, P. Patil, U.T. Uthappa, H.Y. Jung, T. Altalhi, et al., Paper-based microfluidic analytical devices for colorimetric detection of toxic ions: a review, Trends Anal. Chem. 93 (2017) 212–227.

[4] W. Alahmad, N. Tungkijanansin, T. Kaneta, P. Varanusupakul, A colorimetric paper-based analytical device coupled with hollow fiber membrane liquid phase microextraction (HF-LPME) for highly sensitive detection of hexavalent chromium in water samples, Talanta 190 (2018) 78–84.

[5] H. Tabani, F. Dorabadi Zare, W. Alahmad, Pakorn Varanusupakul, Dual-gel electromembrane extraction: an efficient and green method for speciation of chromium species in environmental water samples, Environ. Chem. Lett. (2019). Available from: https://doi.org/10.1007/s10311-01900921-w.

[6] W. Alahmad, P. Varanusupakul, T. Kaneta, P. Varanusupakul, Chromium speciation using paper-based analytical devices by direct determination and with electromembrane microextraction, Anal. Chim. Acta 1085 (2019) 98–106.

[7] S. Karita, T. Kaneta, Acid–base titrations using microfluidic paper-based analytical devices, Anal. Chem. 86 (2014) 12108–12114.

[8] S. Karita, T. Kaneta, Chelate titrations of Ca^{2+} and Mg^{2+} using microfluidic paper-based analytical devices, Anal. Chem. Acta 924 (2016) 60–67.

[9] J. Park, J.K. Park, Pressed region integrated 3D paper-based microfluidic device that enables vertical flow multistep assays for the detection of C-reactive protein based on programmed reagent loading, Sens. Actuators B Chem. 246 (2017) 1049–1055.

[10] R.A.G. de Oliveira, F. Camargo, N.C. Pesquero, R.C. Faria, A simple method to produce 2D and 3D microfluidic paper-based analytical devices for clinical analysis, Anal. Chim. Acta 957 (2017) 40–46.

[11] X. Weng, S.R. Ahmed, S. Neethirajan, A nanocomposite-based biosensor for bovine haptoglobin on a 3D paper-based analytical device, Sens. Actuators B Chem. 265 (2018) 242–248.

[12] L. Fan, Q. Hao, X. Kan, Three-dimensional graphite paper based imprinted electrochemical sensor for tertiary butylhydroquinone selective recognition and sensitive detection, Sens. Actuators B Chem. 256 (2018) 520–527.

[13] F. Li, J. Liu, L. Guo, J. Wang, K. Zhang, J. He, et al., High-resolution temporally resolved chemiluminescence based on double-layered 3D microfluidic paper-based device for multiplexed analysis, Biosens. Bioelectron. 141 (2019) 111472.

[14] L. Ge, S. Wang, X. Song, S. Ge, J. Yu, 3D Origami-based multifunction-integrated immunodevice: low-cost and multiplexed sandwich chemiluminescence immunoassay on microfluidic paper-based analytical device, Lab. Chip 12 (2012) 3150–3158.

[15] A.T. Singh, D. Lantigua, A. Meka, S. Taing, M. Pandher, G. Camci-Unal, Paper based sensors: emerging themes and applications, Sensors 18 (2018) 28–38.

[16] Y. Xia, J. Si, Z. Li, Fabrication techniques for microfluidic paper-based analytical devices and their applications for biological testing: a review, Biosens. Bioelectron. 77 (2016) 774–789.

[17] J. Yu, S. Wang, L. Ge, S. Ge, A novel chemiluminescence paper microfluidic biosensor based on enzymatic reaction for uric acid determination, Biosens. Bioelectron. 26 (2011) 3284–3289.

[18] X. Chen, Y. Luo, B. Shi, X. Liu, Z. Gao, Y. Du, et al., Chemiluminescence diminishment on a paper-based analytical device: high throughput determination of β-agonists in swine hair, Anal. Methods 6 (2014) 9684–9690.

[19] X. Han, M. Cao, M. Wu, Y.J. Wang, C. Yu, C. Zhang, et al., A paper-based chemiluminescence immunoassay device for rapid and high-throughput detection of allergen-specific IgE, Analyst 144 (2019) 2584–2593.

[20] Y. Chen, W. Chu, W. Liu, X. Guo, Y. Jin, B. Li, Paper-based chemiluminescence immunodevice for the carcinoembryonic antigen by employing multi-enzyme carbon nanosphere signal enhancement, Microchim. Acta. 185 (2018) 187.

[21] W. Chu, Y. Chen, W. Liu, L. Zhang, X. Guo, Three-dimensional ring-oven washing technique for a paper based immunodevice, Luminescence (2019) 1–9. Available from: https://doi.org/10.1002/bio.3756.

[22] S. Wang, L. Ge, X. Song, J. Yu, S. Ge, J. Huang, et al., Paper-based chemiluminescence ELISA: Lab-on-paper based on chitosan modified paper device and wax-screen-printing, Biosens. Bioelectron. 31 (2012) 212–218.

[23] M. Zhao, H. Li, W. Liu, Y. Guo, W. Chu, Plasma treatment of paper for protein immobilization on paper-based chemiluminescence immunodevice, Biosens. Bioelectron. 79 (2016) 581–588.

[24] W. Chu, Y. Chen, W. Liu, M. Zhao, H. Li, Paper-based chemiluminescence immunodevice with temporal controls of reagent transport technique, Sens. Actuators B Chem. 250 (2017) 324–332.

[25] R. Yang, F. Li, W. Zhang, W. Shen, D. Yang, Z. Bian, et al., Chemiluminescence immunoassays for simultaneous detection of three heart disease biomarkers using magnetic carbon composites and three-dimensional microfluidic paper-based device, Anal. Chem. 91 (2019) 13006–13013.

[26] X. Guo, Y. Guo, W. Liu, W. Chu, Fabrication of paper-based microfluidic device by recycling foamed plastic and the application for multiplexed measurement of biomarkers, Spectrochim. Acta A 223 (2019) 117341.

[27] F. Li, L. Guo, Y. Hu, Z. Li, J. Liu, J. He, et al., Multiplexed chemiluminescence determination of three acute myocardial infarction biomarkers based on microfluidic paper-based immunodevice dual amplified by multifunctionalized gold nanoparticles, Talanta 207 (2020) 120346.

[28] W. Alahmad, K. Uraisin, D. Nacapricha, T. Kaneta, A miniaturized chemiluminescence detection system for a microfluidic paper-based analytical device and its application to the determination of chromium (III), Anal. Methods 8 (2016) 5414–5420.

[29] Q. Shang, P. Zhang, H. Li, R. Liu, C. Zhang, A flow chemiluminescence paper based microfluidic device for detection of chromium (III) in water, J. Innov. Opt. Health Sci. 6 (2019) 1950016.

[30] F. Jahanpeyma, M. Forouzandeh, M.J. Rasaee, N. Shoaie, An enzymatic paper-based biosensor for ultrasensitive detection of DNA, Front. Biosci 11 (2019) 122–135.

[31] E.M. Spyrou, D.P. Kalogianni, S.S. Tragoulias, P.C. Ioannou, T.K. Christopoulos, Digital camera and smartphone as detectors in paper-based chemiluminometric genotyping of single nucleotide polymorphisms, Anal. Bioanal. Chem. 408 (2016) 7393–7402.

[32] S. Wang, L. Ge, L. Li, M. Yan, S. Ge, J. Yu, Molecularly imprinted polymer grafted paper-based multi-disk micro-disk plate for chemiluminescence detection of pesticide, Biosens. Bioelectron. 50 (2013) 262–268.

[33] F. Liu, C. Zhang, A novel paper-based microfluidic enhanced chemiluminescence biosensor for facile, reliable and highly-sensitive gene detection of *Listeria monocytogenes*, Sens. Actuators B 209 (2015) 399–406.

[34] A. Roda, E. Michelini, L. Cevenini, D. Calabria, M.M. Calabretta, P. Simoni, Integrating biochemiluminescence detection on smartphones: mobile chemistry platform for point-of-need analysis, Anal. Chem. 86 (2014) 7299–7304.

[35] D. Zhang, Q. Liu, Biosensors and bioelectronics on smartphone for portable biochemical detection, Biosens. Bioelectron. 75 (2016) 273–284.

[36] X.-X. Zhang, J.-J. Liu, Y. Cai, S. Zhao, Z.-Y. Wu, A field amplification enhanced paper-based analytical device with a robust chemiluminescence detection module, Analyst 144 (2019) 498–503.

[37] W. Liu, C.L. Cassano, X. Xu, Z.H. Fan, Laminated paper-based analytical devices (LPAD) with origami-enabled chemiluminescence immunoassay for cotinine detection in mouse serum, Anal. Chem. 85 (2013) 10270–10276.

[38] J. Hassanzadeh, H.A.J. Al Lawati, I. Al Lawati, Metal – organic framework loaded by rhodamine B as a novel chemiluminescence system for the paper-based analytical devices and its application for total phenolic content determination in food samples, Anal. Chem. 91 (2019) 10631–10639.

[39] Z. Li, M. Li, F. Li, M. Zhang, Paper-based chemiluminescence enzyme-linked immunosorbent assay enhanced by biotin-streptavidin system for high-sensitivity C-reactive protein detection, Anal. Biochem. 559 (2018) 86–90.

Chapter 8

Fluorescent paper-based analytical devices

Marylyn Setsuko Arai[1], Andrea Simone Stucchi de Camargo[1] and Emanuel Carrilho[2,3]
[1]São Carlos Institute of Physics, University of São Paulo, São Carlos, Brazil, [2]São Carlos Institute of Chemistry, University of São Paulo, São Carlos, Brazil, [3]National Institute of Science and Technology in Bioanalytics, INCTBio, Campinas, Brazil

8.1 Introduction

Fluorescence (FL) is a photoluminescence process defined as the emission of a light quantum by a molecule (fluorophore) or a material (phosphor) after initial electronic excitation from a light-absorption process. After excitation, the material resides for sometime in a so-called excited state, and its FL emission can be observed usually with lower energy (longer wavelengths) than the excitation [2]. This emission, however, can also be of higher energy, as will be discussed in the next sections. In recent years there has been a growing interest in translating fluorescent technology onto paper-based analytical devices (PADs) for chemical and biological sensing, as a disposable and straightforward alternative. Paper, a ubiquitous product, has recently been demonstrated as promising substrate support of fluorescent materials owing to its portability, low-cost, lightweight, flexibility, and biodegradation. Additionally assays performed on paper substrates require only small sample volume [1]. As discussed in this chapter, fluorescent PADs can be easily fabricated by immobilizing fluorescent materials, such as organic dyes, fluorescent polymers, metal nanoclusters (NC), and different types of nanoparticles (NP), on a piece of paper, and the produced platforms can be classified based on their fluorescent response to the presence of the analyte. In this way, the fluorescent PADs overviewed throughout this chapter will be separated into two major categories: the ones with a brightness-based response and the ones with a color-based response.

The fluorescent detection on PADs has been gaining attention due to its significant advantages, such as (1) high sensitivity, which can reach over one hundred million parts; (2) wide wavelength range offered by the fluorescent

Paper-Based Analytical Devices for Chemical Analysis and Diagnostics.
DOI: https://doi.org/10.1016/B978-0-12-820534-1.00002-5
© 2022 Elsevier Inc. All rights reserved.

probes, and spectral bandwidth that meet specific requirements for a variety of applications, attaining high selectivity; (3) wide dynamic linear range of detection; (4) simple operation process; (5) good detection reproducibility; (6) use of small amounts of sample and reagents, and (7) simple equipment for detection [3]. The phosphors/fluorophores employed in PADs should be chosen based on the desired features for a specific target system and analyte. The next section explores the most frequently used fluorescent materials in PADs, presenting their advantages, limitations, and applications.

8.2 Fluorescent materials

Fluorescent materials for paper-based devices are required to have high FL brightness and broad excitation wavelength so that multiple probes can be excited simultaneously at identical conditions. Also, high resistance to photobleaching is critically essential for the system reliability and accuracy. Meanwhile, fluorescent probes must have surface-functionalized flexibility and tunable emissions for creating reactive/interacting sites with analytes and to produce specific spectral responses upon the presence of the target. Some of the most used fluorescent probes are listed in Table 8.1 and are briefly explained as follows.

TABLE 8.1 Comparison of different fluorescent materials usually used in paper-based devices. (Original).

Fluorescent materials	Advantages	Disadvantages
Fluorescent dyes	Small size, biocompatibility, easy surface modification	Photobleaching, small Stokes shift, autofluorescence
Metal nanoclusters	Small size, tunable and intense emission, nontoxicity, and biocompatibility	Difficulties in obtaining monodispersed metal clusters
Carbon dots	Facile functionalization, low toxicity, biocompatibility, good water solubility, strong fluorescence	Low sensitivity, uncertain origins of fluorescence emission
Quantum dots	High quantum yield, tunable emission, high photostability, high Stokes shift	Toxicity, colloidal instability, harsh and complicated synthesis
Upconversion nanoparticles	Zero autofluorescence background, high anti-Stokes shift, biocompatibility, multicolor-emission capability, high photostability	Colloidal instability, low quantum yield

8.2.1 Organic dyes

Fluorescent organic dyes have been employed as fluorescent markers for decades, with a refulgent track record in biological imaging and analysis [4–6]. They are usually polyaromatic or heterocyclic hydrocarbons that absorb light of a specific wavelength and reemit at a longer wavelength; their FL mainly originates from combined aromatic groups or conjugated π-bonds. Some of the most frequently used fluorescent dyes are shown in Fig. 8.1. In general, these molecules exhibit small size, high FL intensity, excellent biocompatibility, and easy chemical modification [7,8]. However, choosing the appropriate fluorescent dye is relatively tricky since many dyes have drawbacks as environmental toxicity, weak resistance to photobleaching, narrow excitation spectra, and broad emission bands [9]. For these reasons, they have been slowly substituted by nanomaterials, which are more photostable. However, there are still a high number of reports on the use of dyes in FL PADs, especially for the detection and quantification of ions [10] such as Fe^{3+} [11], SO_3 [2–12], CN^- [13], Cu^{2+} [14,15], Ag^+ [16] but also in the sensing of phosgene ($COCl_2$) [17], hydrogen peroxide (H_2O_2) [18], and other compounds [19,20].

8.2.2 Metal nanoclusters

Fluorescent metal nanoclusters (NC) have attracted attention due to their unique electronic structures and the following unusual physical and chemical properties. Metal NCs normally consist of a few to a hundred atoms, which endows them an essential role in the missing link between single metal atoms and plasmonic metal NP [21].

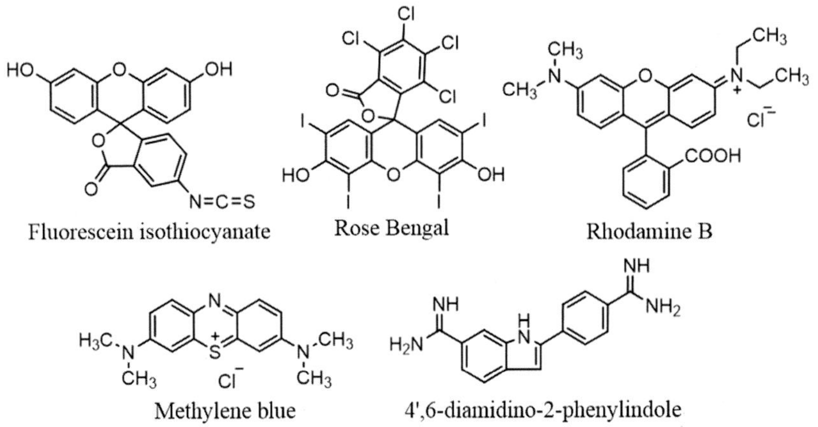

Fluorescein isothiocyanate   Rose Bengal   Rhodamine B

Methylene blue   4',6-diamidino-2-phenylindole

FIGURE 8.1 Structures of fluorescent dye molecules that are most commonly used as fluorescent reporters.

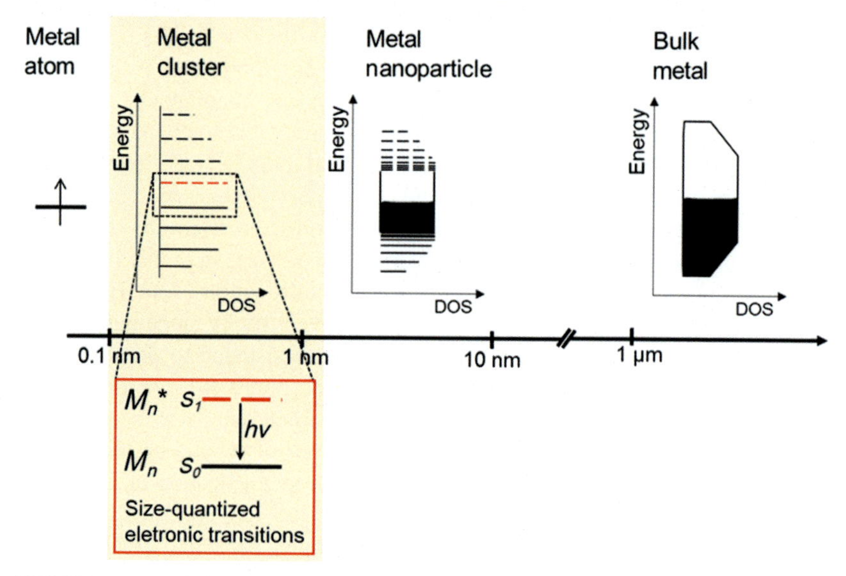

FIGURE 8.2 The effect of size on the electronic transitions in metals. The limited number of atoms in metal nanoclusters results in discrete energy levels, which translates into the unique optical properties of these materials.

Metal NPs exhibit quasi-continuous energy levels and display intense colors due to surface plasmon resonance, which is attributed to the collective oscillation of conduction electrons upon interaction with light. When their dimension is further reduced to the size approaching the Fermi wavelength of electrons, the band structure gives place to discrete energy levels, as shown in Fig. 8.2. Due to the abundance of electrons in metal atoms confined in molecular dimensions and their discrete energy levels, metal NCs exhibit distinct optical, electrical, and chemical properties, including strong photoluminescence, excellent photostability, and good biocompatibility [22,23].

The properties and structures of metal NCs are tunable by changing the core size, the atomic packing mode, and by exposing different lattice planes. Particular features including water-solubility, size-dependent color, the lack of swelling, sharp contrast, and established surface chemistry allow them, mainly AuNCs and CuNCs, to be applied in the development of paper-based devices for detection of Hg^{2+} [24,25], L-cysteine and Ag^{+} [26], nitrite ions (NO_2^{-}) [27], iodine [28], among other analytes.

8.2.3 Carbon dots

Carbon dots (CDs) have also been widely applied as optical nanoprobes in FL sensing. They are defined as clusters of carbon atoms with typical diameters of 2–8 nm, but they also contain substantial fractions of oxygen, hydrogen, or nitrogen [29]. CDs present some advantages over other fluorescent materials like high water-solubility, high chemical resistance, facile functionalization, high resistance to photobleaching, low toxicity, and excellent biocompatibility. Furthermore, CDs can be easily synthesized using a single-step reaction with strong FL without doping or labeling. These are the reasons why there are several reports on their use in PADs [30–32], especially for the detection of metal ions [33–37] and biomolecules [38–41]. Despite their many advantages and widespread use, some shortcomings still limit the application of CDs in fluorescent detection, for example, they are unable to emit intense long-wavelength FL, and the origins of their luminescence is still not fully understood [42,43].

8.2.4 Quantum dots

Quantum dots (QDs), one of the most used luminescent probes, are highly fluorescent inorganic semiconductor nanocrystals in zero dimension with a diameter ranging from 2 to 10 nm [44]. QDs consist of hundreds to thousands of atoms from group II-VI (e.g., CdSe and CdTe), group III-V (e.g., InP and InAs), or group IV-VI (e.g., PbSe). In these materials, after absorption of a photon, an electron is excited from the valence band (VB) to the conduction band (CB), generating a hole in the VB and forming an electron-hole pair so that the combination of electron-hole pairs produces FL.

QDs synthesized from different materials, or with different sizes from the same material, have distinct emission wavelengths (Fig. 8.3) [45]. Due to quantum confinement effects, QDs possess diverse, unique physicochemical characteristics, including high quantum yields (QY), tunable emission wavelength, broad absorption cross-section, and robust photostability. There are several reports in the literature on the use of QDs in fluorescent PADs

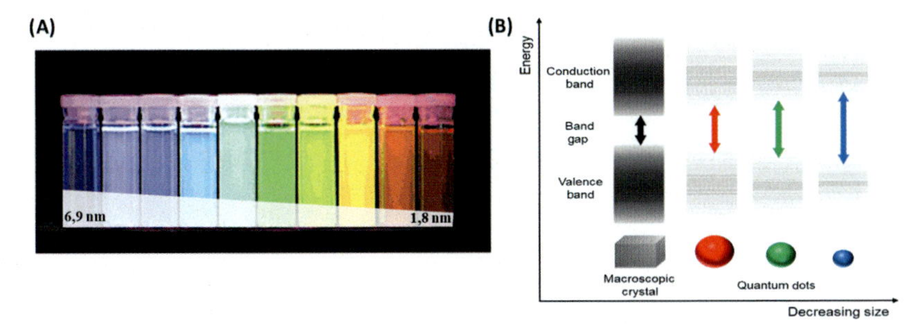

FIGURE 8.3 Characteristics of quantum dots. (A) Ten different emission colors of ZnS-capped CdSe QDs, core diameters ranging from 6.9 to 1.8 nm, excited with a near-UV lamp. From left to right (*blue* to *red*), the emission maxima are located at 443, 473, 481, 500, 518, 543, 565, 587, 610, and 655 nm. *Adapted with permission from M. Han, X. Gao, J.Z. Su, S. Nie. Quantum-dot-tagged microbeads for multiplexed optical coding of biomolecules. Nat. Biotechnol. 19, 631–635 (2001). © 2001 Nature Publishing Group.* (B) QDs energy band diagram compared to a macroscopic crystal - the bandgap energy increases as the nanocrystal size decreases. *QDs*, Quantum dots.

ranging from biosensing of cancer cells [46], bacteria [47], and biomarkers [48–50] for the detection of metal ions [51–54], pesticides, [55] and other analytes [56–59]. Despite the superior optical and electronic properties of QDs, their use suffers from harsh and complicated synthesis procedures, colloidal instability in aqueous media, and toxicity to biological systems [60].

8.2.5 Upconversion nanoparticles

Upconversion nanoparticles (UCNPs) have received much attention in biological applications due to their capability of converting near-infrared light (800–1000 nm) into shorter wavelengths (ultraviolet/visible) fluoresnce (anti-Stokes) via multiphoton mechanisms. As upconversion emission is different from conventional luminescence and does not occur spontaneously in nature, these particles provide unmatched contrast, significantly minimizing the interference of autofluorescence from the sample and the paper, and increasing the signal-to-noise ratio [61,62], especially in biosensing applications.

The size of UCNPs is widely tunable (10–100 nm), and they are usually composed of crystalline inorganic host matrices doped with trivalent lanthanides, such as Er (III), Yb (III), Ho(III), Nd (III), and Tm (III), that confer different emission colors to the particles as shown in Fig. 8.4 [63]. Although the UCNPs present lower FL quantum yield than the other phosphors, they have

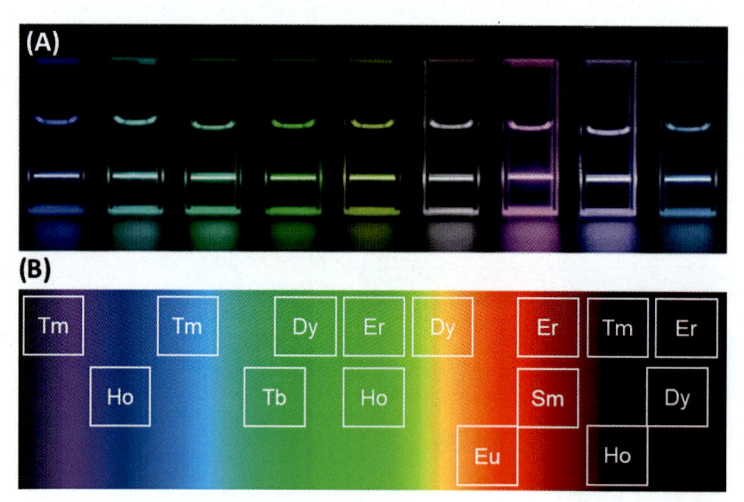

FIGURE 8.4 Characteristic emission colors of lanthanide-doped UCNPs. (A) Photographs of UCNPs emission in colloidal solution under an 800 nm laser irradiation (NPs doped with different amounts of Tm^{3+} and Er^{3+}). *Reprinted with permission from Zhong, Y. et al. Elimination of photon quenching by a transition layer to fabricate a quenching-shield sandwich structure for 800nm excited upconversion luminescence of Nd^{3+}-sensitized nanoparticles. Adv. Mater. 26, 2831–2837 (2014). © 2013 WILEY-VCH Verlag GmbH & Co. KGaA, Weinheim.* (B) Different emission spectral bands of Ln^{3+} ions under NIR excitation, spanning from NIR to VIS and UV regions as each Ln^{3+} has its unique energy level structure. *UCNPs*, Upconversion nanoparticles.

been extensively used in FL paper-based sensors due to their unique characteristics such as high photostability, anti-Stokes shift, biocompatibility and easy surface modification. More specifically, UCNPs have been applied in the development of PADs for sensing of biomarkers [64,65], DNA [66], and peptides [67], as well as cocaine [68], pesticides [69], and other analytes [70,71].

8.3 Fluorescent response

A fluorescent PAD is usually fabricated by linking a recognition element onto a phosphor/fluorophore, which binds explicitly target species and instantly produces spectral responses. These responses should only occur upon analyte presence and will correspond to emission enhancement ("turn-on") or emission quenching ("turn-off") [72]. As discussed in the following examples, in order to meet these requirements, the photoluminescence property and the molecular engineering designs are of equal importance to achieve high sensitivity and selectivity. Based on FL "turn-on" and "turn-off", the fluorescent PADs can be divided into two significant types: "brightness-based" and "color-based," that will be subdivided and explored throughout the next sections.

8.3.1 Brightness-based response

The most typical practice is to use a one signal fluorescent probe for detection via FL "turn-off" or "turn-on" mechanism, which only causes changes in emissions intensities, as shown in Fig. 8.5. Since brightness variations are easily affected by background FL and excitation conditions in real sample assays, this type of probe response is more suitable for yes/no determination of analytes rather than quantitative assays of trace target species. However, with controlled and careful conditions and probe engineering, quantification can also be done, as described in the following examples.

8.3.1.1 Fluorescence "turn-off"

The FL quenching response type, also referred to as the emission "turn-off", has been widely employed in FL PADs. Prabphal et al. reported the use of selective FL quenching of a cationic porphyrin (5,10,15,20-tetrakis(1-methyl-4-pyridinio)porphyrin tetraiodide (TMPyP)) as a reporting method

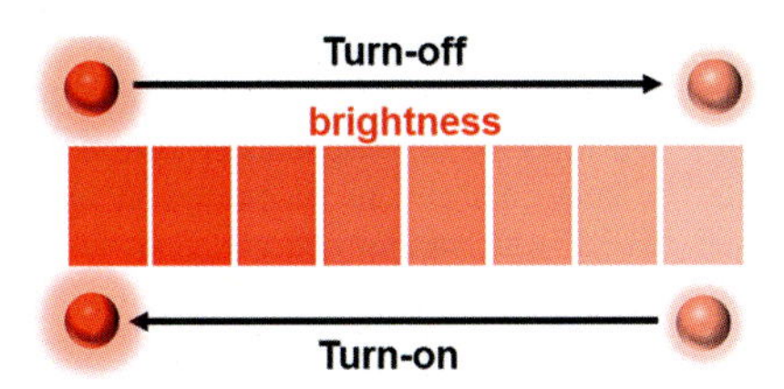

FIGURE 8.5 Single-colored probe with "turn-off" or "turn-on" fluorescent response.

for Cu^{2+} sensing in a paper-based format [15]. The synthesized TMPyP presents emission peaks at 674 and 711 nm which are quenched after the complexation of the dye with Cu^{2+} ions [73,74]. The sensor was fabricated by wax printing, with circular shapes being used as "reaction wells" for depositing all solutions. Twenty micromolar of TMPyP solution was added to the wells and air-dried for 10 min. The change in FL was then observed by simply dropping 5 μL of different metal solutions into each well, as shown in Fig. 8.6A. The data were collected by taking an image of the sensor under commercial black light (365 nm), followed by processing by ImageJ software. Briefly, the image was transformed into a grayscale image, where the grayscale intensity of each pixel was converted into numerical values (ranging from 0 to 255). The response range of FL quenching by different concentrations of Cu^{2+} was determined, along with its limit of detection (LOD) by varying the concentrations of Cu^{2+} in each well and converting the image of these wells to pixel values (Fig. 8.6B and C). The authors constructed a simple "turn-off" PAD for Cu^{2+} with good selectivity and sensitivity, with a LOD at 0.16ppm, which is below the recommended amount of Cu^{2+} in drinking water. The same group had previously reported the use of an anionic porphyrin that also showed FL quenching with Cu^{2+} ion [75].

In general, the "turn-off" FL response has been widely used in PADs for the detection of metal ions [16,76–80] being Cu^{2+} one of the most studied ions. Niammont et al. used four different organic fluorophores to construct a paper-based system for visual detection of picomole amounts of Cu^{2+} in aqueous media [14]. The high FL quenching induced by Cu^{2+} observed in these molecules was attributed to the photoinduced electron transfer (PET)

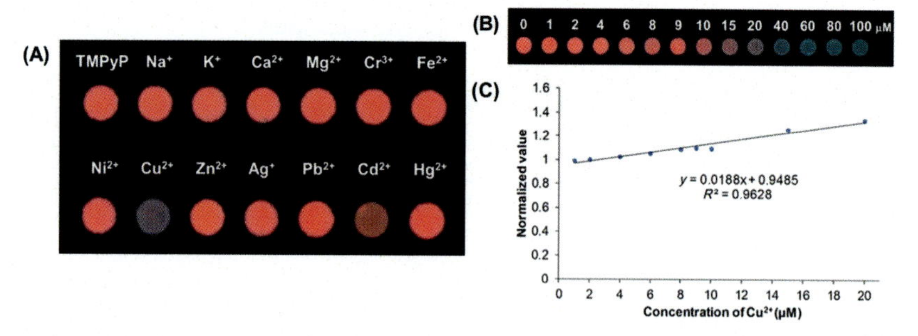

FIGURE 8.6 Example of an fluorescent paper-based analytical device developed by Prabphal et al. for the detection of Cu^{2+}. (A) Fluorescence responses of TMPyP to a variety of metal ions in a paper-based format. (B) The image and (C) the calibration curve from the digitized image data of the sensor containing TMPyP with various concentrations of Cu^{2+} (A: 0–100 μM; B: 0–20 μM). *Reprinted with permission from J. Prabphal, T. Vilaivan, T. Praneenararat. Fabrication of a paper-based turn-off fluorescence sensor for Cu^{2+} Ion from a pyridinium porphyrin. ChemistrySelect 3, 894–899 (2018); Copyright© 2018 Wiley-VCH Verlag GmbH & Co. KGaA, Weinheim.*

process, thus the quenching ability should depend on the energy of the LUMO level of the electron-donating fluorophore matching that of the Cu^{2+} complex, the electron-accepting unit [81].

A paper-based chip device for multiplexed detection of Cu^{2+} and Hg^{2+} ions was constructed by Qi et al. using CdTe QDs with emission at 527 nm. The platform detection was based on QDs' electron transfer FL quenching triggered by the presence of the metal ions [82]. Green-fluorescent CDs covered with carboxyl and amino groups were also applied in Cu^{2+} detection. Oxygen (O) of carboxyl groups and nitrogen (N) of amino groups on CDs' surface have lone-pair electrons, while Cu^{2+} has an empty valence orbital. So, the sensor response was attributed to the coordination of O and N with Cu^{2+}, and the experiments suggested that FL suppression was due to static quenching [35,83]. Based on other quenching mechanisms, CDs were also used in paper-based fluorescent sensing of Al^{3+} [34], Cr^{6+} [36], and Pb^{2+} [33].

FL "turn-off" has also been extensively applied in the sensing of nitroaromatic compounds (NACs). Using dodecyl-functionalized silicon nanocrystals (Si-NCs), Gonzalez et al. [84] produced a PAD for nitrobenzene (NB), nitrotoluene, dinitrotoluene (DNT), and other NACs. The quenching behavior induced by the NACs on the Si-NCs was evaluated using the Stern-Volmer equation ($I_0/I = K_{sv}[Q] + 1$), and a linear relationship of I_0/I *vs* NACs concentration indicated that in this system, the quenching arises from a dynamic process such as electron transfer [85]. Lu and coworkers developed a FL PAD prepared by absorbing hydrophilic pyrene-functionalized polymer into cellulose-based filter papers. The system enables naked-eye detection of low-ppm-level 2,4,6-trinitrotoluene (TNT) and relies on the formation of charge-transfer complexes between the electron acceptor (TNT) and the electron donor (pyrene fluorophores) and consequent FL quenching [86]. Luminescent porous organosilicon polymers (LPOPs) [87], terbium(III)-based coordination polymer [88], and 8-pyrenyl-substituted fluorene dimers [89] have also been also used in FL "turn-off" PADs for NACs.

Paper-based FL "turn-off" devices also find a high number of applications in biosensing. For example, Ruiguo et al. developed a competitive lateral flow immunoassay for multi β-agonist residues by using a single monoclonal antibody labeled with red fluorescent NP [90]. Shan-Wen et al. reported a fluorescent platform for the analysis of Fe^{3+} and colorimetric ELISA for ferritin in blood by using *in situ* CDs and AuNPs sequential patterning techniques [30]. Li and coworkers combined QDs with molecularly imprinted polymers (MIPs) to create microfluidic PADs (μPAD) for specific recognition and sensitive detection of phycocyanin [91]. QDs have also been used for "turn-off" FL sensing of *Staphylococcus aureus* in food, and environmental samples [47], norfloxacin in milk [59], and dopamine detection in serum samples [92].

In pesticide sensing, for specific recognition and sensitive "turn-off" detection of 2,4-dichlorophenoxyacetic acid, Zhang et al. [55] chose yellow-emissive CdTe QDs while Zou et al. [31] used N,S codoped CDs with blue emission at 440 nm to develop a fluorescent sensor to determine fluazinam. For a smartphone-based visual and quantitative sensing of thiram, the group of Mei et al. produced $NaYF_4$:Yb/Tm UCNPs covered with poly (acrylic acid) and decorated with copper ions ($NaYF_4$:Yb/Tm@PAA-Cu), as shown in Fig. 8.7A. The interaction of thiram and the nanoprobe leads to the coordination of the pesticide to the Cu^{2+} ions causing a luminescent resonant energy transfer (LRET) process between the UCNP and the formed complex leading to a consequent decrease in Tm^{3+} emission at 475 nm, as schematically demonstrated in Fig. 8.7B and in the emission spectrum in Fig. 8.7C. The upconversion nanosensor was produced using a piece of ordinary filter paper by immersion and ultrasonic agitating in the aqueous solution of $NaYF_4$: Yb/Tm@PAA-Cu nanoprobe. After drying at 50°C, the prepared test paper was placed into a tailor-made optical accessory. The images in Fig. 8.7D demonstrate the luminescence evolutions on test paper with increasing the amounts of thiram. The authors extracted the blue channel intensities of the RGB (Red-Green-Blue) colored images to quantify the luminescence through a self- written Android program installed on the smartphone, and the reliable detection limit of the system was determined to be 0.1 μM [69].

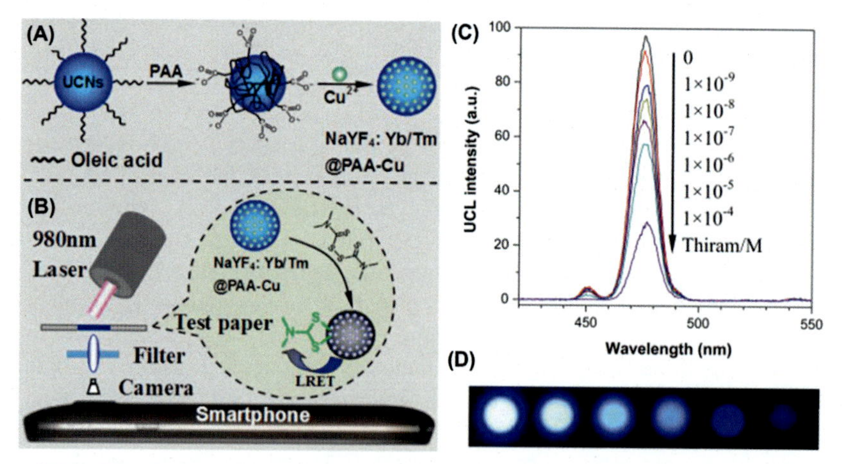

FIGURE 8.7 Fluorescent paper-based analytical device developed by Mei et al. for thiram sensing. (A) Schematic illustration of the preparation procedures of $NaYF_4$:Yb/Tm@PAA-Cu nanoprobes. (B) Inner optical structure of the system and detection principle for pesticide thiram on the upconversion test paper. (C) Luminescence quenching of $NaYF_4$:Yb/Tm@PAA-Cu nanoprobe upon additions of different amounts of thiram. (D) Luminescent images of test paper upon additions of different amounts of thiram, which were taken by the smartphone detection setup. The concentrations of thiram from left to right are 0, 1×10^{-7}, 1×10^{-6}, 1×10^{-5}, 1×10^{-4}, 1×10^{-3} M, respectively. *Reprinted with permission from Mei, Q. et al. Smartphone based visual and quantitative assays on upconversional paper sensor. Biosens. Bioelectron. 75, 427–432 (2016); © 2016 Elsevier B.V.*

Although "turn-off"-based FL PADs have been used for some analytes and have led to good results, environmental factors, like pH and solvents, may cause decrease in the probes emission, causing this response type to usually exhibit low selectivity and reliability. In this way, significant efforts have been devoted to the development of FL "turn-on" sensors by combining higher intrinsic sensitivity with higher chemical selectivity [72].

8.3.1.2 Fluorescence "turn-on"

In fluorescent PADs based on the "turn-on" mechanism, the analyte presence will induce the appearance and/or enhancement of the fluorophore/phosphor FL emission. A FL "turn-on" platform for the detection of cyanide (CN^-) ions in aqueous media was developed by Incel et al. [93] using an organometallic dye, europium tetrakis dibenzoylmethide triethylammonium (EuD_4TEA), and gold nanoparticles (AuNPs). An ordinary filter paper with 6 μm pore size was employed as solid support which facilitates the impregnation of EuD_4TEA and AuNPs and provides durability. EuD_4TEA acts in the specific detection of cyanide while the AuNPs were employed to control FL quenching and recovery. The strategy relies on complete FL quenching of EuD_4TEA impregnated paper by controlled deposition of AuNPs and subsequent dissolution of AuNPs by cyanide presence, causing FL recovery and enhancement, as shown in Fig. 8.8A. The PAD exhibits naked eye distinguishable color transition upon CN^- addition from 10^{-2} to 10^{-12} M. To standardize the methodology, a homemade image processing algorithm has been developed and enabled calibration of color change and quantification of CN^- concentration. Fig. 8.8B shows the digital images of paper substrates: row I. is EuD_4TEA impregnated paper, which is fluorescent, row II. is AuNPs modified EuD_4TEA impregnated paper quenched due to energy transfer from EuD_4TEA to the NPs. The third row shows the paper fluorescence development with varying concentrations of CN^-.

Oligothiophene-benzothiazole receptor (3TBN) [13] and benzylidene derivatives [94] have also been reported for CN^- detection in DMSO and aqueous media, respectively. Fluorescent PADs based on the "turn-on" mechanism using organic dyes were developed for Hg^{2+} [95], F^- in organic phase [96], bisulfite and benzoyl peroxide in food [97], phosgene gas ($COCl_2$) [17], and Promchat et al. reported a macroarray immobilization of fluorophores on filter papers that can synergistically discriminate up to 12 metal ions including Ba^{2+}, Al^{3+}, Cr^{3+}, Fe^{2+}, Co^{2+}, Ni^{2+} among others [10]. "Turn-on" detection of formaldehyde [98], H_2O_2 [99], adrafinil [100], Pb^{2+} and Cd^{2+} [54], and hypochlorous acid [71] have also been reported.

The area in which "turn-on" FL PADs find a significant number of applications is in biosensing [46,101–108]. For instance, a paper-supported aptasensor was constructed by Jiang et al. [109] for immunoglobulin E (IgE) detection using UCNPs and carbon nanoparticles (CNPs) as the energy

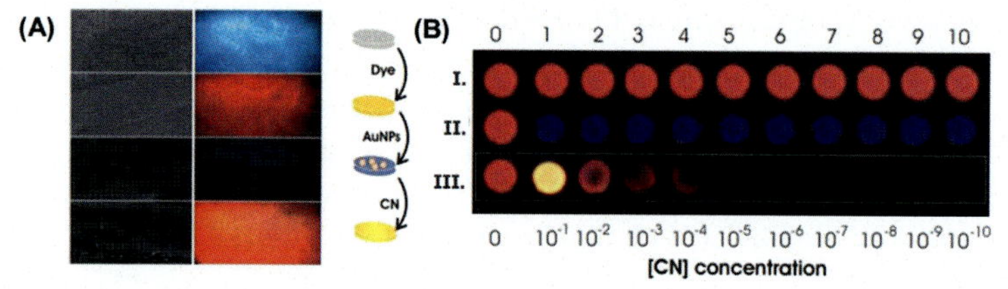

FIGURE 8.8 Example of an fluorescent paper-based analytical device developed by Incel et al. for the detection of CN^-. (A) Scanning electron microscopy and fluorescence microscopy images of pristine paper, EuD_4TEA impregnated paper, EuD_4TEA–Au NPs doped paper, and CN^- treated paper. (B) Digital images of paper sensors upon AuNPs impregnation and cyanide addition with varying concentrations (10^{-1} to 10^{-10} M). Row I. EuD_4TEA impregnated paper, row II. AuNPs modified EuD_4TEA impregnated paper and row III. *Reprinted with permission from İncel, A., Akın, O., Çağır, A., Yıldız, Ü.H. & Demir, M.M. Smart phone assisted detection and quantification of cyanide in drinking water by paper based sensing platform. Sens. Actuators, B Chem. 252, 886–893 (2017); Copyright © 2017 Elsevier B.V.*

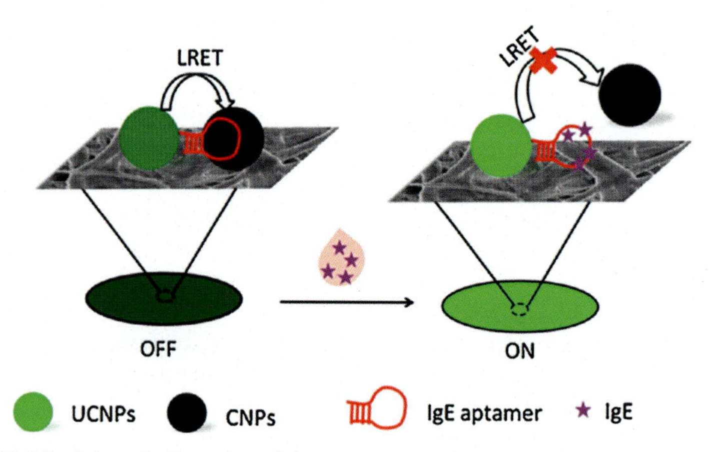

FIGURE 8.9 Schematic illustration of the paper supported aptasensor for IgE detection using upconversion nanoparticles and carbon nanoparticles as energy donor-acceptor pair. *Reprinted with permission from P. Jiang, M. He, L. Shen, A. Shi, Z. Liu. A paper-supported aptasensor for total IgE based on luminescence resonance energy transfer from upconversion nanoparticles to carbon nanoparticles. Sens. Actuators, B Chem. 239, 319–324 (2017); Copyright © 2016 Elsevier B.V.*

donor-acceptor pair. The FL "turn-on" mechanism was based on the conformation change of IgE aptamer before and after interaction with its target. As previously mentioned, UCNPs are suitable and extensively used in bio-applications due to their property of being activated by NIR light and the ability to avoid the biomolecules autofluorescence [110]. As shown in Fig. 8.9, in Jiang's system, in the absence of IgE, the UCNPs-aptamer loaded on the surface of paper interacts with CNPs through π-π stacking interaction, leading to the quenching of luminescence through an energy transfer process. In the presence of IgE, the aptamer explicitly recognizes it and forms the aptamer-IgE complex, which is accompanied by a conformational change. Under this circumstance, the distance between UCNPs and CNPs is enlarged, blocking the energy transfer and, consequently, the emission of UCNPs is restored in an IgE concentration-dependent manner.

Dual-color UCNPs were also used by You et al. in the development of a multiplexed lateral flow strip (LFS) platform for heart failure (HF) prognosis at home, integrating UCNPs-based LFS and a portable smartphone reader. UCNPs with green ($NaYF_4$: Yb,Er) and blue ($NaYF_4$: Yb,Tm) FL were applied for multiplexed detection of brain natriuretic peptide (BNP) and suppression of tumorigenicity 2 (ST2), two target antigens/biomarkers associated with HF. When the sample is added onto the LFS, the target analytes (BNP or ST2 antigens or both) would first specifically bind to UCNPs probes forming the UCNPs-analyte conjugates, as schematically shown in Fig. 8.10, Er^{3+} doped UCNPs respond to ST2, and Tm^{3+} doped NPs relate to BNP. The conjugates would flow through the NC membrane as driven by the capillary force and then be captured by the ST2 and BNP antibodies immobilized in the test lines. The excess UCNPs probes without binding analytes would react with the goat antimouse immunoglobulin G (IgG) and be trapped in the control line. If the analytes are absent in the sample, all UCNPs probes would directly flow through the test lines without reaction and be captured on the control line. In Fig. 8.10D, the emission "turn-on" evolution of both probes with increasing concentration of BNP and ST2 is shown. A smartphone-based portable reader reads the FL signal from the LFS, with the images decomposed to their blue and green channels. As a result, the PAD was able to detect the minimal concentration of 29.92 ng mL^{-1} for ST2 and 17.46 pg mL^{-1} for BNP [111].

He et al. produced a similar LFS using Tm^{3+} and Er^{3+} doped UCNPs as two independent sensors and achieved new records of the LOD with 89 and

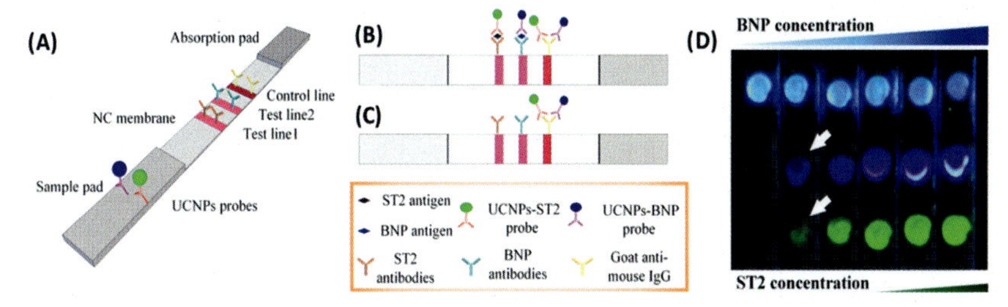

FIGURE 8.10 LFS platform developed by You et al. for heart failure prognosis. (A), (B) and (C) Schematic illustration of the dual-color UCNPs-based LFS platform. (D) Photograph of the LFS for simultaneously detecting BNP and ST2 antigens. *LFS*, lateral flow strip; *UCNPs*, upconversion nanoparticles. *Reprinted with permission from M. You et al. Household fluorescent lateral flow strip platform for sensitive and quantitative prognosis of heart failure using dual-color upconversion nanoparticles. ACS Nano 11, 6261–6270 (2017); Copyright © 2017 American Chemical Society.*

400 pg mL^{-1} for prostate-specific antigen (PSA) and ephrin type-A receptor 2 (EphA2) biomarkers, respectively [65]. Upconversion NPs were also applied in paper-based FL "turn-on" detection of metalloproteinase-2 (MMP-2) in human serum and whole blood samples [67], using a DNA sequence for diagnostic of spinal muscular atrophy [66], and in visceral adipose tissue-derived serpin (Vaspin) for early diagnosis of type-2 diabetes [112].

Other lateral flow assays (FLA) for biosensing and based on the "turn-on" mechanism were reported for different phosphors: blue emissive CDs for DNA detection in the fM range [38], carboxyl-functionalized polystyrene NPs for sensing and quantification of C-reactive protein (a biomarker related to cardiovascular diseases) [48], red emissive QDs for detection of HIV-DNA [49] in human serum and, more recently, QDs for the detection of Zika virus nonstructural protein 1 (ZIKV NS1) [113].

Liang et al. designed a fluorescent μPAD for sensing cancer cells composed of different colored mesoporous silica nanoparticles (MSNs), QDs-labeled aptamers, and graphene oxide (GO) for real-time visual monitoring of multiple cancer cells. The sensing method employed by the authors combines the quenching capability of GO and an aptamer with high specificity and affinity employed as the molecular recognition element. The CdTe QDs FL is quenched via Forster resonance energy transfer (FRET) between them and the GO, with subsequent recovery of the FL upon addition of target cells [46]. GO has also been used with functionalized silver nanoparticles (AgNPs) in paper printed sensors for the detection of biomolecules (peptide, DNA, proteins); in this system, GO acts as the fluorescent probe while the AgNPs act as the quencher [114].

Concerning qualitative and quantitative detection of targets, the use of only emission intensity at a wavelength may be problematic because unavoidable interferences can arise from the target-independent factors, such as light scattering by the sample matrix, excitation source fluctuation, particular microenvironment around probes, and local concentration variations of probes. In this way, PADs in which the detection mechanism relies only on FL "turn-on" or "off" may find some limitations since their response is entirely based on brightness variations. An accurate visual quantification mainly depends on the color variations rather than only brightness. Additionally, the use of color variations enables the development of platforms with a broader sensing range and a lower LOD. [115] Therefore a vibrant color response is critical to improving the quantification capability of PADs, and more about such strategies are discussed in the next sections.

8.3.2 Color-based response

To overcome some of the drawbacks of signal brightness-based FL response, one can construct a ratiometric FL system using dual-emissive probes or a dual-probe configuration. The mixing of two different colored probes/signals may provide optical engineering of fluorescent sensors and visual responses of both color variations and brightness change. Besides that, ratiometric FL avoids the influence of environmental and intrinsic FL, and, owing to the self-calibration function of FL intensities, this method has improved sensitivity and accuracy [116].

There are two general categories to carry out ratiometric FL detection. One is to introduce the second signal as a reference that is target insensitive—ratiometric FL with one reference signal (1RS). The other is to apply target-responsive two signal changes that enable the achievement of dual-emission—ratiometric FL with two reversible signal changes (2RSC) [115].

8.3.2.1 Ratiometric fluorescence with one reference signal

In ratiometric FL with one reference signal, the first signal is target sensitive and responds explicitly to the analyte presence in a "turn-on" or "turn-off" manner. The second signal is target-insensitive and acts as an internal standard, which has a negligible variation caused by target-related events and retains very stable fluorescence. By introducing this reference signal, interferences from microenvironment-related factors are eliminated, providing a built-in correction, which improves the reproducibility of signal transduction during the recognition process of specific targets. [115] In terms of probes, the 1RS sensing system can be designed in two ways, as shown in Fig. 8.11: (A) dual-probe system (1RS-2P) in which one probe provides a target sensitive signal, and the other is inert to the analyte and (B) one dual-emission probe (1RS-1P) with FL signals that respond differently to the analyte presence.

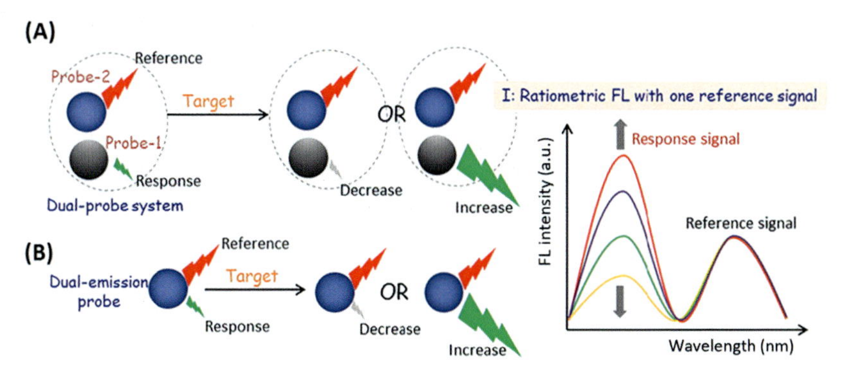

FIGURE 8.11 The general categories for fabricating dual-emissive ratiometric fluorescent probes with one reference signal. Representation of (A) dual-probe system: the reference and the response signals originate from different fluorescent probes, and (B) dual-emission probe in which the same probe generates both reference and response signals. The spectrum represents the FL emission response, one signal is target sensitive, increase or decreasing upon analyte presence, and the other is inert to the target and used as a reference. *Modified with permission from R. Gui et al. Recent advances in dual-emission ratiometric fluorescence probes for chemo/biosensing and bioimaging of biomarkers. Coord. Chem. Rev. 383, 82–103 (2019); Copyright © 2019 Elsevier B.V.*

Usually, the 1SR-2P FL systems are constructed by a simple mixture of the two luminescent probes [37] as in the PAD developed by Zhou et al. for the detection and quantification of arsenic ions, As^{3+} [56]. Red emitting CdTe QDs were modified with glutathione and dithiothreitol (GSH/DTT-QDs) to obtain the sensitivity to As^{3+}, which occurs by the quenching of red FL through the formation of QDs aggregates. A small number of cyan CDs with spectral blue-green components were mixed into the QDs solution and were used for the reference signal, as shown in the emission spectra in Fig. 8.12A. Upon the addition of As^{3+} into the sensing solution, the FL color could gradually be reversed from red to cyan with a detection limit of 1.7 ppb As^{3+}. Fig. 8.12B illustrates the proposed mechanism of detection: It is well-known that arsenic ions have an extremely high binding affinity to

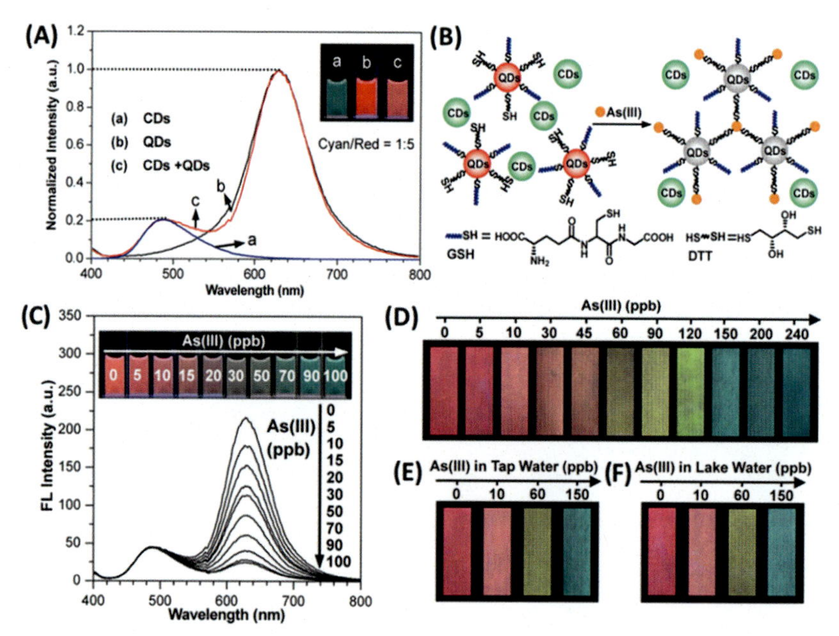

FIGURE 8.12 Ratiometric 1RS-2P fluorescent paper-based analytical device developed by Zhou et al. for As^{3+} sensing. (A) Fluorescent spectra of CDs and GSH/DTT-modified CdTe QDs and the mixing CDs/QDs at the excitation of 350 nm (the inset photos were taken under 365 nm UV lamp). (B) Visualization mechanism of As^{3+} using GSH/DTT-QDs as fluorescent sensing probe and CDs as an internal standard probe. (C) Fluorescent spectra of mixing GSH/DTT-QDs/CDs with the addition of As^{3+}. The inset photos show the evolution of corresponding colors under a 365-nm UV lamp. (D) Visualization of As^{3+} using the fluorescent test papers prepared by printing GSH/DTT-QDs/CDs ink on a piece of filter paper. (E) and (F) Visual detections of As^{3+} in tap water and lake water, respectively. *CDs*, Carbon dots; *DTT*, dithiothreitol; *GSH*, glutathione; *QDs*, quantum dots. *Reprinted with permission from Y. Zhou et al. Color-multiplexing-based fluorescent test paper: dosage-sensitive visualization of arsenic(III) with discernable scale as low as 5 ppb. Anal. Chem. 88, 6105–6109 (2016); Copyright© 2016 American Chemical Society.*

DTT, so the addition of As^{3+} leads to $As-S$ bond formation that induces the aggregation of QDs and consequent red FL quenching. In the presence of cyan CDs as an internal standard, the gradual quenching of red FL will produce a consecutive color variation with the increase of the As^{3+} amount in the system.

Fig. 8.12C shows the evolution of fluorescent spectra of the mixture of CDs and GSH/DTT-QDs with the titration of As^{3+}. The red fluorescence peak at 630 nm was gradually quenched, and the cyan FL peak at 486 nm kept highly steady. When the sensing solution was printed onto a piece of filter paper, a sequential color evolution from peach to pink to orange to khaki to yellowish to yellow-green to final cyan, with the addition of As^{3+}, was displayed and allowed clear discernment of the dosage scale as low as 5 ppb (Fig. 8.12D). The same group developed a similar sensor for As^{3+} using CDs combined with colloidal QD chains [117].

Gold nanoclusters (AuNCs) with an emission band in the red region (600–750 nm) have been used as target-sensitive responders in 1RS-2P PAD in combination with other fluorescent probes. Qi et al. reported a device in which bovine serum albumin capped gold nanoclusters (BSA–AuNCs) acted as the signal indicator and terbium(III) (Tb^{3+}) was used as the build-in reference for visual detection of Hg^{2+} [24]. A mixture of AuNCs and CDs were employed in the detection of Hg^{2+}[25], Ag^{+}, and L-cysteine (Cys) [26]. In these systems, the blue emission of CDs at 450 nm still unchanged, while the AuNCs acts as a recognition unit having its FL quenched by increasing analytes concentration.

Su et al. reported 1RS-2P fluorescent PAD to determine blood glucose in human serum by using AuNCs stabilized by bovine serum albumin (BSA) combined with fluorescent GO. Glucose was oxidized by glucose oxidase and produced hydrogen peroxide (H_2O_2), which can, in turn, oxidize BSA–AuNCs and quench their FL while being inert to GO. A carefully adjusted ratio of red AuNCs and blue fluorescent GO results in a colorimetric sensor accompanied by a color change from pink to blue. The distinct color differences exhibited by the probing system was utilized to measure glucose concentrations from 0.5 to 10 mM, as shown in Fig. 8.13A. Test papers were prepared through an inkjet printer by utilizing the colorimetric probe as ink (Fig. 8.13B). Typically, glucose was incubated with glucose oxidase and was evenly added onto a piece of test paper; with increasing analyte concentration, the pink color turned to blue under a 365 nm UV lamp, as displayed in Fig. 8.13C [118].

The mixing of two different fluorescent probes may cause precipitation and optical instability. In this way, a single dual-emissive probe can be designed alternatively for eliminating the instability and achieve an ideal visual effect. Usually, these 1RS-1P systems are constructed by incorporating two different probes into a core-shell structure or linking them through covalent bonds, especially EDC/NHS coupling [20,50,119].

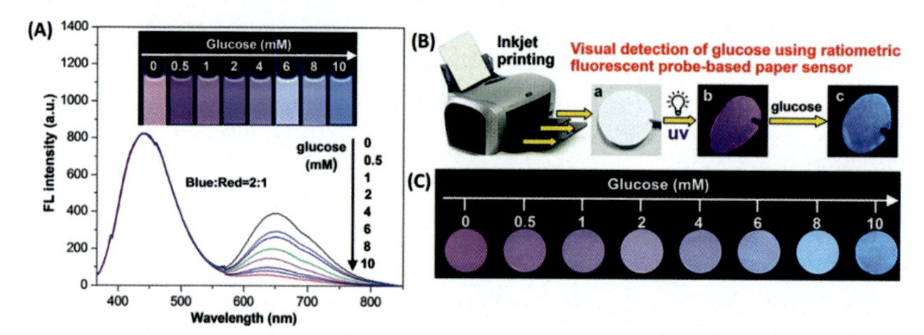

FIGURE 8.13 Ratiometric 1RS-2P fluorescent paper-based analytical device developed by Su et al. for the determination of glucose. (A) The fluorescence (FL) spectra of the ratiometric probe upon the addition of different concentrations of glucose. The inset displays the color variation resulting from FL quenching of red emission from BSA–Au NCs under a 365 nm UV lamp. (B) Visual detection of glucose using a ratiometric fluorescent probe-based paper sensor. The photographs of the ratiometric FL probe-based test strips were taken (a) under natural light, (b) under a 365 nm UV lamp, and (c) under a 365 nm UV lamp in the presence of 10 mM glucose. (C) Fluorescent test papers applied to detect glucose visually. *Reprinted with permission from L. Su et al. A ratiometric fluorescent paper sensor for consecutive color change-based visual determination of blood glucose in serum. N. J. Chem. 42, 6867–6872 (2018); Copyright© 2018 The Royal Society of Chemistry and the Centre National de la Recherche Scientifique.*

Huang et al. developed a single dual-emissive nanofluorophore for quantification of blood glucose. As schematically shown in Fig. 8.14A, red emissive QDs were embedded into silica nanoparticles as a stable internal standard emission, and blue CDs were further covalently linked onto the surface of silica (QDs@SiO2-CDs), in which the ratiometric FL intensity of blue to red was controlled at 5:1. The oxidation of glucose induces the formation of Fe^{3+} ions and the electron transfer to Fe^{3+} quenches the emission of CDs, displaying a series of consecutive color variations with the dosage of glucose, as shown in Fig. 8.14B and C. The fluorescence intensity ratio I_{445}/I_{630} was linearly related to the glucose concentration in the range $15-75$ μM and the calculated detection limit in solution was 3.0 μM. The fluorescent PADs were prepared using QDs@SiO_2-CDs probe as fluorescent ink through an inkjet printer connected with a computer. The repeating 30 times printing led to enough and even coverage of probe on a piece of cellulose acetate membrane. The same samples were tested with the fluorescent test papers and observed under a UV lamp. Fig. 8.14D shows that the colors of test papers gradually changed from the original blue to pink with the increase of glucose concentration in the serum [39].

Based on similar approaches and detection modes, QDs and CDs were also used in visual detection of Cu^{2+} [120], Hg^{2+} [121], and dopamine [122]. CDs combined with Eu(III) complex and CuNCs were applied in paper-based 1RS-1P ratiometric fluorescent sensing of 2, 6-dipicolinic acid (DPA), an important biomarker of *Bacillus anthracis spore* [41] and dopamine [40], respectively. There are also reports on the use of differently colored QDs sensitive to

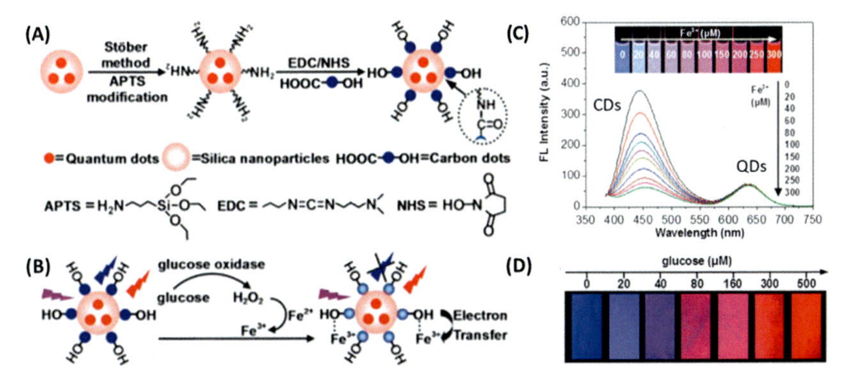

FIGURE 8.14 Ratiometric 1RS-1P fluorescent paper-based analytical device developed by Huang et al. for quantification of blood glucose. Schematic illustrations of (A) the synthesis of the dual-emissive ratiometric fluorescent probe QDs@SiO_2-CDs nanoparticles and (B) the visual detection of glucose by quenching blue emission via Fe^{3+} with the aid of glucose oxidase and Fe^{2+}. (C) Fluorescent spectra ($\lambda_{ex} = 360$ nm) of the ratiometric probe, as a function of the Fe^{3+} concentration. The inset shows the corresponding fluorescence colors. (D) Visual detection of glucose using fluorescent test papers. The photos were taken under a 365 nm UV lamp. *Modified with permission from X. Huang et al. A single dual-emissive nanofluorophore test paper for highly sensitive colorimetry-based quantification of blood glucose. Biosens. Bioelectron. 86, 530–535 (2016); Copyright © 2016 Elsevier B.V.*

tyrosinase [57] and trinitrotoluene particulates [58], QDs mixed with rhodamine derivatives (CRDs) for visual sensing of Fe^{3+} [51], and UCNPs combined with Rhodol [123] for detection of organophosphonate (OP).

8.3.2.2 Ratiometric fluorescence with two reversible signal changes

Another category of ratiometric probes involves the ratiometric FL with two signal changes. Upon the analytes presence, specific binding of probes with targets can perturb the systems to trigger reversible variations of two or more different signals, thus allowing for a versatile ratiometry. As shown in Fig. 8.15, the presence of target-related events induces the disappearance of one signal along with the emergence of a new one. Ratiometric FL probes with two reversible signal changes (2RSC) can distinctly increase the sensitivity of signal responses to the target and avoid interferences from target-independent factors [115]. The 2RSC sensors can also be divided into two subcategories: (1) single-emission probe (2RSC-1P) and (2) dual-probe (2RSC-2P) system. Note that in 1P systems, both signal changes come from a single fluorophore, and even though two fluorophores are combined in a core-shell structure or linked by covalent bonds, differently from 1RS platforms, this system will be considered 1P because, as it will be shown, the changes in emission in one probe depends on the presence of the other and the interactions between them.

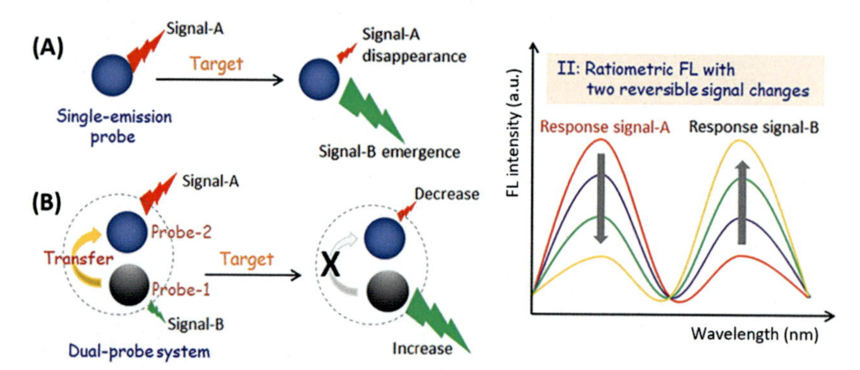

FIGURE 8.15 The general categories for fabricating dual-emissive ratiometric fluorescence (FL) probes with two reversible signal changes. Representation of (A) single emission probe, which is responsible for generating both response signals; and (B) dual-probe system in which each signal change originates from different probes. The spectrum represents the FL emission response; both signals are target sensitive and respond to the analyte presence. *Modified with permission from R. Gui et al. Recent advances in dual-emission ratiometric fluorescence probes for chemo/biosensing and bioimaging of biomarkers. Coord. Chem. Rev. 383, 82–103 (2019); Copyright © 2019 Elsevier B.V.*

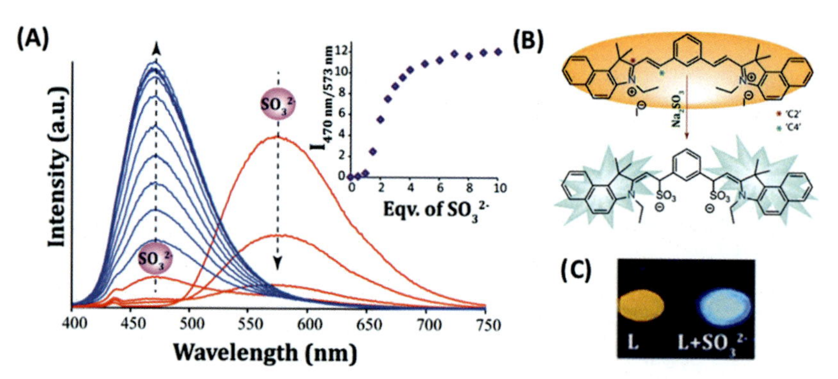

FIGURE 8.16 2RSC-1P fluorescent paper-based analytical device developed by Samanta et al. for ratiometric detection of SO_3^{2-}. (A) Fluorescence spectra of the probe **L** (10 μM) in the presence of varying concentrations of SO_3^{2-}. Inset: Changes in the emission intensity ratio (I_{470}/I_{573}) with the addition of equivalents of SO_3^{2-}. (B) The hypothetic sensing mechanism. (C) **L** drop-cast on cellulose paper with and without the addition of SO_3^{2-} under irradiation of 380 nm UV lamp. *Reprinted with permission from S. Samanta et al. A ratiometric fluorogenic probe for the real-time detection of SO_3^{2-} in aqueous medium: application in a cellulose paper-based device and potential to sense SO_3^{2-} in mitochondria. Analyst 143, 250–257 (2018); Copyright © 2018 The Royal Society of Chemistry.*

Samanta et al. designed a fluorogenic probe (L), shown in Fig. 8.16, that demonstrates a specific ratiometric detection of an SO_2 derivative (SO_3^{2-}) in 100% aqueous medium and living cells. They reported a 2RSC-1P PAD in which the orange-red FL of L turns into an intense blue FL upon interaction with SO_3^{2-}. The free probe L showed a well-defined, intense emission peak at 573 nm, which can be accredited to the uninterrupted extended π conjugation in the system. Increasing the addition of SO_3^{2-} to L results in the generation of a new emission peak at 470 nm, whereas the original emission maximum of L at 573 nm disappears entirely, as observed in Fig. 8.16A. It was observed that L could retain its fluorogenic behavior by exhibiting a visible orange FL when dropcast on a cellulose paper (Whatman 42 filter paper). Subsequent addition of SO_3^{2-} rapidly altered the orange FL of L into a highly blueish-green emitting spot, as shown in Fig. 8.16C. The authors achieved a detection limit for SO_3^{2-} of 8.45 ppb and proposed a sensing mechanism based on nucleophilic addition reaction and formation of the $L-SO_3$ adduct shown in Fig. 8.16B [12].

Lee's group reported two different 2RSC-2P systems based on the use of rhodamine as an energy acceptor in FRET processes. The first probe consists of rhodamine B moieties immobilized on the surface of water-soluble C-dots for the detection of aluminum ions (Al^{3+}). The CDs serve as the energy donors as well as the conjugation site for the rhodamine. The nanohybrid system initially presented only the CDs emission at 410 nm due to the fact that unopened rhodamine could not absorb the CDs' blue emission. Upon exposure to Al^{3+}, the spirolactam ring in the rhodamine moiety opens and generates a shift in its absorption. The absorption of the rhodamine open form overlaps with the CDs emission, leading to FRET and causing a decrease in the blue emission followed by the emergence of rhodamine's yellow emission at 560 nm. In this work, the LOD of the produced PAD was determined as 3.89×10^{-5} M, similar to that of the solution, however, a broader range of detectable concentrations of Al^{3+} ($0-1 \times 10^{-3}$ M) was achieved. This example indicated that the paper-based sensor was more effective and had better performance than the solution-based sensor [124]. The other reported probe has a very similar mechanism, but instead of CDs, the author used blue-emitting conjugated polymers of poly(p-phenylene) as energy donors in the detection of iron ions (Fe^{3+}) [11].

Bu et al. developed a 2RSC-2P CuNC-based ratiometric FL biosensor for choline detection. As shown in the scheme in Fig. 8.17A, in an aqueous suspension containing CuNCs, choline oxidase, horseradish peroxidase (HRP), and o-phenylene-diamine (OPD), the added choline is catalyzed by choline oxidase to yield H_2O_2. The OPD is then oxidized by H_2O_2 and HRP to yield 2,3-diaminophenazine (DAP). DAP—as the oxidation product of OPD, has an emission peak at 550 nm that is gradually enhanced with the increase of choline concentration. However, the FL of the CuNCs at 440 nm is reduced by DAP presence due to the inner filter effect (IFE), inducing FL quenching. By plotting the linear relationship of ratiometric FL intensities (I_{DAP}/I_{CuNCs}) *vs* choline concentrations, a novel ratiometric FL biosensor of choline could be obtained with an excellent linear relationship within the range 0.1–80 μM and a low LOD of 25 nM. The choline-spiked sensing systems were placed

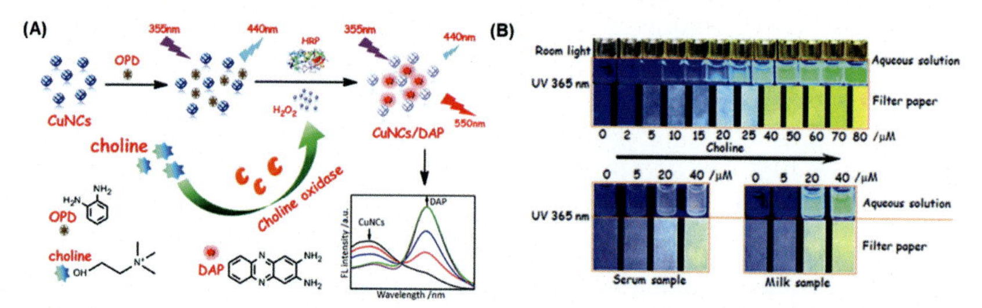

FIGURE 8.17 Ratiometric 2RSC-2P fluorescent paper-based analytical device developed by Bu et al. for choline detection. (A) Schematic illustration of the preparation process and detection mechanism of the biosensor. (B) Ratiometric fluorescence sensing of choline (dispersed in PBS or real samples) at different coexisting concentrations in CuNC-based mixture sensing systems. Visual FL colors of choline-spiked sensing systems in aqueous solutions and on wet filter papers were observed upon excitation using a 365 nm UV lamp. *Reprinted with permission from X. Bu, Y. Fu, H. Jin, R. Gui. Specific enzymatic synthesis of 2,3-diaminophenazine and copper nanoclusters used for dual-emission ratiometric and naked-eye visual fluorescence sensing of choline. N. J. Chem. 42, 17323–17330 (2018); Copyright © 2018 The Royal Society of Chemistry and the Centre National de la Recherche Scientifique.*

onto wet filter papers or into glass vials. With the increase of [choline] in the CuNC-based mixture solution, remarkable FL changes were observed (Fig. 8.17B) [125].

Both 1RS and 2RSC strategies can be applied to the same system in order to achieve more extensive color variations as proposed by Cai et al., which reported a profuse color-evolution-based fluorescent test paper sensor for monitoring Cu^{2+} in human urine by printing tricolor probe onto filter paper. As shown in Fig. 8.18A, the probe consists of blue-emission CDs (bCDs), green-emission QDs (gQDs), and red-emission QDs (rQDs) mixed in with different ratios in one system. As observable in Fig. 8.18B, the tree-probe system ensures the most comprehensive color-varying range when compared to the other ones. This sensor system is based on the principle that the FL of gQDs and rQDs is simultaneously quenched by Cu^{2+}, whereas the bCDs as the photostable internal standard is insensitive to Cu^{2+}. When the tricolor probe solution was printed onto a sheet of filter paper, a dosage scale as low as 6.0 nM could be discriminated [53].

8.4 Conclusions

In this chapter, we have explored the development of FL paper-based analytical devices by focusing on the applied fluorescent probes and the types of FL signal responses. The summarized examples have demonstrated the tremendous potential of FL PADs for qualitative and/or quantitative sensing of a wide range of targets, and their advantages concerning feasibility, simplicity, operation facility, portability, and cost-benefit relations. Although many advances have been achieved in the development of FL PADs, they still

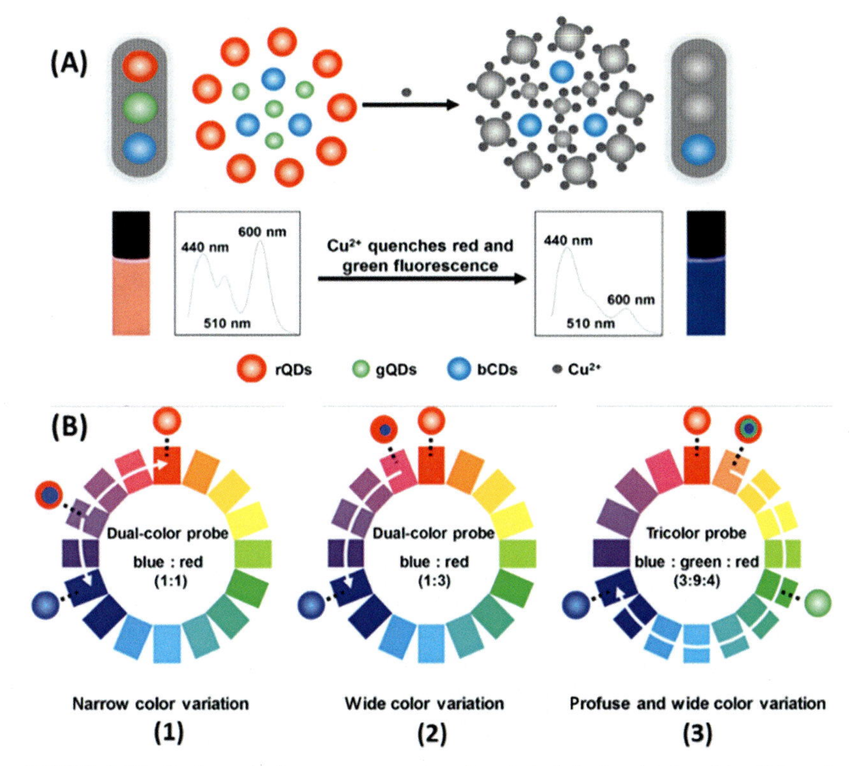

FIGURE 8.18 Ratiometric fluorescent paper-based analytical device developed by Cai et al. for Cu^{2+} sensing. (A) Schematic illustration of the visual detection principle for Cu^{2+} using the tricolor probe. (B) RGB-based colorimetric assay on the range of color variation while the different color probes coexist in a sensing system: (1) 1:1 dual-color probe; (2) 1:3 dual-color probe, and (3) 3:9:4 tricolor probe. The arrows indicate the ranges of color variations. *Reprinted with permission from Y. Cai et al. Profuse color-evolution-based fluorescent test paper sensor for rapid and visual monitoring of endogenous* Cu^{2+} *in human urine. Biosens. Bioelectron. 99, 332–337 (2018); Copyright © 2017 Elsevier B.V.*

suffer from certain limitations in accuracy, sensitivity, and difficulty in multiplex analysis. To improve the performance of these systems, beyond the fabrication method, one needs to be extremely careful when choosing the fluorescent material and the recognition element, which determines the type of sensing mechanism and the FL response of the PAD. Hence, modeling these aspects are critical for the development of FL PADs with improved properties, always focusing on application in real samples.

References

[1] A. Singh, et al., Paper-based sensors: emerging themes and applications, Sensors 18 (2018) 2838.

[2] A.P. Demchenko, Fluorescence detection techniques, Introduction to Fluorescence Sensing, Springer International Publishing, 2015, pp. 69–132.

[3] M. Wu, et al., Paper-based fluorogenic devices for in vitro diagnostics, Biosens. Bioelectron. 102 (2018) 256–266.

[4] C. Chen, J. Wu, A fast and sensitive quantitative lateral flow immunoassay for Cry1Ab based on a novel signal amplification conjugate, Sensors 12 (2012) 11684–11696.

[5] D. Carradori, K. Barreau, J. Eyer, The carbocyanine dye DiD labels in vitro and in vivo neural stem cells of the subventricular zone as well as myelinated structures following in vivo injection in the lateral ventricle, J. Neurosci. Res. 94 (2016) 139–148.

[6] J.-H. Cho, et al., Two-dimensional paper chromatography-based fluorescent immunosensor for detecting acute myocardial infarction markers, J. Chromatogr. B 967 (2014) 139–146.

[7] W. Feng, et al., A 1,8-naphthalimide-derived turn-on fluorescent probe for imaging lysosomal nitric oxide in living cells, Chin. Chem. Lett. 27 (2016) 1554–1558.

[8] Y. Long, J. Zhou, M.-P. Yang, B.-Q. Yang, A selective and sensitive off–on probe for palladium and its application for living cell imaging, Chin. Chem. Lett. 27 (2016) 205–210.

[9] U. Resch-Genger, M. Grabolle, S. Cavaliere-Jaricot, R. Nitschke, T. Nann, Quantum dots versus organic dyes as fluorescent labels, Nat. Methods 5 (2008) 763–775.

[10] A. Promchat, K. Wongravee, M. Sukwattanasinitt, T. Praneenararat, Rapid discovery and structure-property relationships of metal-ion fluorescent sensors via macroarray synthesis, Sci. Rep. 9 (2019) 1–9.

[11] H. Namgung, J. Kim, Y. Gwon, T.S. Lee, Synthesis of poly(: P -phenylene) containing a rhodamine 6G derivative for the detection of Fe(III) in organic and aqueous media, RSC Adv. 7 (2017) 39852–39858.

[12] S. Samanta, et al., A ratiometric fluorogenic probe for the real-time detection of SO 32- in aqueous medium: application in a cellulose paper-based device and potential to sense SO 32- in mitochondria, Analyst 143 (2018) 250–257.

[13] Q. Niu, et al., A highly selective turn-on fluorescent and the naked-eye colorimetric sensor for cyanide detection in food samples and its application in imaging of living cells. *Sensors Actuators*, B Chem. 276 (2018) 13–22.

[14] N. Niamnont, N. Kimpitak, G. Tumcharern, P. Rashatasakhon, M. Sukwattanasinitt, Highly sensitive salicylic fluorophore for visual detection of picomole amounts of Cu2 + , RSC Adv. 3 (2013) 25215–25220.

[15] J. Prabphal, T. Vilaivan, T. Praneenararat, Fabrication of a paper-based turn-off fluorescence sensor for Cu^{2+} Ion from a pyridinium porphyrin, ChemistrySelect 3 (2018) 894–899.

[16] N.H. Kim, M. Won, J.S. Kim, Y. Huh, D. Kim, A highly sensitive and fast responsive fluorescent probe for detection of Gold(III) ions based on the AIEgen disaggregation, Dye Pigment 160 (2019) 647–653.

[17] L. Yang, et al., A novel NBD-based fluorescent turn-on probe for detection of phosgene in solution and the gas phase, Anal. Methods 11 (2019) 4600–4608.

[18] J. Chang, H. Li, T. Hou, W. Duan, F. Li, Paper-based fluorescent sensor via aggregation-induced emission fluorogen for facile and sensitive visual detection of hydrogen peroxide and glucose, Biosens. Bioelectron. 104 (2018) 152–157.

[19] T. Jarangdet, et al., A fluorometric paper-based sensor array for the discrimination of volatile organic compounds (VOCs) with novel salicylidene derivatives, Dye Pigment 159 (2018) 378–383.

[20] L. Jia, et al., A ratiometric fluorescent nano-probe for rapid and specific detection of tetracycline residues based on a dye-doped functionalized nanoscaled metal-organic framework, Nanomaterials 9 (2019).

[21] L. Zhang, E. Wang, Metal nanoclusters: new fluorescent probes for sensors and bioimaging, Nano Today 9 (2014) 132–157.
[22] J.P. Wilcoxon, B.L. Abrams, Synthesis, structure and properties of metal nanoclusters, Chem. Soc. Rev. 35 (2006) 1162.
[23] Y. Lu, W. Chen, Sub-nanometre sized metal clusters: from synthetic challenges to the unique property discoveries, Chem. Soc. Rev. 41 (2012) 3594.
[24] Y.X. Qi, M. Zhang, A. Zhu, G. Shi, Terbium(III)/gold nanocluster conjugates: the development of a novel ratiometric fluorescent probe for mercury(II) and a paper-based visual sensor, Analyst 140 (2015) 5656–5661.
[25] Y. Yan, et al., Dual-emissive nanohybrid of carbon dots and gold nanoclusters for sensitive determination of mercuric ions, Nano Res. 9 (2016) 2088–2096.
[26] B. Han, et al., Paper-based visual detection of silver ions and l-cysteine with a dual-emissive nanosystem of carbon quantum dots and gold nanoclusters, Anal. Methods 10 (2018) 3945–3950.
[27] X.J. Zheng, R.P. Liang, Z.J. Li, L. Zhang, J.D. Qiu, One-step, stabilizer-free and green synthesis of Cu nanoclusters as fluorescent probes for sensitive and selective detection of nitrite ions, Sens. Actuators, B Chem. 230 (2016) 314–319.
[28] F. Pena-Pereira, N. Capón, I. de la Calle, I. Lavilla, C. Bendicho, Fluorescent poly(vinylpyrrolidone)-supported copper nanoclusters in miniaturized analytical systems for iodine sensing, Sens. Actuators, B Chem. 299 (2019) 126979.
[29] S.N. Baker, G.A. Baker, Luminescent carbon nanodots: emergent nanolights, Angew. Chem. Int. (Ed.) 49 (2010) 6726–6744.
[30] S.W. Hu, et al., Dual-functional carbon dots pattern on paper chips for Fe^{3+} and Ferritin analysis in whole blood, Anal. Chem. 89 (2017) 2131–2137.
[31] S. Zou, et al., An efficient fluorescent probe for fluazinam using N, S co-doped carbon dots from L-cysteine, Sens. Actuators B Chem. 239 (2017) 1033–1041.
[32] X. Du, G. Wen, Z. Li, H.W. Li, Paper sensor of curcumin by fluorescence resonance energy transfer on nitrogen-doped carbon quantum dot, Spectrochim. Acta A Mol. Biomol. Spectrosc. 227 (2020) 117538.
[33] A. Gupta, et al., Paper strip based and live cell ultrasensitive lead sensor using carbon dots synthesized from biological media, Sens. Actuators B Chem. 232 (2016) 107–114.
[34] L. Shi, et al., Excitation-independent yellow-fluorescent nitrogen-doped carbon nanodots for biological imaging and paper-based sensing, Sens. Actuators B Chem. 251 (2017) 234–241.
[35] L. Shi, et al., Green-fluorescent nitrogen-doped carbon nanodots for biological imaging and paper-based sensing, Anal. Methods 9 (2017) 2197–2204.
[36] M.R. Pacquiao, M.D.G. de Luna, N. Thongsai, S. Kladsomboon, P. Paoprasert, Highly fluorescent carbon dots from enokitake mushroom as multi-faceted optical nanomaterials for Cr^{6+} and VOC detection and imaging applications, Appl. Surf. Sci. 453 (2018) 192–203.
[37] C. Liu, et al., Dual-colored carbon dot ratiometric fluorescent test paper based on a specific spectral energy transfer for semiquantitative assay of copper ions, ACS Appl. Mater. Interfaces 9 (2017) 18897–18903.
[38] S. Takalkar, K. Baryeh, G. Liu, Fluorescent carbon nanoparticle-based lateral flow biosensor for ultrasensitive detection of DNA, Biosens. Bioelectron. 98 (2017) 147–154.
[39] X. Huang, et al., A single dual-emissive nanofluorophore test paper for highly sensitive colorimetry-based quantification of blood glucose, Biosens. Bioelectron. 86 (2016) 530–535.

[40] W. He, et al., Ratiometric fluorescence and visual imaging detection of dopamine based on carbon dots/copper nanoclusters dual-emitting nanohybrids, Talanta 178 (2018) 109–115.

[41] M. Rong, et al., A ratiometric fluorescence visual test paper for an anthrax biomarker based on functionalized manganese-doped carbon dots, Sens. Actuators B Chem. 265 (2018) 498–505.

[42] M.L. Liu, B.B. Chen, C.M. Li, C.Z. Huang, Carbon dots: synthesis, formation mechanism, fluorescence origin, and sensing applications, Green Chem. 21 (2019) 449–471.

[43] H. Li, Z. Kang, Y. Liu, S.-T. Lee, Carbon nanodots: synthesis, properties, and applications, J. Mater. Chem. 22 (2012) 24230.

[44] W.C. Chan, Quantum dot bioconjugates for ultrasensitive nonisotopic detection, Science 281 (1998) 2016–2018.

[45] M. Han, X. Gao, J.Z. Su, S. Nie, Quantum-dot-tagged microbeads for multiplexed optical coding of biomolecules, Nat. Biotechnol. 19 (2001) 631–635.

[46] L. Liang, et al., Aptamer-based fluorescent and visual biosensor for multiplexed monitoring of cancer cells in microfluidic paper-based analytical devices, Sens. Actuators B Chem. 229 (2016) 347–354.

[47] C.H. R, M. Venkataramana, M.D. Kurkuri, G. Balakrishna R, Simple quantum dot bioprobe/label for sensitive detection of *Staphylococcus aureus* TNase, Sens. Actuators B Chem. 222 (2016) 1201–1208.

[48] Y. Cai, K. Kang, Y. Liu, Y. Wang, X. He, Development of a lateral flow immunoassay of C-reactive protein detection based on red fluorescent nanoparticles, Anal. Biochem. 556 (2018) 129–135.

[49] X. Deng, et al., Applying strand displacement amplification to quantum dots-based fluorescent lateral flow assay strips for HIV-DNA detection, Biosens. Bioelectron. 105 (2018) 211–217.

[50] K. Malhotra, M.O. Noor, U.J. Krull, Detection of cystic fibrosis transmembrane conductance regulator Δf508 gene mutation using a paper-based nucleic acid hybridization assay and a smartphone camera, Analyst 143 (2018) 3049–3058.

[51] H. Wu, L. Yang, L. Chen, F. Xiang, H. Gao, Visual determination of ferric ions in aqueous solution based on a high selectivity and sensitivity ratiometric fluorescent nanosensor, Anal. Methods 9 (2017) 5935–5942.

[52] C. Yuan, K. Zhang, Z. Zhang, S. Wang, Highly selective and sensitive detection of mercuric ion based on a visual fluorescence method, Anal. Chem. 84 (2012) 9792–9801.

[53] Y. Cai, et al., Profuse color-evolution-based fluorescent test paper sensor for rapid and visual monitoring of endogenous Cu^{2+} in human urine, Biosens. Bioelectron. 99 (2018) 332–337.

[54] J. Zhou, et al., ZnSe quantum dot based ion imprinting technology for fluorescence detecting cadmium and lead ions on a three-dimensional rotary paper-based microfluidic chip, Sens. Actuators B Chem. 305 (2020) 1–9.

[55] Z. Zhang, et al., Deposition of CdTe quantum dots on microfluidic paper chips for rapid fluorescence detection of pesticide 2,4-D, Analyst 144 (2019) 1282–1291.

[56] Y. Zhou, et al., Color-multiplexing-based fluorescent test paper: dosage-sensitive visualization of arsenic(III) with discernable scale as low as 5 ppb, Anal. Chem. 88 (2016) 6105–6109.

[57] X. Yan, H. Li, W. Zheng, X. Su, Visual and fluorescent detection of tyrosinase activity by using a dual-emission ratiometric fluorescence probe, Anal. Chem. 87 (2015) 8904–8909.

[58] K. Zhang, et al., Instant visual detection of trinitrotoluene particulates on various surfaces by ratiometric fluorescence of dual-emission quantum dots hybrid, J. Am. Chem. Soc. 133 (2011) 8424–8427.

[59] L. Zong, et al., Paper-based fluorescent immunoassay for highly sensitive and selective detection of norfloxacin in milk at picogram level, Talanta 195 (2019) 333–338.

[60] X. Gong, et al., A review of fluorescent signal-based lateral flow immunochromatographic strips, J. Mater. Chem. B 5 (2017) 5079–5091.

[61] M. Lin, et al., Recent advances in synthesis and surface modification of lanthanide-doped upconversion nanoparticles for biomedical applications, Biotechnol. Adv. 30 (2012) 1551–1561.

[62] S. Wen, et al., Advances in highly doped upconversion nanoparticles, Nat. Commun. 9 (2018) 2415.

[63] Y. Zhong, et al., Elimination of photon quenching by a transition layer to fabricate a quenching-shield sandwich structure for 800nm excited upconversion luminescence of Nd^{3+}-sensitized nanoparticles, Adv. Mater. 26 (2014) 2831–2837.

[64] S. Xu, et al., Paper-based upconversion fluorescence resonance energy transfer biosensor for sensitive detection of multiple cancer biomarkers, Sci. Rep. 6 (2016) 1–9.

[65] H. He, et al., Quantitative lateral flow strip sensor using highly doped upconversion nanoparticles, Anal. Chem. 90 (2018) 12356–12360.

[66] Q. Ju, U. Uddayasankar, U. Krull, Paper-based DNA detection using lanthanide-doped LiYF4 upconversion nanocrystals as bioprobe, Small 10 (2014) 3912–3917.

[67] M. He, Z. Liu, Paper-based microfluidic device with upconversion fluorescence assay, Anal. Chem. 85 (2013) 11691–11694.

[68] M. He, Z. Li, Y. Ge, Z. Liu, Portable upconversion nanoparticles-based paper device for field testing of drug abuse, Anal. Chem. 88 (2016) 1530–1534.

[69] Q. Mei, et al., Smartphone based visual and quantitative assays on upconversional paper sensor, Biosens. Bioelectron. 75 (2016) 427–432.

[70] W. Wang, et al., Highly selective and sensitive sensing of 2,4,6-trinitrophenol in beverages based on guanidine functionalized upconversion fluorescent nanoparticles, Sens. Actuators B Chem. 255 (2018) 1422–1429.

[71] R. Zhang, et al., Responsive upconversion nanoprobe for background-free hypochlorous acid detection and bioimaging, Small 15 (2019) 1–11.

[72] C. Jiang, B. Liu, M.-Y. Han, Z. Zhang, Fluorescent nanomaterials for color-multiplexing test papers toward qualitative/quantitative assays, Small Methods 2 (2018) 1700379.

[73] M. Biesaga, Porphyrins in analytical chemistry. A review, Talanta 51 (2000) 209–224.

[74] Z. Valicsek, O. Horváth, Application of the electronic spectra of porphyrins for analytical purposes: the effects of metal ions and structural distortions, Microchem. J. 107 (2013) 47–62.

[75] J. Prabpal, T. Vilaivan, T. Praneenararat, Paper-based heavy metal sensors from the concise synthesis of an anionic porphyrin: a practical application of organic synthesis to environmental chemistry, J. Chem. Educ. 94 (2017) 1137–1142.

[76] M. Zheng, et al., Fast response and high sensitivity europium metal organic framework fluorescent probe with chelating terpyridine sites for Fe^{3+}, ACS Appl. Mater. Interfaces 5 (2013) 1078–1083.

[77] K. Patir, S.K. Gogoi, Facile synthesis of photoluminescent graphitic carbon nitride quantum dots for Hg^{2+} detection and room temperature phosphorescence, ACS Sustain. Chem. Eng. 6 (2018) 1732–1743.

[78] P. Nath, M. Chatterjee, N. Chanda, Dithiothreitol-facilitated synthesis of bovine serum albumin–gold nanoclusters for Pb(II) ion detection on paper substrates and in live cells, ACS Appl. Nano Mater. 1 (2018) 5108–5118.

[79] C. Wang, et al., Highly selective, rapid-functioning and sensitive fluorescent test paper based on graphene quantum dots for on-line detection of metal ions, Anal. Methods 10 (2018) 1163–1171.

[80] X. Liu, Y. Yang, X. Xing, Y. Wang, Grey level replaces fluorescent intensity: fluorescent paper sensor based on ZnO nanoparticles for quantitative detection of Cu^{2+} without photoluminescence spectrometer, Sens. Actuators B Chem. 255 (2018) 2356–2366.

[81] Bissell, R.A. *et al.* Fluorescent PET (photoinduced electron transfer) sensors. In: Mattay J. (Ed.) Photoinduced Electron Transfer V. Topics in Current Chemistry, vol 168. Springer, Berlin, Heidelberg. 223–264 (1993). doi: 10.1007/3–540-56746–1_12.

[82] J. Qi, et al., Three-dimensional paper-based microfluidic chip device for multiplexed fluorescence detection of Cu^{2+} and Hg^{2+} ions based on ion imprinting technology, Sens. Actuators B Chem. 251 (2017) 224–233.

[83] L.K. Fraiji, D.M. Hayes, T.C. Werner, Static and dynamic fluorescence quenching experiments for the physical chemistry laboratory, J. Chem. Educ. 69 (1992) 424.

[84] C.M. Gonzalez, et al., Detection of high-energy compounds using photoluminescent silicon nanocrystal paper based sensors, Nanoscale 6 (2014) 2608–2612.

[85] I.N. Germanenko, S. Li, M.S. El-Shall, Decay dynamics and quenching of photoluminescence from silicon nanocrystals by aromatic nitro compounds, J. Phys. Chem. B 105 (2001) 59–66.

[86] W. Lu, et al., Self-diffusion driven ultrafast detection of ppm-level nitroaromatic pollutants in aqueous media using a hydrophilic fluorescent paper sensor, ACS Appl. Mater. Interfaces 9 (2017) 23884–23893.

[87] Z. Gou, Y. Zuo, M. Tian, W. Lin, Siloxane-based nanoporous polymers with narrow pore-size distribution for cell imaging and explosive detection, ACS Appl. Mater. Interfaces 10 (2018) 28979–28991.

[88] R. Feyisa Bogale, et al., A terbium(III)-based coordination polymer for selective and sensitive sensing of nitroaromatics and ferric ion: synthesis, crystal structure and photoluminescence properties, N. J. Chem. 41 (2017) 12713–12720.

[89] L. Chen, et al., Femtogram level detection of nitrate ester explosives via an 8-pyrenyl-substituted fluorene dimer bridged by a 1,6-hexanyl unit, ACS Appl. Mater. Interfaces 6 (2014) 8817–8823.

[90] R. Wang, W. Zhang, P. Wang, X. Su, A paper-based competitive lateral flow immunoassay for multi β-agonist residues by using a single monoclonal antibody labelled with red fluorescent nanoparticles, Microchim. Acta 185 (2018).

[91] B. Li, et al., Quantum dot-based molecularly imprinted polymers on three-dimensional origami paper microfluidic chip for fluorescence detection of phycocyanin, ACS Sens. 2 (2017) 243–250.

[92] X. Chen, N. Zheng, S. Chen, Q. Ma, Fluorescence detection of dopamine based on nitrogen-doped graphene quantum dots and visible paper-based test strips, Anal. Methods 9 (2017) 2246–2251.

[93] A. İncel, O. Akın, A. Çağır, Ü.H. Yıldız, M.M. Demir, Smart phone assisted detection and quantification of cyanide in drinking water by paper based sensing platform, Sens. Actuators B Chem. 252 (2017) 886–893.

[94] A. Promchat, P. Rashatasakhon, M. Sukwattanasinitt, A novel indolium salt as a highly sensitive and selective fluorescent sensor for cyanide detection in water, J. Hazard. Mater. 329 (2017) 255–261.

[95] G. Aragay, H. Montón, J. Pons, M. Font-Bardía, A. Merkoi, Rapid and highly sensitive detection of mercury ions using a fluorescence-based paper test strip with an N-alkylaminopyrazole ligand as a receptor, J. Mater. Chem. 22 (2012) 5978–5983.

[96] J. Hu, R. Liu, X. Cai, M. Shu, H. Zhu, Highly efficient and selective probes based on polycyclic aromatic hydrocarbons with trimethylsilylethynyl groups for fluoride anion detection, Tetrahedron 71 (2015) 3838–3843.

[97] L. Zeng, X. Wu, Q. Hu, H.Q. Yuan, G.M. Bao, A single fluorescent chemosensor for discriminative detection of bisulfite and benzoyl peroxide in food with different emission, Sens. Actuators B Chem. 299 (2019) 126994.

[98] J.M.C.C. Guzman, L.L. Tayo, C.C. Liu, Y.N. Wang, L.M. Fu, Rapid microfluidic paper-based platform for low concentration formaldehyde detection, Sens. Actuators B Chem. 255 (2018) 3623–3629.

[99] L. Liang, et al., Fluorescence 'turn-on' determination of H_2O_2 using multilayer porous SiO_2/NGQDs and PdAu mimetics enzymatic/oxidative cleavage of single-stranded DNA, Biosens. Bioelectron. 82 (2016) 204–211.

[100] M.G. Caglayan, S. Sheykhi, L. Mosca, P. Anzenbacher, Fluorescent zinc and copper complexes for detection of adrafinil in paper-based microfluidic devices, Chem. Commun. 52 (2016) 8279–8282.

[101] L. Liang, et al., Metal-enhanced fluorescence/visual bimodal platform for multiplexed ultrasensitive detection of microRNA with reusable paper analytical devices, Biosens. Bioelectron. 95 (2017) 181–188.

[102] N.K. Thom, G.G. Lewis, K. Yeung, S.T. Phillips, Quantitative fluorescence assays using a self-powered paper-based microfluidic device and a camera-equipped cellular phone, RSC Adv. 4 (2014) 1334–1340.

[103] U.H. Yildiz, P. Alagappan, B. Liedberg, Naked eye detection of lung cancer associated miRNA by paper based biosensing platform, Anal. Chem. 85 (2013) 820–824.

[104] K. Yamada, T.G. Henares, K. Suzuki, D. Citterio, Distance-based tear Lactoferrin assay on microfluidic paper device using interfacial interactions on surface-modified cellulose, ACS Appl. Mater. Interfaces 7 (2015) 24864–24875.

[105] J. Chang, et al., Paper-based fluorescent sensor for rapid naked-eye detection of acetylcholinesterase activity and organophosphorus pesticides with high sensitivity and selectivity, Biosens. Bioelectron. 86 (2016) 971–977.

[106] G. Jang, J. Kim, D. Kim, T.S. Lee, An emission color-changeable sensor for biothiol using a water-soluble conjugated polymer in a paper-based strip, J. Nanosci. Nanotechnol. 16 (2016) 8728–8732.

[107] B. Li, et al., Simultaneous detection of antibiotic resistance genes on paper-based chip using [Ru(phen)2dppz]2 + turn-on fluorescence probe, ACS Appl. Mater. Interfaces 10 (2018) 4494–4501.

[108] J. Zhou, et al., A rapid and highly selective paper-based device for high-throughput detection of cysteine with red fluorescence emission and a large Stokes shift, Anal. Methods 11 (2019) 1312–1316.

[109] P. Jiang, M. He, L. Shen, A. Shi, Z. Liu, A paper-supported aptasensor for total IgE based on luminescence resonance energy transfer from upconversion nanoparticles to carbon nanoparticles, Sens. Actuators B Chem. 239 (2017) 319–324.

[110] B.P. Chhetri, A. Karmakar, A. Ghosh, Recent advancements in Ln-Ion-based upconverting nanomaterials and their biological applications, Part. Part. Syst. Charact. 36 (2019) 1900153.

[111] M. You, et al., Household fluorescent lateral flow strip platform for sensitive and quantitative prognosis of heart failure using dual-color upconversion nanoparticles, ACS Nano 11 (2017) 6261–6270.

[112] M. Ali, et al., A fluorescent lateral flow biosensor for the quantitative detection of Vaspin using upconverting nanoparticles, Spectrochim. Acta A Mol. Biomol. Spectrosc. 226 (2020) 117610.

[113] Z. Rong, et al., Smartphone-based fluorescent lateral flow immunoassay platform for highly sensitive point-of-care detection of Zika virus nonstructural protein 1, Anal. Chim. Acta 1055 (2019) 140–147.

[114] Q. Mei, Z. Zhang, Photoluminescent graphene oxide ink to print sensors onto microporous membranes for versatile visualization bioassays, Angew. Chem. Int. (Ed.) 51 (2012) 5602–5606.

[115] R. Gui, et al., Recent advances in dual-emission ratiometric fluorescence probes for chemo/biosensing and bioimaging of biomarkers, Coord. Chem. Rev. 383 (2019) 82–103.

[116] Y.-J. Gong, et al., Through bond energy transfer: a convenient and universal strategy toward efficient ratiometric fluorescent probe for bioimaging applications, Anal. Chem. 84 (2012) 10777–10784.

[117] Q. Sun, et al., Colloidal quantum dot chains: self-assembly mechanism and ratiometric fluorescent sensing, RSC Adv. 7 (2017) 53977–53983.

[118] L. Su, et al., A ratiometric fluorescent paper sensor for consecutive color change-based visual determination of blood glucose in serum, N. J. Chem. 42 (2018) 6867–6872.

[119] Q.X. Wang, et al., Dual lanthanide-doped complexes: the development of a time-resolved ratiometric fluorescent probe for anthrax biomarker and a paper-based visual sensor, Biosens. Bioelectron. 94 (2017) 388–393.

[120] Y. Wang, et al., Ratiometric fluorescent paper sensor utilizing hybrid carbon dots-quantum dots for the visual determination of copper ions, Nanoscale 8 (2016) 5977–5984.

[121] H. Xu, K. Zhang, Q. Liu, Y. Liu, M. Xie, Visual and fluorescent detection of mercury ions by using a dually emissive ratiometric nanohybrid containing carbon dots and CdTe quantum dots, Microchim. Acta 184 (2017) 1199–1206.

[122] J. Wang, et al., Molecularly imprinted fluorescent test strip for direct, rapid, and visual dopamine detection in tiny amount of biofluid, Small 15 (2019) 1–9.

[123] X. Wang, et al., A ratiometric upconversion nanosensor for visualized point-of-care assay of organophosphonate nerve agent, Sens. Actuators B Chem. 241 (2017) 1188–1193.

[124] Y. Kim, G. Jang, T.S. Lee, New fluorescent metal-ion detection using a paper-based sensor strip containing tethered rhodamine carbon nanodots, ACS Appl. Mater. Interfaces 7 (2015) 15649–15657.

[125] X. Bu, Y. Fu, H. Jin, R. Gui, Specific enzymatic synthesis of 2,3-diaminophenazine and copper nanoclusters used for dual-emission ratiometric and naked-eye visual fluorescence sensing of choline, N. J. Chem. 42 (2018) 17323–17330.

Chapter 9

Electrochemiluminescence paper-based analytical devices

Erin M. Gross* **and Samaya Kallepalli**
Creighton University Department of Chemistry and Biochemistry, Omaha, NE, United States
**Corresponding author. e-mail address: eringross@creighton.edu*

9.1 Background

9.1.1 Electrochemiluminescence

The process of electrochemiluminescence (ECL), also known as electrogenerated chemiluminescence [1], results in the production of luminescence in solution without an external light source. ECL generally occurs via two different pathways, an annihilation pathway or a coreactant pathway (Fig. 9.1). In the annihilation pathway (Fig. 9.1A), ECL intermediates are generated heterogeneously at an electrode via potential steps that oxidize an acceptor species (A) and reduce a donor species (D). Upon electron transfer between $A^{+\bullet}$ and $D^{-\bullet}$, a luminescent excited state is produced (A*) which emits photons at a characteristic wavelength range. In the annihilation pathway, A and D may be the same or different species. Alternatively, coreactant pathways are either "oxidative-reductive" or "reductive-oxidative." In the former, both a luminophore and a "coreactant" are oxidized by heterogeneous or both heterogeneous and homogeneous electron transfer processes. For example, Fig. 9.1B shows both

A) Annihilation pathway

$$A \rightarrow A^{+\bullet} + e^- \quad (1)$$
$$D + e^- \rightarrow D^{-\bullet} \quad (2)$$
$$D^{-\bullet} + A^{+\bullet} \rightarrow A^* + D \quad (3)$$
$$A^* \rightarrow A + h\nu \quad (4)$$

B) Coreactant pathway

$$Ru(bpy)_3^{2+} \rightarrow Ru(bpy)_3^{3+} + e^- \quad (1)$$
$$TPrA \rightarrow TPrA^{+\bullet} + e^- \quad (2a)$$
$$Ru(bpy)_3^{3+} + TPrA \rightarrow TPrA^{+\bullet} + e^- \quad (2b)$$
$$TPrA^{+\bullet} \rightarrow TPrA^{\bullet} + H^+ \quad (3)$$
$$TPrA^{\bullet} + Ru(bpy)_3^{3+} \rightarrow {}^*Ru(bpy)_3^{2+} + h\nu \quad (4)$$

FIGURE 9.1 Annihilation (A) and coreactant (B) ECL pathways. The coreactant pathway shown here is oxidative-reductive ECL between tris(2,2′-bipyridyl)ruthenium(II) and tri-*n*-propylamine. *Credit: authors.*

Paper-Based Analytical Devices for Chemical Analysis and Diagnostics.
DOI: https://doi.org/10.1016/B978-0-12-820534-1.00003-7
© 2022 Elsevier Inc. All rights reserved.

a luminophore and coreactant being oxidized at an electrode. The oxidized coreactant is not stable on the time frame of the experiment and forms a "reductant" species, which reacts with the oxidized luminophore to form the emissive state. One of the most common oxidation reduction reactions is that of tris(2,2′-bipyridyl)ruthenium(II) ($Ru(bpy)_3^{2+}$) and tripropylamine [2] and is shown in Fig. 9.1B. In "reductive-oxidative" ECL, both the luminophore and coreactant are reduced via heterogeneous or both heterogeneous and homogeneous electron transfer processes to form the precursors to the formation of the emissive state. One of the first examples of this coreactant pathway was the reaction of $Ru(bpy)_3^{2+}$ and peroxydisulfate ($S_2O_8^{2-}$) [3]. Annihilation pathways generally require aprotic solvents and removal of oxygen to generate the free radical intermediates. Therefore the coreactant pathway is more widely utilized in making chemical and biochemical measurements because it can be performed in aqueous solutions at milder potentials. Numerous small molecules have been developed for ECL detection schemes [1]; however, $Ru(bpy)_3^{2+}$ and its derivatives and luminol remain the most common molecular luminophores. Nanomaterials have become popular ECL materials [4], particularly in the area of bioanalysis. These materials include quantum dots, along with carbon, silica, and metal-based nanomaterials. Nanomaterials have opened up a wide range of possibilities for the improvement of sensor specificity and sensitivity. In the literature reviewed for this chapter, nanomaterials have served a variety of roles in ECL systems, including increasing sensor surface area and conductivity, and enhancing ECL signal, along with quenching ECL from resonance energy transfer.

ECL clinical diagnostic methods, including immunoassays and DNA-probe assays, were first commercialized as Origen technology in the early 1990s by the company Igen [5]. In 2003 Roche Diagnostics acquired Igen and the Origen technology [6] that are currently marketed as Cobas and Elecsys systems.

ECL uniquely combines the miniaturization capabilities and low cost of electrochemical measurements with the high sensitivity of luminescent methods. ECL methods do not require the expensive equipment necessary for photoluminescence methods. As with chemiluminescent methods, ECL theoretically has no background but also has more control over the reactions and increased selectivity via manipulation of the applied potential. These characteristics give it the potential for applications of rapid on-site detection, in which low-cost, simple, and disposable sensors are preferred. For these reasons, ECL is an excellent detection method to combine with microfluidic and microscale detection platforms, including paper-based analytical devices.

9.1.2 Paper devices and electrochemiluminescence

Paper has long been a platform for making chemical measurements, as evidenced by pH paper and home-pregnancy tests. However, a "lab-on-paper" concept was not imagined until an article on that concept was published by the Whitesides group in 2007 [7]. This groundbreaking research patterned chromatography paper using photolithography techniques typically used in the semiconductor industry, and used the patterned paper to perform inexpensive, low-volume, and portable bioassays. This work demonstrated the utility of paper as a low cost, ubiquitous, porous, easily disposable (e.g., combustion) platform for analysis. Additional benefits included its biodegradability, and the fact that paper is easy to use, store, transport, and print. Paper also has a wide range of chemical compatibility, a large surface area and has been shown to be easily patterned. This 2007 report and the versatility and advantages of paper inspired many researchers. Not long after, reports of using patterned paper for applications, such as telemedicine [8]; clinical measurements of urinary ketones, glucose, and salivary nitrite [9]; and other small molecule (glucose, lactate, and uric acid) measurements [10], were published. Soon more devices were patterned with the less expensive and more environmentally-friendly wax [11,12] rather than photoresists.

Just as in pH paper and home-pregnancy tests, initial paper-based analytical devices had outputs based on a visual change, such as a color change. In 2009 electrochemical detection on paper devices was first reported by Dungchai et al [13]. In this work, carbon electrodes were screen printed directly onto patterned chromatography paper. The electrodes were characterized by cyclic voltammetry and used amperometry to detect glucose, lactate, and uric acid in control serum samples. Once electrodes were patterned onto paper, it was only natural that the development of ECL detection systems would follow. In 2011 a paper-microfluidic device with ECL detection was reported by Hogan's group [14]. The catalytic effect of Whitesides' 2007 paper on the development of "lab-on-paper" devices can be visualized in the graph in Fig. 9.2. The graph depicts the number of citations versus year resulting from a search of "paper-based analytical devices" in the Scopus database. The number of reports on paper-based devices appears to have exponentially grown since 2007, with nearly 5000 citations in 2019, and an even larger amount expected for 2020.

With a myriad of reports using paper-based analytical devices, it is helpful to differentiate between the different device types. Numerous acronyms have been used to describe a variety of devices and many of these are summarized in Table 9.1. The devices using patterned paper previously mentioned [7–14], all contained micron-sized channels that guided fluid flow on the device. These are classified as microfluidic paper-based analytical devices (μ-PADs). Paper-based devices that are not microfluidic are generally defined as paper chips (PCs). Electrochemical [15] and ECL-based devices [14,16–18] also have acronyms as denoted in Table 9.1. In reviewing the literature, we have found a wide range of acronyms, some which have been used in multiple articles and some that have not endured. Table 9.1 reports commonly used acronyms for devices and for many of the materials and substances used to fabricate devices.

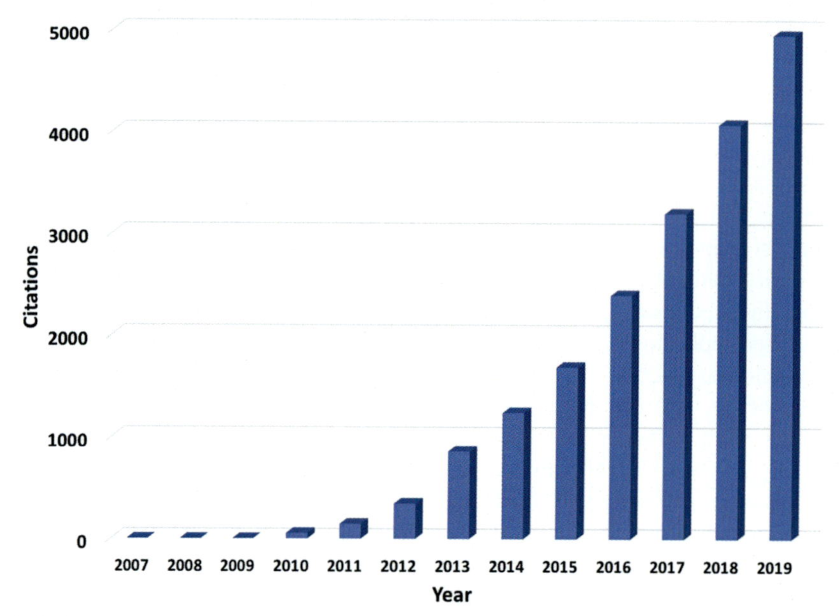

FIGURE 9.2 Graph of citations on a search of "paper-based analytical devices" for the years 2007–2019 using the Scopus database. *Credit: authors.*

As mentioned earlier, the first report of ECL detection on a μ-PAD was by Hogan's group in 2011 [14]. A 2014 review [16] covered 14 reports from the literature on microfluidic paper-based ECL devices (μ-PECLDs). A 2019 [18] review on μ-PECLDs for point-of-care (POC) testing applications reported on 32 studies from 2012 to 2018. In our preparation of this chapter, we found more than 50 manuscripts reporting paper-based analytical devices using ECL detection. We are aware of three review articles (from 2014 to 2019) discussing μ-PECLDs so the work here will focus on devices reported more recently (2016–2020) but will discuss older works (2011–2015) to highlight key studies and for establishing trends.

9.2 Paper-device fabrication methods

ECL-based devices come in a variety of designs. These designs include both microfluidic and nonmicrofluidic PCs. Devices can be planar (2D) or 3D. Three-dimensional devices have been prepared via stacking paper layers [19,20] to fabricate devices or folded, which is more common. The first folded ECL-based devices were reported by Yu's group in 2012 [21] and were given the moniker microfluidic "origami" ECL device (μ-OECLD). Folded devices have the advantage of straightforward alignment between layers. The first devices reported were fairly simple, with one fold to align a sample area/working electrode and the counter and reference electrodes as

TABLE 9.1 Definition of acronyms used in paper-based analytical devices.

Acronym	Definition
ABEI	luminol derivative
AuNP	gold nanoparticles
BPE	bipolar electrode
CMC	carboxymethyl chitosan
CNCs	carbon nanocrystals
CNDs	carbon nanodots
ePAD	electrochemical paper-based analytical device
Gr	graphene
GQDs	graphene quantum dots
HC	hollow channel
HCR	hybridization chain reaction
HRP	horseradish peroxidase
ITO	indium tin oxide
MWCNT	multiwalled carbon nanotubes
μ-OECLD	microfluidic origami ECL device
μ-s-OECL	self-powered 3D origami microfluidic electrochemiluminescence
μ-PAD	microfluidic paper-based analytical device
PECLOD	paper-based ECL origami device
NPS	nanoporous silver
PC	paper-based chip (not microfluidic)
POC	point of care
PoP-ECL	pen-on-paper ECL device
PSA	Prostate-specific antigen
PSSG	poly(sodium 4-styrenesulfonate) modified graphene
PWE	paper working electrode
RCA	rolling circle amplification
Ru, $Ru(bpy)_3^{2+}$	tris(2,2′-bipyridyl)ruthenium(II)
$Ru(phen)_2dppz^{2+}$	phen = 1,10-phenanthroline; dppz = dipyridophenazine
SNC	silica nanochannel
SPCE	screen-printed carbon electrode
SPWPE	screen-printed working paper electrode

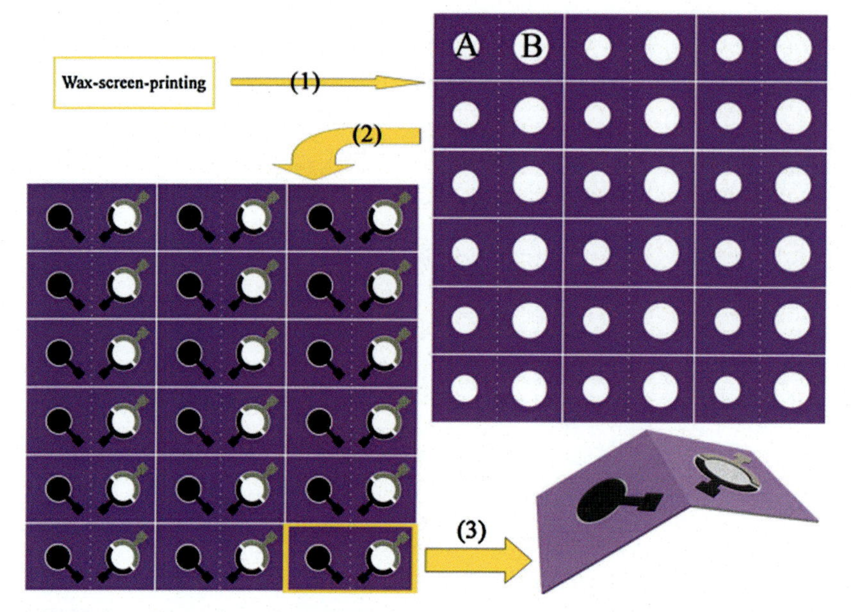

FIGURE 9.3 Schematic representation of the process for fabricating a folded paper-based ECL device. First, cellulose paper sheets were patterned using a wax printer (Procedure 1). After baking, three electrodes were screen printed onto the wax-patterned sheet (Procedure 2). Lastly, the prepared sheet was cut to rectangular paper (60.0 mm × 30.0 mm) and folded (Procedure 3). *Reproduced with permission from S. Wang, L. Ge, M. Yan, J. Yu, X. Song, S. Ge, et al., 3D microfluidic origami electrochemiluminescence immunodevice for sensitive point-of-care testing of carcinoma antigen 125, Sens. Actuators B 176 (2013) 1–8. Copyright (2013) Elsevier.*

shown in Fig. 9.3 part 3 [22]. We will see later in this chapter how increasingly complex origami devices that integrate more laboratory functions have been developed.

9.2.1 Paper types

More details on the chemistry and modification of paper can be found in Chapter 2, Chemistry of paper—properties, modification strategies, and uses in bioanalytical chemistry, but this section will summarize the types of paper and patterning methods used with ECL detection. Paper is generally the hydrophilic part of the device and is used to contain sample and reagents and to fabricate electrodes [e.g., paper working electrodes (PWEs)]. The most common paper used in ECL studies is cellulose. Whatman #1 and #2 chromatography/filter paper primarily have been used as the paper substrate. It is inexpensive and available in most laboratories, and can be purchased in a variety of thicknesses and sizes. Additionally, paper can also be precisely cut to manage device design.

9.2.2 Fabrication of hydrophobic barriers

Hydrophobic barriers are needed to control the volume of reagents contained by the device and also to direct flow, especially as devices become more sophisticated and integrate more functions into a single device. As devices move toward integrating more functions onto the paper (e.g., potential generation, rinsing, reagent mixing, fluid handling), it is important to have efficient, inexpensive, and easy to prototype methods.

Photolithographic methods similar to those used to pattern silica wafers [23,24] were used to pattern paper in the first reports of μ-PADs [7,9,10] and e-PADs [13,25]. These methods were able to obtain channel dimensions as narrow as around 200 μm [24] to direct fluid flow. Fig. 9.4 shows a photograph and diagram of a device fabricated via photolithography. After patterning, the paper is hydrophobic in places where the photoresist was patterned, leaving a hydrophilic space in the center here for fluid flow and reagent deposition. The photoresist also hardens the paper, making is less susceptible to tearing. The first ECL devices that patterned paper using photolithography, including both nonmicrofluidic PCs [26–28] and a microfluidic device [29], were reported by Wang's group around 2013–2015.

Wax is the most commonly used material for patterning the hydrophobic barrier. Filter paper was first patterned with wax in 2009 by Lin's group [11]. They demonstrated three different wax-patterning methods: (1) painting with a wax pen, (2) printing with an ink-jet printer followed by painting with a wax pen, and (3) direct printing with a wax printer. After wax is applied to the paper in the desired pattern, the wax/paper undergoes a heating step that embeds the paper with the wax. Although wax can be applied to paper as simply as by writing on the paper through a stencil with a wax pen crayon (screen printing), direct wax printing is the most common method for patterning paper. Compared with photolithography, this process is simple to perform with fabrication taking around 5–10 min, without equipment and materials such as a clean room, UV lamp, spin coater, or organic solvent. A wax printer is not

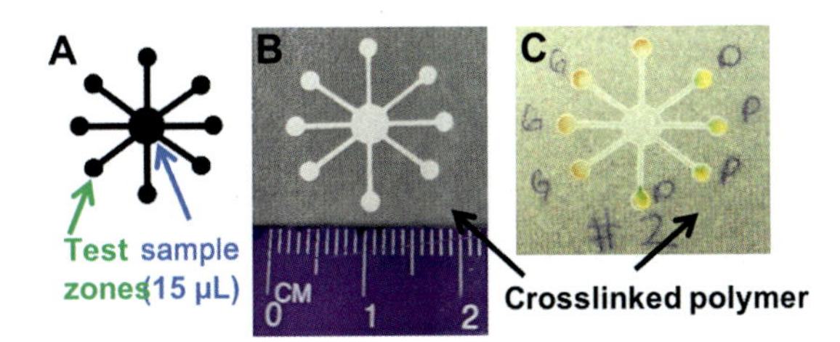

FIGURE 9.4 Diagram (A) and photograph (B) of a paper-based microfluidic device patterned using photolithography. This device was patterned so that sample was introduced in the center and traveled via capillary action to test zones containing colorimetric reagents. Reagents for both glucose (G) and protein (P) were spotted in the test zones and changed color if the analyte was present (C). *Credit: authors.*

necessary to fabricate paper ECL devices. For example, Yu's group fabricated an inexpensive pen-on-paper (PoP) ECL device to detect a tumor biomarker using crayon to pattern wax on cellulose paper [30]. The working electrode for this device was fabricated using a graphite pencil, and the device was run using a rechargeable battery. Rusling's group (2013) reported patterning microfluidic devices by heat transfer of wax paper onto filter paper [31]. These devices were subsequently used for genotoxicity screenings.

Although most ECL-based devices use wax-patterning methods to design microfluidic channels and other fluid-containing reservoirs, other methods have been successfully implemented. Bao and Gu used a laser printer [32] and chromatography paper to fabricate ECL paper-chip "hybrid" devices [33]. The device was hybrid in that it contained a cellulose paper layer that defined the sample volume that would be in contact with an electrode. They optimized the postprinting heating time so that the hydrophobic barriers penetrated the paper sufficiently to contain solution; they determined optimal conditions of 150°C for a duration of 300 min. It is important to note that the electrode was not constructed on paper, making this a hybrid device. It was constructed from a piece of double-sided carbon-tape mounted on indium tin oxide (ITO) glass. The electrode was modified with CdS quantum dots (QDs) and was used to detect dopamine. Here, we are denoting hybrid devices as those where the entire device is not fabricated from paper, such as in the example earlier. Additional devices and electrode materials will be discussed in the next section on electrodes.

9.3 Fabrication of electrodes onto paper

9.3.1 Screen printing

Zhang and Zing's groups [34–36] have used screen printing in two ways: (1) wax screen printing the hydrophobic pattern that defines the device barriers and (2) carbon ink screen printing. Screen printing the electrodes along with contact pads is a popular method due to its simplicity and cost-effectiveness. Generally carbon-based materials (ink or paste) are used for the working and counter electrodes, while an AgCl ink can be used for the reference electrode. Fig. 9.3 shows devices in which all three electrodes and contacts were screen printed onto the paper. In general, a stencil or screen is placed over the patterned paper. The electrode material is applied onto the paper and when the stencil is removed, the ink is in the pattern of the screen.

ECL is generated at the working electrode and much research has been done developing working electrodes that can be modified to both fabricate biosensors and increase performance. Modifications allow for attachment of antibodies, aptamers, and even ECL luminophores. Screen-printed carbon electrodes (SPCEs), such as the one shown in Fig. 9.3, were some of the first

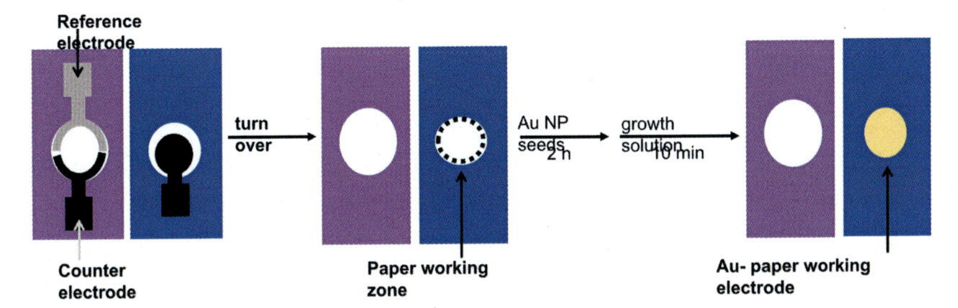

FIGURE 9.5 Fabrication of the process of growing AuNPs on the paper working zone on a 3D origami device. *Adapted from C. Ma, H. Liu, L. Zhang, L. Li, M. Yan, J. Yu, et al., Microfluidic paper-based analytical device for sensitive detection of peptides based on specific recognition of aptamer and amplification strategy of hybridization chain reaction, ChemElectroChem 4 (2017) 1744–1749. Figure S2. Copyright (2017) Wiley.*

working electrodes fabricated onto paper, modified and used as the working electrode. Yu's group fabricated both single electrode [20–22] and electrode array [19,37] devices. As research continued in this area, the porous nature of paper was utilized to fabricate conductive, porous, high surface area PWEs. This novel type of electrode was first developed by Yu's group for ECL paper devices and used to detect adenosine triphosphate (ATP) [38]. The steps in fabricating a PWE are shown in Fig. 9.5 [39]. On each wax-patterned (blue and purple areas) origami device, a carbon working electrode is screen printed onto a paper working zone, and a carbon ink counter electrode and Ag/AgCl ink reference electrode are screen printed on a separate paper auxiliary zone, respectively. Metal nanoparticles are synthesized or grown in the paper working zone on the other side of the working electrode. This area, shown in yellow, is in contact with the SPCE and serves as the working electrode. Growing nanoparticles on the high surface area cellulose network creates an electrode with a high surface area. The porosity of these materials also increases the surface area, which accomplishes two things: (1) increases the "loading" of the biorecognition agent and (2) increases electron transfer rates. Modification can also increase the biocompatibility, which is important when making measurements on real samples. Before the device is used, it is folded at the center and placed in a holder for electrical connection.

The device in Fig. 9.5 was modified with Au nanoparticles (AuNPs) and fabricated into a biosensor to detect the peptide Mucin-1. The paper platform allows for a plethora of modification schemes, opening up many possibilities of analytes to detect with high specificity and sensitivity. Devices have been fabricated with AuNPs and DNA to detect a wide range of analytes including heavy metals [40–43], tumor markers [30,44–49], DNA [50], peptides [39], and proteins [51]. More recently, with more sophisticated modification using nanomaterials, Yu's group has fabricated cytodevices that measure cells [52–55]. Cytodevices have also been applied to measuring the release of H_2S from cells [56] and Ca153 antigens at MCF-7 cell surfaces [57].

As mentioned earlier, the first PWEs were fabricated using AuNPs. Soon, other materials were investigated for fabricating biosensors. A graphene oxide-chitosan (GCA)/gold nanoparticle PWE was used to detect the tumor markers prostate-specific antigen (PSA) and carcinoembryonic antigen (CEA) via an immunoassay [44]. Utilizing GCA improved the biocompatibility, facilitated electron transfer and increased the ECL signal from a phenyleneethynylene derivative-TPrA ECL system. In addition to modifying PWEs with carbon-based materials such as graphene [45,50,58], they also have been used with Au and other metal nanomaterials such as silver [41,46,48,59] and dual metal systems [45,53,55]. The work in this area has been summarized in Tables 9.2 and 9.3.

9.3.2 Bipolar electrodes

Most traditional screen printed and PWEs (Figs. 9.3 and 9.5) are three electrode systems, which require wired contacts to electrochemical workstations or batteries. Bipolar or "floating" electrodes [73] allow for wireless contact and can controlled by small batteries. Bipolar electrodes were first introduced by the Manz group in 2001 [74] for ECL detection after an electrophoretic separation. The Pt working electrode was not connected to any external power source, and the voltage required to carry out the ECL reaction on this floating electrode arose from the electric field in the separation channel. In 2014 Xu's group first reported the use of a BPE on a paper-based ECL device [62]. Fig. 9.6 shows a diagram of a paper-based BPE used to generate ECL [35]. BPE−ECL only requires one conductor with opposite polarity on each of two ends to allow for direct coupling of the electrochemical and ECL reactions. Fig. 9.6 depicts an open BPE−ECL system, in which the BPE is in a single channel immersed in a single solution. A closed BPE−ECL system contains two separate solutions, and each pole of the BPE is in contact with a different solution. Electrical connection between the two half cells is only through the BPE.

The device in Fig. 9.6 was fabricated by Zhang's group using screen printing of both the wax hydrophobic barriers (blue) and the carbon ink electrodes (black). It has the format of an "open" BPE system in which the conductor (thin black rectangle) is in a paper fluidic channel (white area). A voltage is applied to a pair of driving electrodes via a battery, and two coupled electrochemical and ECL reactions occur at each pole of the BPE. In this case, luminol and hydrogen peroxide are oxidized at the anode and a species (O′) is reduced at the cathode. The luminol ECL signal in this particular device was visible to the naked eye and images were captured with a smartphone detector (detection will be discussed in the next section). This device was used to detect hydrogen peroxide and glucose in both buffer and artificial urine. Zhang and coworkers have fabricated similar devices to detect pathogenic bacteria [36] and choline, lactate, and cholesterol [67].

TABLE 9.2 Summary of ECL at paper-based devices 2011–2015.

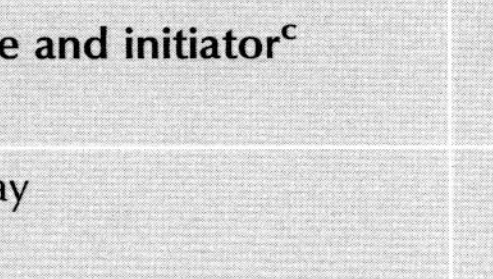

Analyte and assay type	Detection limit and detector[a]	ECL system and amplification[b]	Fabrication method and device type	Electrode and initiator[c]	References
AFP CA125 CA199 CEA Immunoassay	0.15 ng mL^{-1} 0.6 U mL^{-1} 0.17 U mL^{-1} 0.5 ng mL^{-1}	$Ru(bpy)_3^{2+}$ + TPrA	Wax printing/stacked 3D μ-PAD	SPCE array	[19]
CEA *Solutions* *Serum* Immunoassay	 0.001 ng mL^{-1} 0.008 ng mL^{-1}	$Ru(bpy)_3^{2+}$ + TPrA	Wax printing/stacked 3D μ-PAD	SPCE	[20]
AFP CA 153 CA 199 CEA Immunoassay	0.02 ng mL^{-1} 6.0 mU mL^{-1} 5.0 mU mL^{-1} 4.0 pg mL^{-1}	CND, W1 CND, W2 Ru, W2 Ru, W1	Wax printing/stacked 3D μ-PAD	SPCE/MWCNT array	[37]
HCG antigen Immunoassay	0.0019 mIU mL^{-1}	$Ru(bpy)_3^{2+}$ + TPrA	Wax printing/3D origami μ-PAD	SPCE/nanoporous gold	[21]
Carcinoma antigen 125 Immunoassay	0.0074 U mL^{-1}	Luminol + H_2O_2 AuNPs	Wax printing/3D origami μ-PAD	AuNP-modified SPCE	[22]
ATP Aptasensor	0.1 pM	P-acid + TPrA Pt–Ag NPs	Wax printing/3D origami μ-PAD	Porous Au-PWE	[38]

(*Continued*)

TABLE 9.2 (Continued)

Analyte and assay type	Detection limit and detector[a]	ECL system and amplification[b]	Fabrication method and device type	Electrode and initiator[c]	References
Pb^{2+} Hg^{2+} Aptasensor	10 pM 0.2 nM	Si@CNCs Ru@AuNPs	Wax printing/3D origami μ-PAD	SPWPE w/ DNA-labeled NPs	[40]
PSA CEA Immunoassay	1.0 pg mL^{-1} 0.8 pg mL^{-1}	P-acid/TPrA NPS	Wax printing/3D μ-PAD	SPWPE GrO-AuNP array Battery triggered	[44]
Carcinoma antigen 125 Immunoassay	2.5 mU mL^{-1}	CdTe QD @carbon microspheres	Wax printing/3D μ-PAD	SPCE w/ Au–Ag nanocomposite-functionalized graphene	[45]
CEA Immunoassay	0.12 pg mL^{-1}	CdTe QD's + $S_2O_8^{2-}$ NPS	Wax printing/3D origami μ-PAD	Ag-PWE	[46]
CEA Immunoassay	0.6 pg mL^{-1}	GQDs GOx/Au@Pt	Wax printing/3D origami μ-PAD	Au-PWE	[47]
DNA Aptasensor	8.5×10^{-18} M	AgNPs $CaCO_3$/CMC@AgNPs	Wax printing/3D folded μ-PAD	GR/Au-PWE	[50]
Carbohydrate antigen 199 Immunoassay	0.0055 U mL^{-1}	$Ru(bpy)_3^{2+}$ Ru@AuNPs	Crayon/pen on cellulose paper	PoP carbon Battery triggered	[30]
CEA Immunoassay	0.0007 ng mL^{-1}	$Ru(bpy)_3^{2+}$ Ru@Au nanocages	Wax printing/3D origami μ-PAD	Ag-PWE nanospheres	[48]
MCF-7 cancer cells Aptasensor	250 cells mL^{-1}	${}^1(O_2)_2^* + S_2O_8^{2-}$ AuPdNPs	Wax printing/μ-PECLOD	Au-PWE (3D macroporous)	[52]

IgG protein Aptasensor	0.15 fM	Carbon dots RCA	Wax printing/μ-PECLOD	Au-PWE (3D macroporous)	[51]
H_2S from tumor cells	Visualization of release	GrQDs + $S_2O_8^{2-}$ Au@PtNPs	Wax printing/HC-μ-PECLOD	Au@Pt-PWE array	[56]
CA153 antigens at MCF-7 cell surface	0.0014 U mL^{-1}	GQDs@Au nanocages	Wax printing/μ-PECLOD	Au-PWE (nanoflower)	[57]
DBAE NADH	0.9 μM 72 μM Smartphone	$Ru(bpy)_3^{2+}$	Ink-jet printed μ-PAD	SPCE	[14]
DBAE Proline	100 μM 100 μM Smartphone	$Ru(bpy)_3^{2+}$	Paper punch/filter paper	SPCE	[60]
TPrA DNA	5.0 nM Single-base mismatch	$Ru(bpy)_3^{2+}$	Photolithography/solid-state PC	PSSG-Nafion	[26]
TPrA Glucose in blood	0.11 μM NA	$Ru(bpy)_3^{2+}$ Luminol	Photolithography/μ-OECLD	SPCE On-chip battery	[29]
TPrA (direct) Tetracycline	5.0 nm 2.22 nm	$Ru(bpy)_3^{2+}$	Photolithography/solid-state PC	Carbon paste, functionalized graphene	[27]
NA	NA	$Ru(bpy)_3^{2+}$ Nafion@CNDs	Photolithography/solid-state PC	SPCE/CNDs	[28]
Dopamine	NA	CdS QD's + H_2O_2	Laser-printed hybrid device	Carbon tape/ITO	[33]

(*Continued*)

TABLE 9.2 (Continued)

Analyte and assay type	Detection limit and detector[a]	ECL system and amplification[b]	Fabrication method and device type	Electrode and initiator[c]	References
Benzo[a]pyrene	150 nM CCD camera	Ru metallopolymeric film	Wax paper heat transfer/layered	SPCE	[31]
HL-60 cancer cells Aptasensor	56 cells mL^{-1}	RuSi@ AuNPs + DBAE	Paper thin-layer electrochemical cell/ hybrid	ITO/AuNP/Gr	[61]
PSA Immunoassay	1 pg mL^{-1}	$Ru(bpy)_3^{2+}$ + TPrA $MWCNT/H_2O_2$	Wax printing/screen-printed PC	Screen-printed BPE	[62]
Pathogenic DNAs Aptasensor	0.1 fmol L^{-1} 0.5 fmol L^{-1} 0.2 fmol L^{-1}	$Ru(bpy)_3^{2+}$ + TPrA PtNPs	Wax printing/screen-printed PC	Screen-printed BPE array	[63]
TPrA H_2O_2	8.7 μM 46.6 μM CCD camera	$Ru(bpy)_3^{2+}$	Wax screen printing/ μ-PAD	Screen-printed BPE array	[34]
α-Fetoprotein antigen Immunoassay	1.2 pg mL^{-1}	N-GrQDs + DBAE MWCNT	Wax printing/folded μ-PAD	Au-PWE (nanoflowers)	[64]

[a] *If ECL was detected with equipment other than a PMT, the equipment is noted.*
[b] *If a material was used to amplify the ECL signal, it is noted second in the box.*
[c] *If ECL was initiated with equipment other than a conventional electrochemical workstation, the equipment is noted.*

TABLE 9.3 Summary of ECL at paper-based devices 2016–2020.

Analyte and assay type	Detection limit and detector[a]	ECL system and amplification[b]	Fabrication method and device type	Electrode and initiator[c]	References
AFP CEA Immunosensor	NA	Luminol@AuPd CdTe@AuPd H_2O_2 coreactant	Wax printing/3D origami μ-PAD	Au-PWE (nanoflowers)	[47]
Cancer cells Aptasensor	300 cells mL^{-1}	PtNi@CNDs + $S_2O_8^{2-}$ Nanoporous PtNi alloy	Wax printing/3D origami μ-PAD	Au@Pd-PWE (3D macroporous)	[53]
Mucin-1 peptide Aptasensor	8.33 pM	$Ru(phen)_3^{2+}$ + TPrA HCR		AuNP-PWE	[39]
Cancer cells MCF-7 CCRF-CEM HeLa K562 Aptasensor	 38 cells mL^{-1} 53 cells mL^{-1} 67 cells mL^{-1} 42 cells mL^{-1}	Graphene QDs (GQDs) + $S_2O_8^{2-}$ AgNPs	Wax printing/3D origami μ-PAD (multiplexed)	AuNP-PWE	[54]
Pb^{2+} Aptasensor	0.016 nM	Luminol + H_2O_2 CeO_2NPs	Wax printing/3D μ-PAD (multiplexed)	AgNP-PWE	[41]
"Power Paper"	NA	$Ru(bpy)_3^{2+}$ + TPrA	Cellulose/wax printing μ-PAD	Porous gold on-chip battery	[65]
Pb^{2+}	0.14 nM	Luminol + H_2O_2	Wax printing/3D origami μ-PAD	AuNP-PWE	[42]

(Continued)

TABLE 9.3 (Continued)

Analyte and assay type	Detection limit and detector[a]	ECL system and amplification[b]	Fabrication method and device type	Electrode and initiator[c]	References
H_2O_2 released from MCF-7 tumor cells Aptameric	40 cells mL^{-1}	Luminol + H_2O_2	Wax printing/3D origami μ-PAD	Screen-printed BPE	[66]
Ni^{2+} Hg^{2+} Aptasensor	3.1 pM 3.8 pM	ABEI + H_2O_2 Cu_2O-Au and Ag NPs	Wax printing/ multichannel μ-PAD	AgNP-PWE	[43]
H_2O_2 released from MCF-7 tumor cells Aptasensor	30 cells mL^{-1}	$Ru(bpy)_3{}^{2+}$ + TPA AuPdNPs	Wax printing/3D origami μ-PAD	AuPdNP-PWE	[55]
Pb^{2+} Aptasensor	0.003 nM	CdS QDs and luminol + H_2O_2 Au/HRP nanocubes	Wax printing/3D origami μ-PAD	SPCE w/ dendritic Au and Gr	[58]
miRNA-107 Aptasensor	0.04 pM	Black phosphorus nanosheets + $S_2O_8{}^{2-}$ Target recycling signal	Wax printing/PC	Ag-modified SPCE	[59]
H_2O_2 Glucose *in PBS* Glucose *in art. urin*	1.75 μM 0.017 mM 0.030 mM	Luminol + H_2O_2 Smartphone	Wax screen printing/ μ-PAD	Screen-printed BPE Battery power	[35]
Listeria monocytogenes Aptasensor	10 copies μL^{-1}	$Ru(phen)_2dpp]^{2+}$	Wax-screen printing/ μ-PAD	Screen-printed BPE	[36]

H_2O_2 Choline Lactate Cholesterol	0.424 μM 0.573 μM 3.132 μM 7.418 μM	Luminol	Wax-screen printing PC	Screen-printed BPEs	[67]
Alkaloid drugs	1.799–11.43 nM	$Ru(bpy)_3^{2+}$	SNC with hydrophobic paper cover μ-chip	ITO	[68]
H_2O_2	0.5 μM	$Ru(bpy)_3^{2+} + C_2O_4^{2-}$	PDMS-graphite paper BPE/PC	Pt NPs@Cathode	[69]
CEA Immunoassay	5 pg mL^{-1}			Au@Pt@cathode	
Hepatitis B antigen Immunoassay	34.2 pg mL^{-1}	$Ru(bpy)_3^{2+}$-NHS ester + *N*-butyldiethanolamine	Cut paper disks PC	SPCE + magnetic beads	[70]
CEA PSA	0.07 ng mL^{-1} 0.03 ng mL^{-1}	$Ru(bpy)_3^{2+}$ + TPrA	Paper disks with punched holes/3D rotational	SPCE + MWCNT	[71]
TPrA Licocaine	2.8 nM 7.8 nM	$Ru(bpy)_3^{2+}$	Hybrid with PWE	3D graphene porous paper electrode	[72]

[a]If ECL was detected with equipment other than a PMT, the equipment is noted.
[b]If a material was used to amplify the ECL signal, it is noted second in the box.
[c]If ECL was initiated with equipment other than a conventional electrochemical workstation, the equipment is noted.

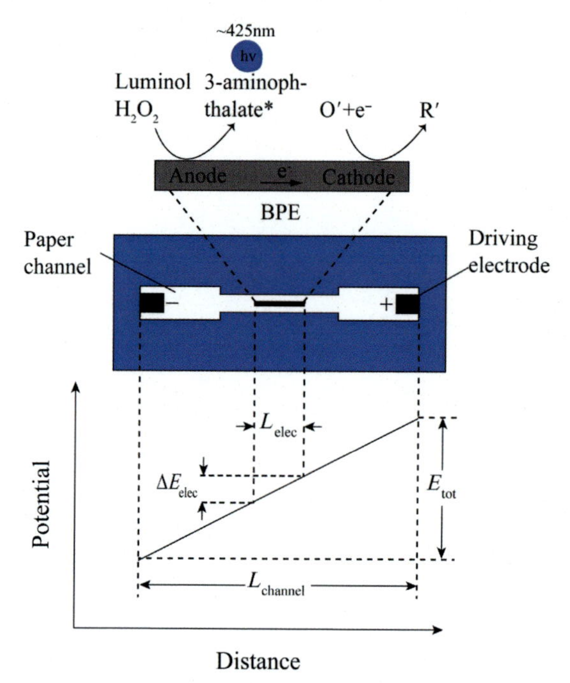

FIGURE 9.6 Schematic of the design and principle of an open bipolar electrode developed in Ref. [35] for biochemical analysis. Blue = wax, black = carbon ink, and white = cellulose paper. L = length and E = potential. *Reproduced with permission from L. Chen, C. Zhang, D. Xing, Paper-based bipolar electrode-electrochemiluminescence (BPE-ECL) device with battery energy supply and smartphone read-out: a handheld ECL system for biochemical analysis at the point-of-care level, Sens. Actuators B: Chem. 237 (2016) 308–317. Copyright (2016) Elsevier.*

Much of their work in the area of BPE–ECL biosensors has been to develop low-cost systems controlled with inexpensive batteries and using inexpensive detectors such as smartphones [35] and charge-coupled device (CCD) cameras [67]. Using these imaging methods, detection can be performed on array devices [34].

Xu's group has used wax printing on cellulose paper to fabricate devices that utilize carbon screen-printed BPEs. They fabricated both immunosensors and aptamer-based sensors (see the next section for more details on immunosensors and aptasensors) to detect a PSA cancer biomarker [63] and pathogenic DNA [62]. The pathogenic DNA device also utilized an array of BPE's. Ding's group has also worked to develop an inexpensive and easy to implement closed BPE for ECL detection. As an alternative to SPCE's, they used graphite paper (foil) to construct a hybrid device that also used poly (dimethylsiloxane) (PDMS) and a paper cover.

To summarize, conventional BPE–ECL systems generally use more sophisticated and expensive materials and instrumentation. The work by many groups on paper devices seeks to develop inexpensive, easy to use, and portable devices that are still sensitive and accurate, with the goal of developing POC diagnostics.

9.3.3 Hybrid devices

The previous example of a graphite paper electrode demonstrated that paper can be used in combination with other materials to construct "hybrid" devices. Xu's group coated graphene and AuNPs onto an ITO glass electrode then combined with paper to form a thin-layer electrochemical cell [61]. They attached capture DNA to the electrode to make an aptasensor which was used to detect HL-60 cancer cells.

9.4 Electrochemiluminescence sensing strategies

Most paper-based ECL devices have been used to detect clinical analytes such as cancer biomarkers and cells, with the goal of developing POC devices. In the development of novel paper-based devices, some studies have used the coreactant reactions shown in Fig. 9.1 to detect a coreactant such as TPrA or H_2O_2, with $Ru(bpy)_3^{2+}$ and luminol as luminophores, respectively. Most of these studies use these as model reactions to validate ECL detection on a novel device. The first report of ECL detection on a microfluidic device used the ECL reaction between $Ru(bpy)_3^{2+}$ and 2-(dibutylamino)-ethanol (DBAE) to demonstrate a mobile phone detector [14]. They also used nicotinamide adenine dinucleotide (NADH) as a coreactant and detected it with a limit of detection of 72-μM. In 2015 Zhang's group used these reactions to demonstrate the utility of imaging ECL at a BPE using an inexpensive CCD camera [34].

Most recent studies have been focused on detecting clinically relevant analytes such as tumor markers and cancer cells. Although coreactant ECL has some inherent specificity, it lacks the specificity it would have if coupled to a separation method (e.g., capillary electrophoresis-ECL or high-performance liquid chromatography-ECL). Therefore most recent ECL applications involve either ECL-based immunoassays or aptamer-based ECL assays. Developments in this area include integration of nanomaterials into the detection methods in three ways: (1) modification of the working electrode for immobilization support and increased electron transfer, (2) modification of the detection antibody or probe, and (3) use as an ECL luminophore. Each of these steps serves to increase the sensitivity. Much work has gone into investigating materials for points (1) and (2) to enhance the ECL signal and increase sensitivity.

9.4.1 Electrochemiluminescence immunoassays

As mentioned earlier, ECL-based immunoassays were first developed and commercialized in the 1990s but many are still in development and reported in the literature [75]. Conducting immunoassays on paper has a variety of advantages. Paper is easy to modify, inexpensive, and easily disposable. For example, many devices can be burned as a disposal method. Fig. 9.7 shows

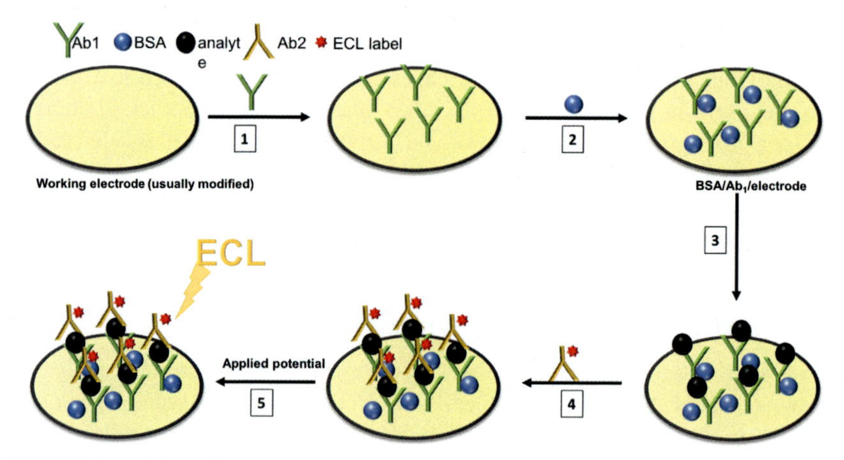

FIGURE 9.7 Schematic representation of the general procedure for the fabrication of an ECL immunosensor and corresponding assay. *Credit: authors.*

the general procedure for the fabrication of an ECL immunosensor and performing the assay. The typical steps in an ECL immunoassay are:

1. Capture antibodies (Ab1) are coated onto the working electrode.
2. The electrode is washed with buffer and bovine serum albumin to block active sites against nonspecific adsorption. After washing and drying, the device can be stored for later use.
3. The biosensor is incubated with sample solution for a specified time then washed.
4. The biosensor is incubated with the ECL labeled-antibody ($* - Ab_2$) and washed with buffer.
5. ECL signal of the immunosensor is triggered under the applied potential in the presence of coreactant (not shown).

Immunoassays have been developed on many of the electrodes discussed in this chapter, particularly those listed in Table 9.2. For example, Yu's group developed an Au-nanoflower-PWE origami immunodevice for the detection of alphafetoprotein (AFP) and human carcinoembryonic (CEA) antigens [47]. Luminol and CdTe QDs served as the luminophores, for AFP and CEA, respectively. Selectivity was achieved via the different ECL potential of each luminophore. The ECL signal was amplified via deposition of the luminophores on bimetallic AuPd nanoparticles. The details of additional immunosensors are summarized in Tables 9.2 and 9.3.

Bipolar electrodes have also been fabricated as immunosensors. Xu's group reported on a dual-channel closed BPE to measure the cancer marker PSA [62]. The device contained two cells that were patterned with wax and electrically connected through a screen-printed carbon ink electrode. The driving electrodes, a carbon ink anode and a Ag, AgCl cathode, were also screen printed. The ECL reaction between $Ru(bpy)_3^{2+}$ and TPrA occurred at the anode and served as the "reporting cell," while the reaction in the cathodic "sensing cell" was the reduction of hydrogen peroxide to water. The capture antibody (Ab1) was immobilized on the cathode in the sensing cell for a sandwich immunoassay. The device was incubated in sample containing PSA, then with Ab2—glucose oxidase (GOx) on silica nanoparticles forming SiO_2-GOx-Ab_2/PSA/Ab_1/BPE. The GOx catalyzed the production of H_2O_2 in the presence of glucose. The more PSA that was captured, the more H_2O_2 was reduced at the sensing cell, causing an increase in ECL intensity at the reporting cell, as $Ru(bpy)_3^{2+}$ and TPrA were oxidized at the reporting cell. They showed that the ECL signal could be enhanced by modification of the cathode with multiwalled carbon nanotubes (MWCNTs). They served to increase Ab1 loading at the cathode and facilitated electron transfer. Highlights of additional paper-based ECL immunodevices can be found in Table 9.2.

9.4.2 Aptamer-based electrochemiluminescence assays

Most of the paper-based ECL biosensors fabricated more recently (2015 and later) have been aptamer based. Aptamers are short chains of nucleic acids that can be custom synthesized and modified. Aptamer-based sensors offer specificity, just like immunosensors but additionally have more versatility because of the fact that the chains can be customized for a wider range of analytes, such as small molecules, other nucleic acid chains and cells.

Just as with immunoassays, ECL-based DNA-probe assays were commercialized in the 1990s. However, research into aptamer-based ECL assays continues. In fact, as shown in Table 9.3, most of the paper-based ECL devices reported in the past 5 years use aptamers in their detection schemes. Aptamer-based ECL assays are more versatile than ECL immunoassays, thus have a variety of designs. Fig. 9.8 shows the general procedure for the fabrication of an ECL aptasensor and corresponding assay. Keeping in mind that specific designs can differ while the common theme is that a change in hybridization changes the signal, sample steps are summarized below:

1. Capture DNA aptamer is immobilized on the Au electrode via a thiol linkage.
2. The electrode is washed with buffer, and mercaptohexanol (MCH) is added to block active sites against nonspecific adsorption. The device is washed and can be stored at 4°C.
3. The aptamer-modified electrode is placed in a solution containing another aptamer specific for the analyte. For example, cells are the analyte in Fig. 9.8 and the aptamer for the cells, which is labeled with the ECL luminophore, is hybridized with the capture DNA. In the presence of coreactant and an applied potential, the ECL signal is strong.

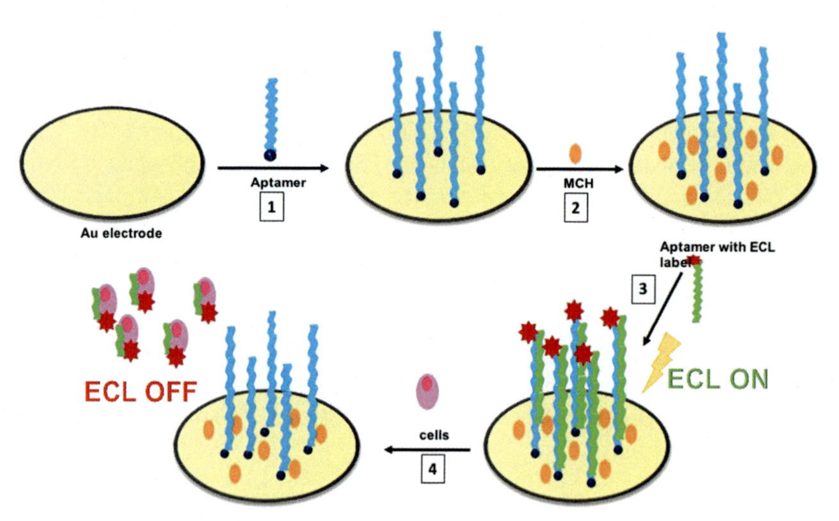

FIGURE 9.8 Schematic illustration of paper-based aptasensor for ECL detection of HL-60 cells. *Adapted from Q.-M. Feng, Z. Liu, H.-Y. Chen, J.-J. Xu, Paper-based electrochemiluminescence biosensor for cancer cell detection, Electrochem. Commun. 49 (2014) 88–92. Fig. 9.1, Copyright 2014 Elsevier.*

4. The electrode is rinsed and in the presence of the analyte (cells), the aptamer attaches to the cell surface and as luminophore-aptamer is released from the electrode surface the ECL signal decreases.

Note that this is a "signal off" aptasensor, where the ECL signal decreases proportionately with an increase in analyte.

A hybrid paper-ITO device [61] used a similar strategy to detect HL-60 cancer cells with a limit of detection of 56 cells mL^{-1}. The ITO electrode was coated with graphene and AuNPs, which functioned to immobilize the capture DNA and accelerated the rate of the electron transfer. Silica nanoparticles loaded with $Ru(bpy)_3{}^{2+}$ served as the ECL luminophore and DBAE as the coreactant.

Aptasensors for tumor cells [66] and pathogenic bacteria [36] have also recently been developed. In the former device, MCF-tumor cells were detected by their release of H_2O_2. The microfluidic device was a closed BPE fabricated via wax printing and screen printing of carbon ink on cellulose paper. The BPE was modified with AuPd NPs which served as (1) a carrier of the capture aptamer and (2) the catalyst for the ECL reaction of luminol and cellular H_2O_2. Luminol and AuNPs were attached to the surface of the captured cells via a hybridization chain reaction between the labeled aptamer and the capture aptamer, as shown in step 3 of Fig. 9.8. Target cells were stimulated with phorbol myristate acetate and released H_2O_2 (the ECL coreactant). The ECL response of the luminol–H_2O_2 system correlated with the number of cancer cells in the testing buffer, which were detected with a limit of detection of 40 cells mL^{-1}. A summary of reported aptasensors can be found in Tables 9.2 and 9.3. Devices such as these have potential value for studying cellular biology and pathophysiology.

9.4.3 Signal amplification

Signal, particularly in biosensors, is usually amplified. Work in nanomaterials [18,76,77] has especially promoted investigations into using the materials in signal amplification. The specific examples for immunosensors and aptasensors have described how nanomaterials have been used to amplify ECL biosensors. Signal amplification strategies are included in Tables 9.2 and 9.3. To summarize, the biosensors described previously have three areas where sensitivity can be enhanced

1. Increasing the surface area of the electrode with deposition of a nanomaterial area to increase the amount of antibody or aptamer immobilized onto the electrode surface.
2. The nanomaterial also increases the electron transfer rate and more species are oxidized or reduced.
3. Direct amplification of the ECL signal by addition of nanomaterial to the ECL probe.

9.5 Potential control and detection: work toward portable devices

9.5.1 Potential control

Most of the method development work has used a benchtop ECL analyzer or conventional electrochemical workstation. Cyclic voltammetry and electrochemical impedance spectroscopy techniques have been used to characterize electrodes on paper-based devices and to determine the optimal applied potential to operate a device. One benefit of aqueous ECL techniques is that only a low DC potential is required to drive an ECL reaction. To develop portable and inexpensive POC devices, small portable batteries and on-chip power sources have been developed.

9.5.2 Detection

Most reports of paper-based ECL devices use photomultiplier tubes (PMTs) as the detector. PMTs are highly sensitive but they are not robust for many field and POC applications. Another advantage of ECL detection is its visual nature. CCD cameras and smartphones can be coupled with paper ECL devices and used as inexpensive, portable, and robust detectors. These are particularly helpful for imaging or array/multiplexed detection. Tables 2 and 3 have included when a detector other than a PMT was used.

Wang's group used screen printing to fabricate a nontoxic battery on a microfluidic origami ECL device [29]. The battery was used to supply power to a red LED and also ECL reactions involving $Ru(bpy)_3^{2+}$/TPrA and luminol/H_2O_2. Zhang and Yu's groups also developed a microfluidic device that fabricated a battery on the paper that they named "power paper." They fabricated a porous gold working electrode along with wires on the cellulose paper device. The power paper was flexible and the battery contained a magnesium anode and a prussian blue cathode. It was used to drive the electropolymerization of a polypyrrole film and a polyaniline network. It also powered electrochemiluminescence of which they acquired images. These "on-chip" power supplies have the potential for developing POC paper-based ECL devices with nonhazardous "green" materials. Zhang and Xing developed a portable paper-based BPE–ECL device that was run with a rechargeable lithium battery and used a smartphone detector [35]. The entire system was encased in a small plastic hand-held case. The battery drove the luminol–H_2O_2 ECL reaction and the device was applied to the detection of glucose in artificial urine samples. This device also has the capacity of high throughput. These are the most recent examples of steps taken to developing portable devices. Other studies using batteries [37] and CCD cameras [34,68,69] are listed in Tables 9.2 and 9.3.

9.6 Integrated devices

The small dimensions of microfluidic and paper-chip devices can speed up analysis time and minimize sample and reagent consumption. These devices also use less power and can be automated, which can increase reliability, functionality, and sensitivity. Additionally, they have multiplexing capabilities and offer portability. All of these factors improve analytical performance and can lead to POC applications. One of the goals of a lab-on-paper application is to integrate multiple lab functions on a single device. For example, we have discussed two devices in which the battery to run the ECL reaction was fabricated right on the paper device [29,35]. This section highlights additional recent advances in this area, while many others are noted in Tables 9.2 and 9.3.

Integrated devices combine multiple functions (e.g., reagent storage and reaction, residual washing, liquid flowing, and dual-mode signal output) on a single paper device. Fluid flow on paper is typically achieved via capillary action. However, sometimes more control or additional washing or reagent introduction steps are needed, particularly in immunoassays and aptamer-based assays. Incorporating washing steps into paper devices can improve both efficiency and sensitivity. For example, Li and Chen's groups developed a novel rotational paper-based analytical device and were able to run multistep ECL immunoassays [71]. The device contained integrated paper-based stacked rotational valves that were easily controlled by rotating cut paper disks manually. These cut disks had holes punched in them and they could be rotated so the holes lined up or not. The response time of the valves could be as short as several seconds. This device was user friendly for carrying out multistep assays. They also noted that these rotational valves were reusable. They applied these devices to multiplexed detection of CEA and PSA.

Yu's group also developed a microfluidic device that integrated autocleaning steps with an aptamer-based assay [43]. They named the device a 3D collapsible autocleaning paper-based ECL biosensor and noted that its design was inspired by "pop-up" greeting cards. The device was folded and collapsible, with two functions: autocleaning and signal collection. Multiple fluidic paths and a hollow-channel structure allowed for repetitive autocleaning of the two working electrodes. This device was applied to the detection of Ni^{2+} and Hg^{2+} with detection limits of 3.1 nM and 3.8 pM, respectively.

Various groups have reported on devices that contain electrode arrays for imaging or the detection of multiple analytes [19,34,37,63]; however, these use a single detection method. Yu's group developed a device that integrated two different detection methods, ECL and colorimetric, for the detection of lead [42]. The aptamer-based assay used the recognition between Pb^{2+} and specific DNAzyme. They designed a cross-shaped origami-based paper-microfluidic device with a AuNP-based detection electrode modified with an aptamer containing the DNAzyme (note: DNAzymes break up complementary strands). The DNAzyme hybridized with reduced graphene oxide−PdAu−GOx labeled oligonucleotide. In the presence of lead(II) ion, Pb^{2+}-dependent DNAzyme was activated and catalyzed the ribonucleotide site and cleaved the substrate strand. The cleaved strand traversed the channel to a visualization bar loaded with H_2O_2 and 3,3′,5,5′-tetramethylbenzidine (TMB). Here, a color change allowed estimation of the Pb^{2+} content. Simultaneously, the uncleaved probe participated in an ECL reaction. The more cleavage, the deeper the color of the TMB oxidation product and the smaller the ECL signal. The colorimetric test was semiquantitative while the ECL measurement was quantitative.

An integrated function aptasensor for microRNA recently was developed by Zhang and Ge's groups [59] and implemented a novel ECL emitter, black phosphorus nanosheets. The ECL power source was a novel paper-based perovskite solar cell device that was designed to act as the power source and external recharging was not required. The paper-chip device was fabricated using wax printing on cellulose and the working electrode was screen-printed carbon modified with AuNPs. They achieved highly sensitive detection of miRNA-107 with a detection limit of 0.04 pM.

9.7 Conclusions and future directions

It has been around 10 years since the first paper-based ECL device was reported. Since then, paper-based ECL devices have been shown to have much promise for developing POC devices. Paper is an excellent substrate for biosensing because of its biocompatibility and passive liquid transport. POC devices also require low power consumption, integration of multiple functions on a single device, ease of use, and a robust detector. ECL-based biosensors offer all of these. Devices have been designed with rechargeable power sources on the chip and have been coupled with inexpensive CCD camera or smartphone detectors. Most of the bioassays performed on paper devices require rinsing steps and the adding of reagents. Recent work has shown that these steps can be integrated onto a device and even automated. We expect to see additional work in this area in the future. To develop POC diagnostics, the devices must be tested using real-life samples, such as saliva or blood. To date, many of the assays reported here have been performed on serum and saliva samples and we expect further work on applying methods to real samples and comparison against reference methods.

During the past 10 + years, a great deal of research has been performed looking into novel ECL emitters and ECL enhancing materials. Nanomaterials are now integral to increasing the sensitivity of ECL-based devices. They are versatile and can function as ECL labels, porous electrode materials, and can be used to modify an electrode to obtain higher coverage of probe molecules. There are a wide range of inorganic and organic nanomaterials, and work will continue to study their roles in ECL bioassays. Overall, we expect to see further work in developing paper-based ECL biosensors using nanomaterials with multiple functions integrated on a single device.

Acknowledgments

The authors would like to acknowledge grants from the National Institute for General Medical Science (NIGMS) (2P20GM103427), a component of the National Institutes of Health (NIH), and its contents are the sole responsibility of the authors and do not necessarily represent the official views of NIGMS or NIH.

References

[1] M. Richter, Ectrochemiluminescence (ECL), Chem. Rev. 104 (2004) 3003–3036.

[2] J.K. Leland, M.J. Powell, Electrogenerated chemiluminescence: an oxidative-reduction type ECL reaction sequence using tripropylamine, J. Electrochem. Soc. 137 (1990) 3127–3131.

[3] H.S. White, A.J. Bard, Electrogenerated chemiluminescence. 41. Electrogenerated chemiluminescence and chemiluminescence of the $Ru(2,2'\text{-}bpy)_3^{2+}$-$S_2O_8^{2-}$ system in acetonitrile-water solutions, J. Am. Chem. Soc. 104 (1981) 6891–6985.

[4] S. Deng, H. Ju, Electrogenerated chemiluminescence of nanomaterials for bioanalysis, Analyst 138 (2013) 43–61.

[5] G. Blackburn, H. Shah, J. Kenten, J. Leland, R. Kamin, J. Link, et al., Electrochemiluminescence detection for development of immunoassays and DNA probe assays for clinical diagnostics, Clin. Chem. 37 (1991) 1534–1539.

[6] A. Bouchie, Roche and Igen in shotgun wedding, Nat. Biotechnol. 21 (2003) 958.

[7] A. Martinez, S. Phillips, M. Butte, G. Whitesides, Patterned paper as a platform for inexpensive, low-volume, portable bioassays, Angew. Chem. Int. Ed. Engl. 46 (2007) 1318–1320.

[8] A. Martinez, S. Phillips, E. Carrilho, S. Thomas, H. Sindi, G. Whitesides, Simple telemedicine for developing regions: camera phones and paper-based microfluidic devices for real-time, off-site diagnosis, Anal. Chem. 80 (2008) 3699–3707.
[9] S. Klasner, A. Price, K. Hoeman, R. Wilson, K. Bell, C. Culbertson, Paper-based microfluidic devices for analysis of clinically relevant analytes present in urine and saliva, Anal. Bioanal. Chem. 397 (2010) 1821–1829.
[10] W. Dungchai, O. Chailapakul, C. Henry, Use of multiple colorimetric indicators for paper-based microfluidic devices, Anal. Chim. Acta 674 (2010) 227–233.
[11] Y. Lu, W. Shi, L. Jiang, J. Qin, B. Lin, Rapid prototyping of paper-based microfluidics with wax for low-cost, portable bioassay, Electrophoresis 30 (2009) 1497–1500.
[12] E. Carrilho, A. Martinez, G. Whitesides, Understanding wax printing: a simple micropatterning process for paper-based microfluidics, Anal. Chem. 81 (2009) 7091–7095.
[13] W. Dungchai, O. Chailapakul, C. Henry, Electrochemical detection for paper-based microfluidics, Anal. Chem. 81 (2009) 5821–5826.
[14] J. Delaney, C. Hogan, J. Tian, W. Shen, Electrogenerated chemiluminescence detection in paper-based microfluidic sensors, Anal. Chem. 83 (2011) 1300–1306.
[15] E. Noviana, C.P. McCord, K.M. Clark, I. Jang, C.S. Henry, Electrochemical paper-based devices: sensing approaches and progress toward practical applications, Lab. Chip 20 (2020) 9–34.
[16] L. Ge, J. Yu, S. Ge, M. Yan, Lab-on-paper-based devices using chemiluminescence and electrogenerated chemiluminescence detection, Anal. Bioanal. Chem. 406 (2014) 5613–5630.
[17] E.M. Gross, H.E. Durant, K.N. Hipp, R.Y. Lai, Electrochemiluminescence detection in paper-based and other inexpensive microfluidic devices, ChemElectroChem 4 (2017) 1594–1603.
[18] S. Chinnadayyala, J. Park, H. Le, M. Santhosh, A. Kadam, S. Cho, Recent advances in microfluidic paper-based electrochemiluminescence analytical devices for point-of-care testing applications, Biosens. Bioelectron. 126 (2019) 68–81.
[19] L. Ge, J. Yan, X. Song, M. Yan, S. Ge, J. Yu, Three-dimensional paper-based electrochemiluminescence immunodevice for multiplexed measurement of biomarkers and point-of-care testing, Biomaterials 33 (2012) 1024–1031.
[20] J. Yan, L. Ge, X. Song, M. Yan, S. Ge, J. Yu, Paper-based electrochemiluminescent 3D immunodevice for lab-on-paper, specific, and sensitive point-of-care testing, Chem. Eur. J. 2012 (18) (2012) 4938–4945.
[21] S. Wang, W. Dai, L. Ge, M. Yan, J. Yu, X. Song, et al., Rechargeable battery-triggered electrochemiluminescence detection on microfluidic origami immunodevice based on two electrodes, Chem. Commun. 48 (2012) 9971–9973.
[22] S. Wang, L. Ge, M. Yan, J. Yu, X. Song, S. Ge, et al., 3D microfluidic origami electrochemiluminescence immunodevice for sensitive point-of-care testing of carcinoma antigen 125, Sens. Actuators B 176 (2013) 1–8.
[23] D. Duffy, J. McDonald, O. Schueller, G. Whitesides, Rapid prototyping of microfluidic systems in poly(dimethylsiloxane), Anal. Chem. 70 (1998) 4974–4984.
[24] A. Martinez, S.T. Phillips, B.J. Wiley, M. Gupta, G.M. Whitesides, FLASH: a rapid method for prototyping paper-based microfluidic devices, Lab. Chip 8 (2008) 2146–2150.
[25] Z. Nie, C. Nijhuis, J. Gong, X. Chen, A. Kumachev, A. Martinez, et al., Electrochemical sensing in paper-based microfluidic devices, Lab. Chip 10 (2010) 477–483.
[26] Y. Xu, B. Lou, Z. Lv, Z. Zhou, L. Zhang, E. Wang, Paper-based solid-state electrochemiluminescence sensor using poly(sodium 4-styrenesulfonate) functionalized graphene/nafion composite film, Anal. Chim. Acta 763 (2013) 20–27.

[27] W. Gu, Y. Xu, B. Lou, Z. Lyu, E. Wang, One-step process for fabricating paper-based solid-state electrochemiluminescence sensor based on functionalized graphene, Electrochem. Commun. 38 (2014) 57–60.

[28] Y. Xu, J. Liu, J. Zhang, X. Zong, X. Jia, D. Li, et al., Chip-based generation of carbon nanodots via electrochemical oxidation of screen printed carbon electrodes and the applications for efficient cell imaging and electrochemiluminescence enhancement, Nanoscale 7 (2015) 9421–9426.

[29] X. Zhang, J. Li, C. Chen, B. Lou, L. Zhang, E. Wang, A self-powered microfluidic origami electrochemiluminescence biosensing platform, Chem. Commun. 49 (2013) 3866–3868.

[30] H. Yang, Q. Kong, S. Wang, J. Xu, Z. Bian, X. Zheng, et al., Hand-drawn&written pen-on-paper electrochemiluminescence immunodevice powered by rechargeable battery for low-cost point-of-care testing, Biosens. Bioelectron. 61 (2014) 21–27.

[31] V. Mani, K. Kadimisetty, S. Malla, A.A. Joshi, J.F. Rusling, Paper-based electrochemiluminescent screening for genotoxic activity in the environment, Environ. Sci. Technol. 47 (2013) 1937–1944.

[32] W.K.T. Coltro, D.P. de Jesus, J.A.F. da Silva, C.L. do Lago, E. Carrilho, Toner and paper-based fabrication techniques for microfluidic applications, Electrophoresis 31 (2010) 2487–2498.

[33] C.-G. Shi, X. Shan, Z.-Q. Pan, J.-J. Xu, C. Lu, N. Bao, et al., Quantum dot (QD)-modified carbon tape electrodes for reproducible electrochemiluminescence (ECL) emission on a paper-based platform, Anal. Chem. 84 (2012) 3033–3038.

[34] R. Liu, C. Zhang, M. Liu, Open bipolar electrode-electrochemiluminescence imaging sensing using paper-based microfluidics, Sens. Actuators B: Chem. (2015) 255–262.

[35] L. Chen, C. Zhang, D. Xing, Paper-based bipolar electrode-electrochemiluminescence (BPE-ECL) device with battery energy supply and smartphone read-out: a handheld ECL system for biochemical analysis at the point-of-care level, Sens. Actuators B: Chem. 237 (2016) 308–317.

[36] H. Liu, X. Zhou, W. Liu, X. Yang, D. Xing, Paper-based bipolar electrode electrochemiluminescence switch for label-free and sensitive genetic detection of pathogenic bacteria, Anal. Chem. 88 (2016) 10191–10197.

[37] S. Wang, L. Ge, Y. Zhang, X. Song, N. Li, S. Ge, et al., Battery-triggered microfluidic paper-based multiplex electrochemiluminescence immunodevice based on potential-resolution strategy, Lab. Chip 12 (2012) 4489–4498.

[38] J. Yan, M. Yan, L. Ge, J. Yu, S. Ge, J. Huang, A microfluidic origami electrochemiluminescence aptamer-device based on a porous Au-paper electrode and a phenyleneethynylene derivative, Chem. Commun. 49 (2013) 1383–1385.

[39] C. Ma, H. Liu, L. Zhang, L. Li, M. Yan, J. Yu, et al., Microfluidic paper-based analytical device for sensitive detection of peptides based on specific recognition of aptamer and amplification strategy of hybridization chain reaction, ChemElectroChem 4 (2017) 1744–1749.

[40] M. Zhang, L. Ge, S. Ge, M. Yan, J. Yu, J. Huang, et al., Three-dimensional paper-based electrochemiluminescence device for simultaneous detection of Pb^{2+} and Hg^{2+} based on potential-control technique, Biosens. Bioelectron. 41 (2013) 544–550.

[41] Y. Huang, L. Li, Y. Zhang, L. Zhang, S. Ge, H. Li, et al., Cerium dioxide-mediated signal "on–off" by resonance energy transfer on a lab-on-paper device for ultrasensitive detection of lead ions, ACS Appl. Mater. Interfaces 9 (38) (2017) 32591–32598.

[42] J. Xu, Y. Zhang, L. Li, Q. Kong, L. Zhang, S. Ge, et al., Colorimetric and electrochemiluminescence dual-mode sensing of lead ion based on integrated lab-on-paper device, ACS Appl. Mater. Interfaces 10 (4) (2018) 3431–3440.

[43] Y. Huang, L. Li, Y. Zhang, L. Zhang, S. Ge, J. Yu, Auto-cleaning paper-based electrochemiluminescence biosensor coupled with binary catalysis of cubic Cu_2O-Au and polyethyleneimine for quantification of Ni^{2+} and Hg^{2+}, Biosens. Bioelectron. 126 (2019) 339–345.

[44] W. Li, M. Li, S. Ge, M. Yan, J. Huang, J. Yu, Battery-triggered ultrasensitive electrochemiluminescence detection on microfluidic paper-based immunodevice based on dual-signal amplification strategy, Anal. Chim. Acta 67 (2013) 66–74.

[45] Y. Zhang, L. Li, H. Yang, Y.-N. Ding, M. Su, J. Zhu, et al., Gold-silver nanocomposite-functionalized graphene sensing platform for an electrochemiluminescent immunoassay of a tumor marker, RSC Adv. 3 (2013) 14701–14709.

[46] W. Li, L. Li, S. Li, X. Wang, M. Li, S. Wang, et al., 3D origami electrochemiluminescence immunodevice based on porous silver-paper electrode and nanoporous silver double-assisted signal amplification, Sens. Actuators B: Chem. 188 (2013) 417–424.

[47] L. Li, W. Li, C. Ma, H. Yang, S. Ge, J. Yu, Paper-based electrochemiluminescence immunodevice for carcinoembryonic antigen using nanoporous gold-chitosan hybrids and graphene quantum dots functionalized Au@Pt, Sens. Actuators B: Chem. 202 (2014) 314–322.

[48] C. Gao, M. Su, Y. Wang, S. Ge, J. Yu, A disposable paper-based electrochemiluminescence device for ultrasensitive monitoring of CEA based on $Ru(bpy)_3^{2+}$@Au nanocages, RSC Adv. 5 (2015) 28324–28331.

[49] M. Su, H. Liu, S. Ge, N. Ren, L. Ding, J. Yu, et al., An electrochemiluminescence lab-on-paper device for sensitive detection of two antigens at the MCF-7 cell surface based on porous bimetallic AuPd nanoparticles, RSC Adv. 6 (2016) 16500–16506.

[50] M. Li, Y. Wang, Y. Zhang, J. Yu, S. Ge, M. Yan, Graphene functionalized porous Au-paper based electrochemiluminescence device for detection of DNA using luminescent silver nanoparticles coated calcium carbonate/carboxymethyl chitosan hybrid microspheres as labels, Biosens. Bioelectron. 59 (2014) 307–313.

[51] L. Wu, C. Ma, X. Zheng, H. Liu, J. Yu, Paper-based electrochemiluminescence origami device for protein detection using assembled cascade DNA–carbon dots nanotags based on rolling circle amplification, Biosens. Bioelectron. 68 (2015) 413–420.

[52] L. Wu, C. Ma, L. Ge, Q. Kong, M. Yan, S. Ge, et al., Paper-based electrochemiluminescence origami cyto-device for multiple cancer cells detection using porous AuPd alloy as catalytically promoted nanolabels, Biosens. Bioelectron. 63 (2015) 450–457.

[53] L. Wu, Y. Zhang, Y. Wang, S. Ge, H. Liu, M. Yan, et al., A paper-based electrochemiluminescence electrode as an aptamer-based cytosensor using PtNi@carbon dots as nanolabels for detection of cancer cells and for in-situ screening of anticancer drugs, Microchim. Acta 183 (2016) 1873–1880.

[54] H. Yang, Y. Zhang, L. Li, L. Zhang, F. Lan, J. Yu, Sudoku-like lab-on-paper cyto-device with dual enhancement of electrochemiluminescence intermediates strategy, Anal. Chem. 89 (14) (2017) 7511–7519.

[55] Y. Jian, H. Wang, X. Sun, L. Zhang, K. Cui, S. Ge, et al., Electrochemiluminescence cytosensing platform based on $Ru(bpy)_3^{2+}$@silica-Au nanocomposite as luminophore and AuPd nanoparticles as coreaction accelerator for in situ evaluation of intracellular H_2O_2, Talanta 199 (2019) 485–490.

[56] L. Li, Y. Zhang, F. Liu, M. Su, L. Liang, S. Ge, et al., Real-time visual determination of the flux of hydrogen sulphide using a hollow-channel paper electrode, Chem. Commun. 51 (2015) 14030–14033.

[57] F. Liu, S. Ge, M. Su, X. Song, M. Yan, J. Yu, Electrochemiluminescence device for in-situ and accurate determination of CA153 at the MCF-7 cell surface based on graphene quantum dots loaded surface villous Au nanocage, Biosens. Bioelectron. 71 (2015) 286–293.

[58] Y. Zhang, J. Xu, S. Zhou, L. Zhu, X. Lv, J. Zhang, et al., DNAzyme-triggered visual and ratiometric electrochemiluminescence dual-readout assay for Pb(II) based on an assembled paper device, Anal. Chem. 92 (2020) 3874–3881.

[59] C. Gao, H. Yu, Y. Wang, D. Liu, T. Wen, L. Zhang, et al., Paper-based constant potential electrochemiluminescence sensing platform with black phosphorus as a luminophore enabled by a perovskite solar cell, Anal. Chem. 92 (10) (2020) 6822–6826.

[60] J.L. Delaney, E.H. Doeven, A.J. Harsant, C.F. Hogan, Use of a mobile phone for potentiostatic control with low cost paper-based microfluidic sensors, Anal. Chim. Acta 790 (2013) 56–60.

[61] Q.-M. Feng, Z. Liu, H.-Y. Chen, J.-J. Xu, Paper-based electrochemiluminescence biosensor for cancer cell detection, Electrochem. Commun. 49 (2014) 88–92.

[62] Q.-M. Feng, J.-B. Pan, H.-R. Zhang, J.-J. Xu, H.-Y. Chen, Disposable paper-based bipolar electrode for sensitive electrochemiluminescence detection of a cancer biomarker, Chem. Commun. 50 (2014) 10949–10951.

[63] Q. Feng, H. Chen, J. Xu, Disposable paper-based bipolar electrode array for multiplexed electrochemiluminescence detection of pathogenic DNAs, Sci. China Chem. 58 (2015) 810–818.

[64] L. Zhang, L. Li, C. Ma, S. Ge, M. Yan, C. Bian, Detection of α-fetoprotein with an ultrasensitive electrochemiluminescence paper device based on green-luminescent nitrogen-doped graphene quantum dots, Sens. Actuators B: Chem. 221 (2015) 799–806.

[65] Y. Zhang, H. Yang, K. Cui, L. Zhang, J. Xu, H. Liu, et al., Highly conductive and bendable gold networks attached on intertwined cellulose fibers for output controllable power paper, J. Mater. Chem. A 6 (2018) 19611–19620.

[66] S. Ge, J. Zhao, S. Wang, F. Lan, M. Yan, J. Yu, Ultrasensitive electrochemiluminescence assay of tumor cells and evaluation of H_2O_2 on a paper-based closed-bipolar electrode by in-situ hybridization chain reaction amplification, Biosens. Bioelectron. 102 (2018) 411–417.

[67] D. Wang, C. Liu, Y. Liang, Y. Su, Q. Shang, C. Zhang, A simple and sensitive paper-based bipolar electrochemiluminescence biosensor for detection of oxidase-substrate biomarkers in serum, J. Electrochem. Soc. 165 (2018) B361–B369.

[68] Y. Xiao, L. Xu, P. Li, X.-C. Tang, L.-W. Qi, A simple microdroplet chip consisting of silica nanochannel-assisted electrode and paper cover for highly sensitive electrochemiluminescent detection of drugs in human serum, Anal. Chim. Acta 983 (2017) 96–102.

[69] X. Zhang, S.-N. Ding, Graphite paper-based bipolar electrode electrochemiluminescence sensing platform, Biosens. Bioelectron. 94 (2017) 47–55.

[70] Y. Chen, J. Wang, Z. Liu, X. Wang, X. Li, G. Shan, A simple and versatile paper-based electrochemiluminescence biosensing platform for hepatitis B virus surface antigen detection, Biochem. Eng. J. 129 (2018) 1–6.

[71] X. Sun, B. Li, C. Tian, F. Yu, N. Zhou, Y. Zhan, et al., Rotational paper-based electrochemiluminescence immunodevices for sensitive and multiplexed detection of cancer biomarkers, Anal. Chim. Acta 1007 (2018) 33–39.

[72] Y. Han, Y. Fang, X. Ding, J. Liu, Z. Jin, Y. Xu, A simple and effective flexible electrochemiluminescence sensor for lidocaine detection, Electrochem. Commun. 116 (106760) (2020). Available from: https://doi.org/10.1016/j.elecom.2020.106760. art.

[73] J.D. Picket, Electrochemical Reactor Design, 2nd ed., Elsevier Scientific Publishing Co., New York, 1979, p. 34.

[74] A. Arora, J.C.T. Eijkel, W.E. Morf, A. Manz, A wireless electrochemiluminescence detector applied to direct and indirect detection for electrophoresis on a microfabricated glass device, Anal. Chem. 73 (2001) 3282–3288.

[75] K. Muzyka, Current trends in the development of the electrochemiluminescent immunosensors, Biosens. Bioelectron. 54 (2014) 393–407.

[76] J. Sun, H. Sun, Z. Liang, Nanomaterials in electrochemiluminescence sensors, ChemElectroChem 4 (2017) 1651–1662.

[77] A. Fiorani, J.P. Merino, A. Zanut, A. Criado, G. Valenti, M. Prato, et al., Advanced carbon nanomaterials forelectrochemiluminescent biosensor applications, Curr. Opin. Electrochem. 16 (2019) 66–74.

Chapter 10

Paper-based immunoassays for mobile healthcare: strategies, challenges, and future applications

Yao-Hung Tsai[1], Ting Yang[1], Ching-Fen Shen[2] and Chao-Min Cheng[1,*]
[1]Institute of Biomedical Engineering, National Tsing Hua University, Hsinchu, Taiwan,
[2]Department of Pediatrics, National Cheng Kung University Hospital, College of Medicine, National Cheng Kung University, Tainan, Taiwan
**Corresponding author. e-mail address: chaomin@mx.nthu.edu.tw*

10.1 Clinical value of paper-based immunoassays

Point-of-care (POC) medical tools are innovative diagnostics that offer new analytical methods for monitoring patient health; they are often praised for their rapidity, low-cost, user-friendliness, small clinical sample requirement, and for the fact that they do not require a trained specialist to complete them. These medical technologies include the use of nanomaterials within portable diagnostics for superior sensitivity and specificity, paper-based, lignocellulose-based, or even cotton-based analytical devices, integrated smartphone devices, and other wearable biosensors [1–4]. A number of well-known POC devices have been successfully commercialized, including diagnostic kits to test for pregnancy, glucose level, HIV, malaria, and other infectious disease tests [5,6]. Among all of these POC applications, paper-based immunoassay devices, which includes lateral flow assay (LFA), lateral flow immunoassay (LFIA), micro-fluid paper-based analytical devices, and dipsticks, not only provide the advantages mentioned earlier due to their inherent capillary action characteristics, they can be used without a power source, thus offering great convenience in a handy, disposable healthcare product [1,7].

Traditional immunoassays such as enzyme-linked immunosorbent assays (ELISA) are highly informative and well-suited for use in laboratories and medical centers. They can be used to detect certain disease and disease-state

Paper-Based Analytical Devices for Chemical Analysis and Diagnostics.
DOI: https://doi.org/10.1016/B978-0-12-820534-1.00007-4
© 2022 Elsevier Inc. All rights reserved.

biomarkers. Despite their high sensitivity and specificity, they are not suitable for use in places without adequate equipment or well-trained people. In order to bring this technology, the ability to detect disease and disease-state biomarkers, to more people who need it, paper-based immunoassays have been developed that provide easy-to-use, inexpensive diagnostics to developing and resource-limited environments. This broadens the potential scope and impact of POC diagnostics.

Paper-based immunoassays detect biomarkers via antibody-antigen recognition. Detecting antibodies conjugated with enzymes, nanoparticles, or fluorescent beads act as reporters to deliver readable and interpretable signals [8,9]. Colorimetric, chemiluminescent, electrochemical, and fluorescent signals may be used to provide qualitative or quantitative analysis [10–12]. Compared to traditional analytical methods, paper-based immunoassays provide the advantages of low turnaround time, low cost, and smaller sample volume requirement. Collectively, these features make paper-based immunoassays powerfully viable tools for the development of analytical devices for disease screening, diagnostics, and healthcare monitoring.

Paper-based ELISA (P-ELISA) is a particular type of paper-based immunoassay, it was first developed in 2010 by G.M. Whitesides and group [13]. P-ELISA relies on the creation of cellulose platform microplates containing a pattern of hydrophobic and hydrophilic boundaries that mimic the wells and well patterns of traditional ELISA plates. These boundaries effectively produce restricted area test zones that can be pre-impregnated with reactive reagents to facilitate quantitative analyses. This technology was a novel departure from existing qualitative dot-immunobinding assay/LFIA technology of the time. P-ELISA was later applied to develop analytical devices for a variety of clinical issues, many of which were focused on providing clinicians with tools for early diagnosis.

P-ELISA has been used to develop diagnostic tools in a variety of clinical areas. One such area is ophthalmology. The level of vascular endothelial growth factor (VEGF) in aqueous humor has been used as an indicator to monitor specific diseases associated with retinal ischemia. However, routinely monitoring VEGF concentration within the eye is hampered by insufficient sample sizes and difficult sampling techniques. For clinical diagnosis, only about 200 μL of aqueous humor can be collected from the anterior chamber before the threat of anterior chamber collapse. P-ELISA only requires 2 μL of aqueous humor and has ~fg/mL-level sensitivity, making it a very useful, inexpensive, and easy-to-handle approach for monitoring VEGF in aqueous humor [14].

Detecting bacteriuria, which is often due to *Escherichia coli*, is critical before certain surgical procedures or in cases of nosocomial infection to prevent further adverse events such as postoperative infection or sepsis. In low-and middle-income countries, where insufficient equipment and facilities preclude modern methods of detection, P-ELISA could be used to provide a valuable, inexpensive, easily used POC diagnostic tool. While P-ELISA analysis for *E. coli* requires hours to complete, it is still considerably faster than routine, clinical culture procedures that require anywhere from 3 to 5 days [15]. Furthermore, P-ELISA can be used to detect a whole variety of other pathogenic bacteria in addition to *E. coli.* Rapid diagnostics allow physicians to quickly identify infecting pathogens and prescribe suitable antibiotics in a timely manner.

In clinical settings, POC devices can be used to conduct in vitro tests and reduce the incidence of invasive patient procedures. Analytes like urine, saliva, body fluid, or fingertip blood samples can be used with POC devices, and they can be easily obtained with less harm. The detection of autoimmune antibodies in body fluids has demonstrated high sensitivity and specificity for diagnosing bullous pemphigoid (BP), which is an increasingly common autoimmune blistering disease with a high mortality rate [16]. Diagnostic criteria for BP include a combination of clinical manifestation, direct immunofluorescence microscopy of a perilesional specimen, and serology testing [17]. Analytical devices based on P-ELISA could be used for initial BP screening; they would be highly suitable for use in clinics or nursing homes [18]. Moreover, P-ELISA for blister fluid detection could also be used in conjunction with burn wound healing to monitor therapeutic efforts. Because wound healing involves multiple cytokines or growth factors, they could be used as analytical biomarkers to determine the effectiveness of care and the state of healing. Secondary burns, for example, can be divided into deep partial thickness burn or superficial partial thickness burns, and the former requires early debridement to ensure wound recovery quality. Because angiogenin, a cytokine that promotes angiogenesis, differs between these two types of secondary burns, P-ELISA targeting angiogenin would provide diagnostic information to guide therapeutic efforts [19,20].

10.2 The strategies of paper-based immunoassay for mobile healthcare

Despite the fact that paper-based immunoassays show great diagnostic promise, assay signal quality, readability, reliability, and accuracy are critical to functionality, whether assays are developed for clinical or at-home use becomes an important issue. Popular paper-based immunoassay products such as pregnancy tests and influenza A/B test strips are qualitative tests and are therefore easy-to-use [21,22]. On the other hand, semiquantitative and quantitative tests such as P-ELISA have signal interpretation issues. Color intensity and color spectrum may vary with target antigen concentration, which makes quantification problematic. In laboratory settings, highly trained researchers and clinicians can record results with readers, cameras, scanners, and smartphones and interpret signals with Image J software, but this process may not be practical for the general public [23–25].

Smartphone technology is continually evolving so that today's devices include high-resolution cameras, larger internal memories, fast processors, and easy-to-use applications (apps) and access to these devices has expanded so that almost everyone has access to one. These technological advances can be leveraged to capture and interpret analytical images that can, in the hands of clinicians and at-home patients, be converted to health-related information to drive and support therapeutic efforts. These data, as well as data from wearable sensors employing digital biomarkers, can also be uploaded to cloud storage, and the Internet of Medical Things (IoMT) can be used to develop platforms for advising and directing diagnostics and therapy at home and in clinical settings [26]. Paper-based POC devices have naturally been integrated with mobile healthcare technologies. Smith et al. developed an inexpensive paper-based wireless colorimetric diagnostic tool that could transmit information to a remote printer for review by clinicians. Similar devices have been successfully used in South African clinics to diagnose HIV or TB, offering inexpensive, easy-to-use, and rapid infectious disease detection [27]. In another effort, Kim's group created saliva-alcohol test strips that provided colorimetric results suitable for capture and analysis with a smartphone camera; these results could be analyzed using machine learning algorithms for improved performance as well [28].

Incorporating paper-based immunoassays with mobile devices like smartphones requires the development of apps for automatic analysis. To support quality image collection and subsequent analysis, a standardized light source that eliminates the influences of environmental light is sometimes necessary. Some aspects of paper-based diagnostic methodology may vary, but one approach is to use sample solution flow length on paper preloaded with one or more agglutinating reagents. In this process, which relies on the principles of immunoagglutination, flow length can then be correlated to the approximate quantity of targeted compound. Other approaches, such as one developed by Park et al., employed paper-based microfluidics and examinations of scattered light to detect the presence of *Salmonella*. In this device, paper was preloaded with antibody-conjugated submicroparticles, test samples were applied, and the light scattering from angle-optimized, standardized-distance digital images was evaluated with a smartphone. If a paper-based device was created with multiple microfluidics channels, a smartphone camera could simultaneously upload and analyze each channel to provide multiplexed assay results [29]. Wu et al. developed a three-channel paper-based microfluidics device to detect C-reactive protein (CRP). In this device, channels were coated with different amounts of latex-antibody reagent. When whole blood samples were applied to the center of the device, the blood flowed in three separate directions and provided three separate flow lengths for analysis with a smartphone [30]. In additional serologically based work, Guan et al. developed a barcode-like paper-based device for blood type testing. In this device, blood type could be determined by flow length and the results, calculated with a smartphone, could be transferred and shared via the internet. This particular study included the capacity for adding text messages along with transmitted results to expand the overall application of the device [31].

Another way of integrating paper-based analytical devices with smartphones is by including readable coding on the paper platform. The coding can be read by a smartphone, which can recognize changing signal output throughout a detection process. Yang et al. have reported on a multiplexed, inkjet-printed, barcoded paper-based assay based on LFIA techniques that is compatible with mobile devices. The color development on the paper device changes the width of the black bars within the originally printed barcodes to provide result signal information that can be interpreted by a smartphone app. This device was used to simultaneously detect hepatitis B virus, hepatitis C virus, human immunodeficiency virus type 1, and *Treponema pallidum* without cross-reaction, indicating that such an approach could be used to create assay systems with multiple targets [32]. Russell et al. developed a paper-based device printed with a QR code. The immunoassay test result signal, the result of changes to nanoparticles or enzyme reactions, disrupted the QR code pattern, which was then evaluated with a smartphone app. This idea has also been applied for *E. coli* detection, in which programmed software could provide an easy-to-interpret "hazard" or "no hazard" test outcome depending on analyte concentration [33].

Paper-based immunoassays with fluorescent reporters usually require integration with an additional device for signal excitation. Joung et al. created an inexpensive, easy to use, paper-based multiantigen assay for early-stage, POC diagnosis of Lyme disease (LD) by combining a vertical flow assay and an optical reader to capture colorimetric results, with a neural network analysis system to provide rapid assay results (less than 15 minutes) [34]. Chi et al. reported on a technique for conducting ELISA on a nitrocellulose platform coated with pseudopaper, a composite of nanoparticles and polymers. The closely packed hexagonal structure of this construct creates a slow-photon effect that greatly enhanced fluorescence emission for ELISA. Detection with a smartphone required the use of a homemade device designed to carry out on-site detection. This device incorporated an LED light source and two filters for excitation and emission, respectively [35]. Additionally, Shah et al. developed low-cost ultraviolet light-emitting diodes with long Stokes-shift quantum dots to enable ratiometric mobile phone fluorescence measurements without optical filters [36].

Paper-based immunoassays using colorimetric signals can be more easily integrated with smartphone applications than those using fluorescent signals because no exciting light source and filters are required. Smartphone apps can easily calculate change in RGB value to provide colorimetric results. An example of such a process was successfully completed by Wang et al., who developed a wax-printed, multilayered μPAD for the colorimetric detection of carcinoembryonic antigen (CEA). This device contained a movable and rotatable detection layer to allow the μPAD to switch the state of the sample solutions [37]. Although there is no need for any external light source or filters, some colorimetric paper-based immunoassay devices have employed an extra device for the purpose of fixing the paper-based tool. A paper-based microfluidics device for CRP detection developed by Dong et al. used a smartphone with an attached microlens and a 3D-printed chip–phone interface to produce diagnostic results [38]. Lee et al. demonstrated an electronics-enabled analytical microfluidic paper-based tool that used a LFA strip, a handheld reader, and a battery. This device was capable of transmitting information from the device to a near-field communication-enabled phone [39].

Combining and building upon these newly developed paper-based POC analytical devices and continuing to integrate them with smartphone applications, and electronic sensors might provide great improvements to the overall digital medical system. Compared to traditional paper-based diagnostics, the integration of mobile devices enables remote and immediate communication with clinicians, rapid readout and quantification, and the use of handheld detection devices that can easily be operated by customers without special training. Further, the addition and expansion of integrated computational tools will improve the sensitivity and specificity of diagnostic results [40,41] (Table 10.1).

10.3 The challenges of paper-based immunoassay approaches for mobile healthcare

Despite the many benefits of integrating paper-based diagnostics with mobile healthcare, there are some challenges to be considered and overcome. As with all paper-based immunoassays, the selected biomarker should directly relate to the disease or disease state being detected in order to provide the best relationship of readout values to clinical meanings. The discovery and study of these digital biomarkers has always been a difficult issue for POC device development. Biomarkers typically demonstrate diagnostic, predictive, and/or monitoring capacities and they have been used to analyze a variety of compounds including CRP, prostate-specific antigen, and infectious disease-related proteins. It is important to note that, digital biomarker discovery must be followed by biomarker validation before effective POC devices can be fully developed. Analytical biomarker results may vary between individuals being tested due to individual heterogeneity as well as other genetic and environmental issues, and the algorithms applied to examine results may alter and influence their accuracy [26,42]. Furthermore, to establish digital biomarker databases to interpret signals and provide clinical diagnosis, a vast amount of data must be collected, analyzed, and controlled. Such work relies on constant communication and feedback from patients, which involves a large tracking effort and active engagement.

TABLE 10.1 Application of POC devices based on paper-based immunoassays compatible with smartphone analysis. (Bb: *Borrelia burgdorferi*).

Application	Target biomarker	Methodology of detection	Assay types	Detection limit	Extra device besides smartphone	Reference
Salmonella	Salmonella	Length	Quantification	10^2 CFU mL^{-1}	Yes	Park et al. [29]
Blood typing	Red cell antigen	Length	Qualification	N/A	No	Guan et al. [31]
Drug residues in milk	Enrofloxacin (ENR)	Paper coding (barcode)	Quantification	8 ng mL^{-1}	No	Yang et al. [32]
Mouse IgG	Mouse IgG	Paper coding (QR code)	Quantification	0.3 μg mL^{-1}	Yes	Russell et al. [33]
Lyme disease	Bb-specific antigens (OspC, BmpA, P41, DbpB, Crasp1, P35, Erpd/Arp37)	Fluorescent signal	Qualification	N/A	Yes	Joung et al. [34]
Human IgG	Human IgG	Fluorescent signal	Quantification	3.8 fg mL^{-1}	Yes	Chi et al. [35]
Influenza	Influenza A Nucleocapsid protein	Fluorescent signal	Quantification	1.5 fmole	Yes	Shah et al. [36]
Gastric cancer	Carcinoembryonic antigen (CEA)	Colorimetric signal	Quantification	0.015 ng mL^{-1}.	No	Wang et al. [37]
Chronic diseases	C-reactive protein(CRP)	Colorimetric signal	Quantification	54 ng mL^{-1}	Yes	Dong et al. [38]
Iron deficiency, vitamin A deficiency, and inflammation	Ferritin, retinol binding protein (RBP), CRP	Colorimetric signal	Quantification	Ferritin:15 ng mL^{-1} RBP:16.3 μg mL^{-1} CRP: 410 ng mL^{-1}	Yes	Lee et al. [39]

After a suitable biomarker is selected, the design of the paper-based diagnostic tool must be carefully planned. Device methodology will greatly influence the possibilities for mobile device integration. Many paper-based diagnostics can only be used in a laboratory, because of the difficulty in converting them to a viable product in the hands of users [43,44]. In addition to following the WHO's ASSURED criteria (affordable, sensitive, specific, user-friendly, robust, equipment-free, delivered), device stability and reagent storage considerations must be considered [45].

Regarding mobile device integration itself, differences in spectral sensitivity and colorimetric metrics between mobile phones may result in arithmetic errors [46]. For example, camera color channels (i.e., red, green, and blue pixel sensitivities) differ significantly between models. Fluorescence-based assays are highly sensitive but would further complicate mobile device analysis using paper-based devices. First of all, the autoluminescence of paper substrate reduces the signal-to-noise ratios of fluorescence-based assays [47]. Secondly, external UV light sources and optical filters are needed for excitation and to distinguish emitted fluorescent light, which while possible, would increase the cost of a diagnostic device [36].

Another problem confronted by such technology is the complexity of clinical samples. Because most paper-based immunoassays rely on specific colorimetric reactions followed by image analysis with software such as ImageJ, the matrix effect from patient samples (i.e., serum, urine, other body fluids) may interfere with the accuracy and reliability of the results. In order to reduce the assay workflow complexity, some researchers have employed various machine learning algorithms to improve reliability and accuracy [48]. However, the lack of existing commercialized paper-based mobile health devices has made the vast data collection process even more difficult.

Finally, although diagnostics with mobile devices could be a convenient technology for both patients and doctors, it carries some social issue burdens. Because all of the patient health records would be uploaded through the internet and/or stored in a "cloud system," patient privacy, related ethics, and data ownership issues would be a concern [49].

10.4 Future applications of paper-based immunoassays for mobile healthcare

Paper-based immunoassays provide rapid, convenient, inexpensive, and easy-to-use diagnostic tools that can be used clinically and at home. Similar POC devices have already been widely applied in resource- or energy-limited areas and for personalized health care [13,50]. Thanks to the recent advances in technology including, digital therapeutics, smartphone devices, computational analysis tools (i.e., machine learning), and the IoMT, integration of paper-based POC devices with mobile health technologies has become easier and more robust and has begun to change the healthcare paradigm. The concept of the IoMT is the extended version of the Internet of Things; it primarily focuses on providing a healthcare platform for connecting medical devices with healthcare providers via the internet. As shown in Fig. 10.1, when applied to personalized healthcare, it could be used to establish a digital medical platform to help patients obtain personalized healthcare without going to a clinic and it could be used to establish a thorough disease database collection of clinical information in cloud storage format. Patients with chronic illnesses, mild symptoms, or long-term care at home could conduct self-sampled diagnosis using a paper-based detection device, and then use a smartphone camera, optical sensor, or smartphone-based colorimetric reader to easily obtain corresponding results. Their individual outcomes can then contribute to the database by being uploaded for storage and data mining on the IoMT platform. These data can be further analyzed and combined by

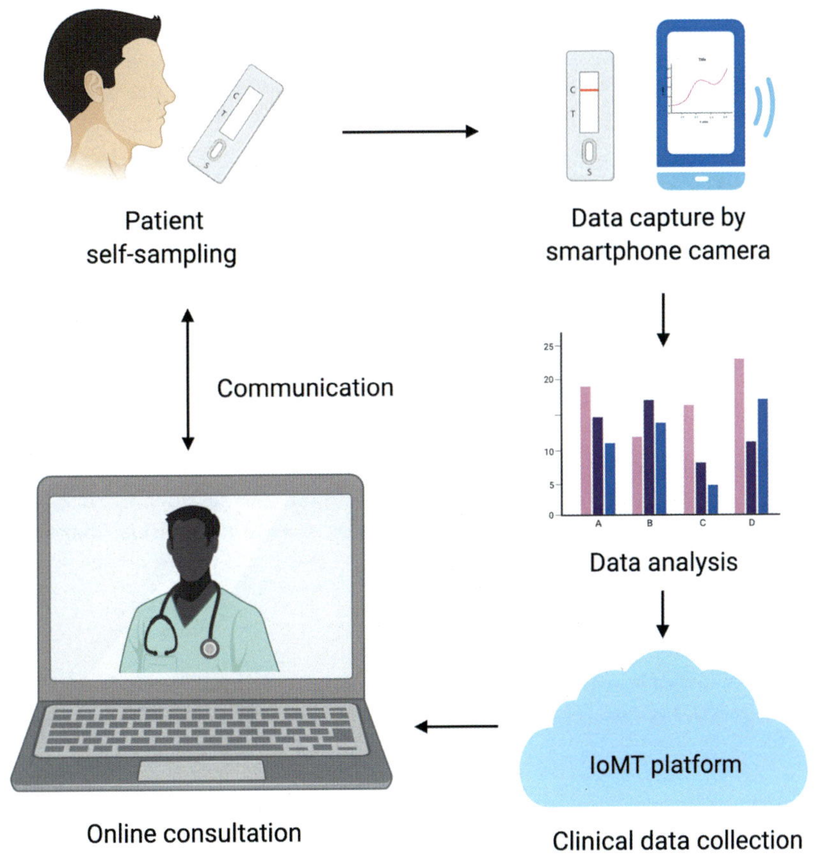

FIGURE 10.1 Schematic showing the integrated workflow for paper-based POC immunoassay techniques, mobile healthcare, and the Internet of Medical Things (IoMT).

researchers or health providers using computational tools for online health consultations as well as clinical studies.

The evolution of paper-based immunoassays combined with digital healthcare advances could benefit patients in the following ways: (1) reduced risk and time required to visit hospitals or healthcare centers, especially for patients with chronic diseases such as diabetics or the elderly who have problems traveling between clinics and home; (2) lowered costs for diagnosis and follow-up care resulting from ease and convenience of diagnosis, a result that would be especially beneficial for resource-deprived areas; (3) shortened overall diagnostic and follow-up care time; (4) uncomplicated and convenient diagnostic process without complex sample preparation; and, (5) protection from the spread of contagious disease facilitated by remote diagnostic capacities.

With ongoing developments in wireless technologies, there is no doubt that the interconnection between POC devices and mobile healthcare will continue to flourish. The future of integrated approaches employing paper-based immunoassay techniques and mobile technology lies in the discovery and validation of new and better digital biomarkers, the establishment of an authorized central clinical database, and most importantly, the successful transformation of these device from academic study to the user market.

10.5 Glossary

1. Internet of Medical Things (IoMT): The combination of medical devices and digital applications that can connect health care information with caregivers via the internet.
2. Digital biomarkers: Physiological and behavioral information or outcomes obtained by portable biosensors, mobile health related tools, and other digital devices.
3. Point-of-care (POC) diagnostic: Medical diagnostic testing that can be performed near the patient or at the bedside and provide rapid results.
4. Mobile healthcare: The application of electronic devices such as wearable biosensors and smartphone applications for disease monitoring, diagnostics, or prevention.

Acknowledgments

This work was supported by grants from the Ministry of Science and Technology in Taiwan (MOST 109–2622-E-007–009-CC3 & MOST 109–2623-E-007–002-D to C.-M. C.). The figure was created with BioRender.com.

References

[1] M.M. Calabretta, M. Zangheri, A. Lopreside, E. Marchegiani, L. Montali, P. Simoni, et al., Precision medicine, bioanalytics and nanomaterials: toward a new generation of personalized portable diagnostics, Analyst 145 (8) (2020) 2841–2853.

[2] C.-M. Kuan, R.L. York, C.-M. Cheng, Lignocellulose-based analytical devices: bamboo as an analytical platform for chemical detection, Sci. Rep. 5 (2015) 18570.
[3] S.-C. Lin, M.-Y. Hsu, C.-M. Kuan, H.-K. Wang, C.-L. Chang, F.-G. Tseng, et al., Cotton-based diagnostic devices, Sci. Rep. 4 (2014) 6976.
[4] S.-C. Lin, C.-Y. Tzeng, P.-L. Lai, M.-Y. Hsu, H.-Y. Chu, F.-G. Tseng, et al., Based CRP monitoring devices, Sci. Rep. 6 (2016) 38171.
[5] V. Gubala, L.F. Harris, A.J. Ricco, M.X. Tan, D.E. Williams, Point of care diagnostics: status and future, Anal. Chem. 84 (2) (2012) 487–515.
[6] R. Peeling, D. Mabey, Point-of-care tests for diagnosing infections in the developing world, Clin. Microbiol. Infect. 16 (8) (2010) 1062–1069.
[7] K.J. Land, Paper-based Diagnostics: Current Status and Future Applications, Springer, 2018.
[8] S. Ramachandran, E. Fu, B. Lutz, P. Yager, Long-term dry storage of an enzyme-based reagent system for ELISA in point-of-care devices, Analyst 139 (6) (2014) 1456–1462.
[9] H. Kim, D.-R. Chung, M. Kang, A new point-of-care test for the diagnosis of infectious diseases based on multiplex lateral flow immunoassays, Analyst 144 (8) (2019) 2460–2466.
[10] Z. Nie, C.A. Nijhuis, J. Gong, X. Chen, A. Kumachev, A.W. Martinez, et al., Electrochemical sensing in paper-based microfluidic devices, Lab. Chip 10 (4) (2010) 477–483.
[11] J.C. Cunningham, N.J. Brenes, R.M. Crooks, Paper electrochemical device for detection of DNA and thrombin by target-induced conformational switching, Anal. Chem. 86 (12) (2014) 6166–6170.
[12] L.-W. Song, Y.-B. Wang, L.-L. Fang, Y. Wu, L. Yang, J.-Y. Chen, et al., Rapid fluorescent lateral-flow immunoassay for hepatitis B virus genotyping, Anal. Chem. 87 (10) (2015) 5173–5180.
[13] C.M. Cheng, A.W. Martinez, J. Gong, C.R. Mace, S.T. Phillips, E. Carrilho, et al., Paper-based ELISA, Angew. Chem. Int. (Ed.) 49 (28) (2010) 4771–4774.
[14] M.-Y. Hsu, C.-Y. Yang, W.-H. Hsu, K.-H. Lin, C.-Y. Wang, Y.-C. Shen, et al., Monitoring the VEGF level in aqueous humor of patients with ophthalmologically relevant diseases via ultrahigh sensitive paper-based ELISA, Biomaterials 35 (12) (2014) 3729–3735.
[15] C.-M. Shih, C.-L. Chang, M.-Y. Hsu, J.-Y. Lin, C.-M. Kuan, H.-K. Wang, et al., Based ELISA to rapidly detect *Escherichia coli*, Talanta 145 (2015) 2–5.
[16] P. Joly, S. Baricault, A. Sparsa, P. Bernard, C. Bédane, S. Duvert-Lehembre, et al., Incidence and mortality of bullous pemphigoid in France, J. Investig. Dermatol. 132 (8) (2012) 1998–2004.
[17] E. Schmidt, D. Zillikens, Pemphigoid diseases, Lancet 381 (9863) (2013) 320–332.
[18] C.-K. Hsu, H.-Y. Huang, W.-R. Chen, W. Nishie, H. Ujiie, K. Natsuga, et al., Based ELISA for the detection of autoimmune antibodies in body fluid· the case of Bullous Pemphigoid, Anal. Chem. 86 (9) (2014) 4605–4610.
[19] S.C. Pan, L.W. Wu, C.L. Chen, S.J. Shieh, H.Y. Chiu, Angiogenin expression in burn blister fluid: implications for its role in burn wound neovascularization, Wound Repair. Regen. 20 (5) (2012) 731–739.
[20] S.-C. Pan, Y.-H. Tsai, C.-C. Chuang, C.-M. Cheng, Preliminary assessment of burn depth by paper-based ELISA for the detection of angiogenin in burn blister fluid—a proof of concept, Diagnostics 10 (3) (2020) 127.
[21] L.A. Bastian, K. Nanda, V. Hasselblad, D.L. Simel, Diagnostic efficiency of home pregnancy test kits: a *meta*-analysis, Arch. Family Med. 7 (5) (1998) 465.

[22] A. Ruest, S. Michaud, S. Deslandes, E.H. Frost, Comparison of the Directigen flu A + B test, the QuickVue influenza test, and clinical case definition to viral culture and reverse transcription-PCR for rapid diagnosis of influenza virus infection, J. Clin. Microbiol. 41 (8) (2003) 3487–3493.

[23] E. Aydindogan, A.E. Ceylan, S. Timur, Based colorimetric spot test utilizing smartphone sensing for detection of biomarkers, Talanta 208 (2020) 120446.

[24] X. Guo, L. Zong, Y. Jiao, Y. Han, X. Zhang, J. Xu, et al., Signal-enhanced detection of multiplexed cardiac biomarkers by a paper-based fluorogenic immunodevice integrated with zinc oxide nanowires, Anal. Chem. 91 (14) (2019) 9300–9307.

[25] R.C. Murdock, L. Shen, D.K. Griffin, N. Kelley-Loughnane, I. Papautsky, J.A. Hagen, Optimization of a paper-based ELISA for a human performance biomarker, Anal. Chem. 85 (23) (2013) 11634–11642.

[26] I. Sim, Mobile devices and health, N. Engl. J. Med. 381 (10) (2019) 956–968.

[27] S. Smith, A. Oberholzer, J.G. Korvink, D. Mager, K. Land, Wireless colorimetric readout to enable resource-limited point-of-care, Lab. Chip 19 (19) (2019) 3344–3353.

[28] H. Kim, O. Awofeso, S. Choi, Y. Jung, E. Bae, Colorimetric analysis of saliva–alcohol test strips by smartphone-based instruments using machine-learning algorithms, Appl. Opt. 56 (1) (2017) 84–92.

[29] T. San Park, W. Li, K.E. McCracken, J.-Y. Yoon, Smartphone quantifies Salmonella from paper microfluidics, Lab. Chip 13 (24) (2013) 4832–4840.

[30] X.-F. Wu, C.-F. Shen, C.-M. Cheng, Integration of mobile devices and point-of-care diagnostic devices—the case of C-reactive protein diagnosis, Diagnostics 9 (4) (2019) 181.

[31] L. Guan, J. Tian, R. Cao, M. Li, Z. Cai, W. Shen, Barcode-like paper sensor for smartphone diagnostics: an application of blood typing, Anal. Chem. 86 (22) (2014) 11362–11367.

[32] M. Yang, W. Zhang, W. Zheng, F. Cao, X. Jiang, Inkjet-printed barcodes for a rapid and multiplexed paper-based assay compatible with mobile devices, Lab. Chip 17 (22) (2017) 3874–3882.

[33] S.M. Russell, A. Doménech-Sánchez, R. de la Rica, Augmented reality for real-time detection and interpretation of colorimetric signals generated by paper-based biosensors, ACS Sens. 2 (6) (2017) 848–853.

[34] H.-A. Joung, Z.S. Ballard, J. Wu, D.K. Tseng, H. Teshome, L. Zhang, et al., Point-of-care serodiagnostic test for early-stage lyme disease using a multiplexed paper-based immunoassay and machine learning, ACS Nano 14 (1) (2019) 229–240.

[35] J. Chi, B. Gao, M. Sun, F. Zhang, E. Su, H. Liu, et al., Patterned photonic nitrocellulose for pseudopaper ELISA, Anal. Chem. 89 (14) (2017) 7727–7733.

[36] K.G. Shah, V. Singh, P.C. Kauffman, K. Abe, P. Yager, Mobile phone ratiometric imaging enables highly sensitive fluorescence lateral flow immunoassays without external optical filters, Anal. Chem. 90 (11) (2018) 6967–6974.

[37] K. Wang, J. Yang, H. Xu, B. Cao, Q. Qin, X. Liao, et al., Smartphone-imaged multilayered paper-based analytical device for colorimetric analysis of carcinoembryonic antigen, Anal. Bioanal. Chem. (2020) 1–12.

[38] M. Dong, J. Wu, Z. Ma, H. Peretz-Soroka, M. Zhang, P. Komenda, et al., Rapid and low-cost CRP measurement by integrating a paper-based microfluidic immunoassay with smartphone (CRP-Chip), Sensors 17 (4) (2017) 684.

[39] S. Lee, A. Aranyosi, M.D. Wong, J.H. Hong, J. Lowe, C. Chan, et al., Flexible opto-electronics enabled microfluidics systems with cloud connectivity for point-of-care micronutrient analysis, Biosens. Bioelectron. 78 (2016) 290–299.

[40] Q. Qin, K. Wang, J. Yang, H. Xu, B. Cao, Y. Wo, et al., Algorithms for immunochromatographic assay: review and impact on future application, Analyst 144 (19) (2019) 5659–5676.

[41] W. Yan, K. Wang, H. Xu, X. Huo, Q. Jin, D. Cui, Machine learning approach to enhance the performance of MNP-labeled lateral flow immunoassay, Nano-Micro Lett. 11 (1) (2019) 7.

[42] A. Coravos, S. Khozin, K.D. Mandl, Developing and adopting safe and effective digital biomarkers to improve patient outcomes, NPJ Digital Med. 2 (1) (2019) 1–5.

[43] A.A. Kumar, J.W. Hennek, B.S. Smith, S. Kumar, P. Beattie, S. Jain, et al., From the bench to the field in low-cost diagnostics: two case studies, Angew. Chem. Int. (Ed.) 54 (20) (2015) 5836–5853.

[44] F.W. Kimani, S.M. Mwangi, B.J. Kwasa, A.M. Kusow, B.K. Ngugi, J. Chen, et al., Rethinking the design of low-cost point-of-care diagnostic devices, Micromachines 8 (11) (2017) 317.

[45] T. Tian, Y. Bi, X. Xu, Z. Zhu, C. Yang, Integrated paper-based microfluidic devices for point-of-care testing, Anal. Methods 10 (29) (2018) 3567–3581.

[46] Matasaru, C., Mobile phone camera possibilities for spectral imaging. in School of Computing Department (University of Eastern Finland, 2014).

[47] K.G. Shah, P. Yager, Wavelengths and lifetimes of paper autofluorescence: a simple substrate screening process to enhance the sensitivity of fluorescence-based assays in paper, Anal. Chem. 89 (22) (2017) 12023–12029.

[48] Z.S. Ballard, H.-A. Joung, A. Goncharov, J. Liang, K. Nugroho, D. Di Carlo, et al., Deep learning-enabled point-of-care sensing using multiplexed paper-based sensors, NPJ Digit. Med. 3 (1) (2020) 1–8.

[49] S.K. Vashist, P.B. Luppa, L.Y. Yeo, A. Ozcan, J.H. Luong, Emerging technologies for next-generation point-of-care testing, Trends Biotechnol. 33 (11) (2015) 692–705.

[50] K. Mahato, A. Srivastava, P. Chandra, Paper based diagnostics for personalized health care: Emerging technologies and commercial aspects, Biosens. Bioelectron. 96 (2017) 246–259.

Chapter 11

Conclusions, challenges, and next steps

Iana V.S. Arantes[1], Letícia F. Mendes[1], Vanessa N. Ataide[1], William R. de Araujo[2] and Thiago R.L.C. Paixão[3,4,*]

[1]Departamento de Química Fundamental, Instituto de Química, Universidade de São Paulo, São Paulo, Brazil, [2]Portable Chemical Sensors Lab, Department of Analytical Chemistry, Institute of Chemistry, State University of Campinas—UNICAMP, Campinas, Brazil, [3]Department of Fundamental Chemistry, Institute of Chemistry, University of Sao Paulo, Sao Paulo, Brazil, [4]National Institute of Bioanalitical Science and Technology, Campinas, Brazil
**Corresponding author. e-mail address: trlcp@iq.usp.br*

11.1 Introduction

Since Whitesides's group reintroduced the use of microfluidic paper-based analytical devices (μPADs) in 2007, creating hydrophobic barriers patterns into the paper matrix using a photolithography technique and wax, enabling *fluid* transport across the hydrophilic channel to the colorimetric detection zones without pumping and external energy [1], the field of paper-based microfluidics emerged as a promising, affordable, portable, and biodegradable alternative to the conventional microfluidic analytical devices, with the possibility of use in several applications, such as disease diagnostics, drug analysis, environmental monitoring, food quality control, among others [2,3]. Hence, one first challenge for this field is directly correlated with fluid transportation on the paper matrix.

Additionally, to detect these several target molecules in different matrices, the paper substrate incorporates different detection strategies. A couple of primary techniques used were color detection and electrochemistry, as shown in chapter 1, Introduction remarks for paper-based analytical devices and timeline. The challenges for these detections on paper will be further discussed in this chapter.

Another critical point in this discussion is related to the application of PADs. Researchers from different parts of the globe created clever approaches for "lab detection" in different sample matrices and using various detection techniques, as discussed throughout this book. However, some drawbacks

Paper-Based Analytical Devices for Chemical Analysis and Diagnostics.
DOI: https://doi.org/10.1016/B978-0-12-820534-1.00010-4
© 2022 Elsevier Inc. All rights reserved.

regarding the mass production of these devices to commercialization and the limited testing directly in the field (outside the lab conditions) hinder their translatability for the real world. Additionally, the usual lack of an integrated circuit and user-friendly interface to interpret the results could be obstacles for nonskilled researchers' applications, i.e., to be used for everyone out of the lab. Hence, challenges need to be examined from a different perspective about the current position of PADs and the possibility to improve characteristics for enhanced commercialization in the future. Thus, in this chapter, we will highlight the importance of fluid transport control on paper, the most used low-cost techniques for in-field applications, and some tips for improvements in future works and next steps to advance in this exciting field.

11.1.1 Challenges in fluid transport

The inherent advantage in using paper to transport fluids lies in its capillary action generated from the hydrophilic network of cellulose fibers to absorb solution into the device, which results in an effortless fluid flow without the requirement of any external pumping facility, as discussed in chapter 2, Chemistry of paper - properties, modification strategies and uses in bioanalytical chemistry. The flow mechanism in such a porous medium is entirely different from any other capillary-based systems like glass, silicon, or polydimethylsiloxanes (PDMS). Flow occurs through paper pores spreads randomly through the capillary network of the porous substrate [4].

By taking advantage of this passive flow, μPADs can analyze simple lateral flow tests. Also perform complex analysis that requires multiple-stage reactions, sample mixing, and washing steps that would otherwise necessitate the use of external pumps and several pipetting steps [5]. These characteristics allow the development of devices capable of performing multiple analyses using only small amounts of samples, aside from the low cost and ease of use, compared to conventional microfluidic apparatus.

Over the past decade, the fabrication techniques, fluid manipulation, and detection methods involving μPADs have considerably improved. Some of the fabrication techniques for creating microfluidic paper-based channels for analytical devices include photolithography [1], wax printing [6], stamping [7], screen-printing [8], and laser cutting [9]. After fabrication, fluid control is the most crucial parameter in the operation of μPADs, because it dictates the efficiency of the interaction between the samples and the reagents in the paper matrix, or simply how fast and continuous the analytes are going to reach the detection zone, generating a stable and reproducible analytical signal.

Fluid transport on paper depends on the quality of cellulose fibers and its degree of randomness, often dictated by the manufacturing process. Consequently, these processes determine the physicochemical properties, the homogeneity and uniformity of the paper, and thus the flow rate on the paper surface. Additionally, transport characteristics on paper-based systems are also dependent on pore radius, average porosity, environmental conditions such as temperature and humidity, evaporation rates, the geometry of the channel, reservoir volume, the viscosity of the fluid, among others [4,10].

Furthermore, there is another critical point associated with μPADs fabrication techniques and flow characteristics. During the fabrication, except in the fabrication methods involving masks or digital printing, the entire paper surface is exposed to chemicals. Therefore, although these different chemical treatments in other fabrication methods may not significantly impact chemical-based assays, these will affect the fundamental transport phenomena on such substrates, as flow characteristics can be involved in different manners [4].

In summary, the paper itself offers limited control over fluid transport since flow is based on the inherent pore structure of the specific paper, and the differences in the substrate manufacturing process and fabrication techniques can lead to a lack of reproducibility in the flow rate [11].

Therefore, while paper has shown great potential as a substrate for analytical purposes, μPADs still do not have the range of capabilities of conventional microfluidic devices, such as rigorous control over flow rates, mixing, the timing of sequential reactions, etc. [11]. It is also known that flow rate decreases over time within capillaries of constant cross-sectional area, such as in lateral flow assays in paper-based channels. Consequently, fluid flow within cellulose channels is intrinsically slow. For example, it takes approximately 15 to 20 minutes for the water to wick through a typical 5 cm long paper channel [12]. This flow rate decay can increase the assay time and change detection response as a function of time [5]. Lengthy analysis times are not desirable, leading to secondary problems such as solvent evaporation, loss of reactive species generated, and paper swelling [13].

In addition, since the flow in the microfluidic channels in the paper is considered a laminar flow, the mixture of reagents only occurs by diffusion in the fluid interface. Thus, promoting a rapid and efficient mixture in these microfluidic platforms is still a challenge to overcome to enable complex reactions in paper substrates [11].

As seen so far, the control over the flow rate is considered a critical parameter to achieve high-performance PADs. Therefore, understanding the flow behavior and the available mechanisms to interfere with it is extremely important for developing accurate, predictable, and programmable microfluidic devices.

Recently, several advancements were achieved in the field of microfluidics that affected fluid transport in paper substrates, either promoting a delay, allowing complex multistep reactions to occur before sample detection, or accelerating the flow to perform rapid and sequential analyses. These fluid manipulation technologies are categorized by some authors into passive and active methods, according to the operation mode [14], and also divided into geometry, chemical, and mechanical-based methods [3].

In general, passive fluid control methods are used to manipulate the flow rate within a paper channel without the need for external inputs or additional equipment, simply by changing the paper's geometrical structure or chemically treating the substrate. Geometry-based fluid manipulation methods can achieve a flow rate control by altering the structural shape of the paper itself, including variations in channel length and width, or by adding additional structures to the channels [14]. About this topic, Yager's group addressed significant contributions by showing that changing channel lengths of the inlets enable reagents and samples to be delivered to the corresponding outlet regions in the proper assay time. Their experiments also showed faster fluid transport in smaller channel widths [15].

Fast flow is another desirable characteristic to reduce the diagnostic time. One way to achieve this is by sandwiching paper channels between two flexible films, preventing evaporation of the sample and accelerating the flow rate compared with channels without films [16]. An alternative method to manipulate flow rate uses a blade to carve the paper surface without cutting it through. When carved in the flow direction, an acceleration occurs, but it acts as a barrier when carved perpendicularly to the flow, causing a flow delay in fluid transport [10].

Another geometry-based method first used by Mendez et al. provided a quasistationary flow in the fluid channels to occur by only adding a fan-shaped structure at the end of the paper channel [17]. This fan geometry is used to compensate for the decay in flow rate within the analysis channel that coincides with the distance a fluid front travels through a capillary network [5].

Chemical-based methods, on the other hand, consists of changing the wicking properties of paper through chemical treatments such as using dissoluble materials, like sucrose, to promote a delay in the flow rate, caused by the viscosity changes in the fluid when the sample reaches the sucrose modified zone [18].

By controlling the flow rate using these methods, chemical signal amplification and higher sensitivity were achieved, otherwise not possible in paper channels. Nevertheless, while passive methods are structurally simple and easy to manufacture, they are still limited to precise fluid control. They sometimes consume more reagents and samples because of more complicated and longer paper channels [14].

On the other hand, active fluid control typically relies on external inputs, including mechanical or electrical methods, which sometimes require manual operation. Fluid manipulation methods using mechanical movement include expandable materials [19], magnetic valves [20], pressure valves [21], and reconfigurable flow switches [22]. Active methods have the advantages of versatile flow manipulation, fast response, reproducibility, compatibility with biological samples by no use of surfactants and chemicals; thus, they have more potential to satisfy the requirements of multistep assays [14]. However, in most cases, the additional mechanical or external systems required are against the simplicity and cost-effectiveness, which are the core values of paper-based diagnostic platforms.

Despite the advances in flow control, there is an agreement among authors in the literature about the future of paper microfluidics. The main challenge in fluid transport in paper devices still lies in the development of an efficient design directed to specific applications, exploring material properties and fluid transport mechanisms to obtain functional devices with predictable behavior [3].

Microfluidics must be able to solve problems for users who are not experts in fluid physics. Therefore, the focus should be on the development of μPADs with integrated flow control methods, which have advanced characterization with accuracy and precision of flow manipulation in a compact device with automated operation and instant analytical response, which is compatible with high-volume fabrication [14].

In conclusion, much research is yet to be done to develop μPADs for use in more complex fully, multistep assays with the advanced capabilities to perform multiple analyses, quantification with low limits of detection, high sensitivity, and selectivity to become an accurate and reliable point-of-care test. The flow control systems either lack sensitivity and reproducibility based on the random arrangement of paper fibers or are complicated and expensive to produce on a mass scale. Nevertheless, one thing is sure, much of the world's technology requires fluid manipulation, and extending those manipulations to small volumes using inexpensive and abundant materials like paper, is undoubtedly very promising [23].

11.1.2 Challenges of colorimetric detection

Chemical analysis based on colorimetric reactions is present since the early uses of paper as a platform for analytical tests [24]. Colorimetric detection methods are most commonly combined with PADs due to their simplicity (it is possible to correlate the color intensity with a particular analyte concentration). They are also compatible with portable reporting systems, such as smartphones, handheld readers, among others [2,25].

Paper is an excellent substrate for colorimetric detection because it has a white background, contrasting with the color appearance. Moreover, it has a high surface-to-volume ratio, facilitating the evaporation of solutions and could promote the concentration of analytes [26]. Besides, its cellulosic-based fibers (made of D-glucose polymer chains) enable reagents storage, filtration, separation, and passive fluid transport [27]. All of these features are important because they allow the creation of increasingly complete analytical devices, in which all stages of the analysis can be performed directly on paper.

There are several strategies for color formation on paper-based devices, such as enzymatic assays, redox indicators, nanoparticles; pH indicators; complexes, dyes, among others [28]. One of the critical aspects of colorimetric analysis on paper is the color formation homogeneity, thus improving the measurement readout, allowing greater accuracy, precision, and reproducibility to the analytical method. Signal uniformity can be affected by the following factors: device design, type of modifiers onto the paper substrate, assay implementation, and consequently, the data processing [25].

Processing the data obtained from the colorimetric assays is challenging because the analytical information acquisition approach depends on the analysis proposal (cost, time, application, and precision). The next step immediately after performing the colorimetric test is to read the color formed to be analyzed. One way to process this information is by digitizing the images using cell phone cameras (smartphones) or scanners [29,30]. It is essential to note that this is a crucial step when the colorimetric test aims to obtain quantitative data.

The advantage of using a scanner to digitize colorimetric data is that the focal distance and illumination are the same (the light is not affected by external conditions) [30], which means that the image processing is homogeneous. This aspect can provide better reproducibility of the measurements [25]. The use of smartphones to acquire colorimetric data is even more attractive because they are portable, easy to use, and economically viable. It is possible to install programs or develop applications that assist in data analysis [31]. In this case, the smartphone offers a complete analysis tool when compared to scanners.

On the other hand, the image's digitalization using smartphones brings some issues, such as variations in ambient light (external conditions), image angle, and focal distance [25]. Some of the problems mentioned above have been overcome due to the efforts of many research groups working with colorimetric PADs. For example, Salles et al. [29] developed a closed PMMA platform with LEDs and a coupled smartphone to control the illumination and focal distance images. Mathematical approaches could be used to circumvent ambient light interference, too, based on an LED light source combined with background and color rescaling [32], but increase the data processor.

In this context, after carrying out the colorimetric tests and digitizing them, it is necessary to analyze the images obtained to extract the analytical information. There are two main ways to do the image analysis: manual and automated. The first one consists of using commercially available software to insert the images and perform the analysis [33]. This type of approach is time-consuming and depends on the operator's skills. The second one deals with software that automatically receives, interprets, and processes the images (called computer vision systems) [34]. Thus, this strategy could enhance the image analysis reproducibility and make them faster.

Additionally, some other freeware software is being available like Photometrix [35] as good options to circumvent the cost and with a friendly interface.

The colorimetric results can be read out three different methods: (1) qualitative (YES/NO output originated by a color change due to the analyte presence), (2) semiquantitative (the colorimetric result is compared with an external calibration curve-used as a reference- in which an analyte concentration range can be estimated), and (3) quantitative (a calibration curve is performed on-device, being possible to obtain an external calibration or standard addition method) [2,25].

From a commercial point of view and considering a broader application of colorimetric paper-based devices, a qualitative readout for rapid diagnosis, which is equipment-free, has excellent potential because it is user-friendly (no training is required) and is a good fit for point-of-care analysis, especially in places with limited resources and in-development countries [25]. Home pregnancy tests for human chorionic gonadotropin (hCG) are examples of a colorimetric PAD of the YES/NO type [36], demonstrating quick diagnosis in a simple way and yet affordable. The development of sensors of this type meets the ASSURED challenge of the World Health Organization (WHO) [28], opening new opportunities for scientific research to promote even more benefits to society. Although the characteristics mentioned above of qualitative colorimetric tests are attractive, there is also a limitation that they are only suitable when the answer is YES/NO. There is no need to know the concentration or concentration range of the analyte.

The semiquantitative colorimetric analysis can also be equipment-free, which, as already discussed, facilitates the results readout and makes it quick. The most well-known practical application of this type of sensor is pH measuring tapes, where the color of the paper strip changes according to the change in the pH of the medium. The result is read by comparing the paper strip's color with a color chart (which establishes a pH range) printed on their box. It is important to note that the H^+/OH^- concentration is estimated according to the color range; it is impossible to know their concentration in the sample [25]. Another critical aspect of this type of readout is that the interpretation of colors varies from person to person, which generates much uncertainty associated with the analysis result.

Specifically dealing with semiquantitative analysis using PADs, some research groups have proposed readings of colorimetric measurements based on the count, distance, and time. In the first approach, the number of colored bars, tabs or, spots on the paper device is counted. In the second, the measurement is based on the distance that the developed color moves along the paper platform. The third analysis strategy is based on the track of the time for color formation and its correlation with the analyte concentration [37–39].

Quantitative colorimetric analysis is necessary when the other approaches discussed above are not sufficient, and the application requires that the concentrations of the analytes be determined. The quantitative aspect of the analysis carried out on portable devices, such as those on paper, is even more relevant when considering those related to medical diagnoses, the results of which may lead to some decision-making in terms of medical treatments [40].

Quantitative analysis is performed using calibration curves, and these can be external calibration or standard addition methods. External calibration provides more accurate results when compared to those obtained through semiquantitative analysis. Some considerations must be made depending on the analysis purposes. First, if the paper-based device's purpose is to perform in-field analysis, the calibration curve must be made in the laboratory. In contrast, the analysis of the samples is carried out outside the laboratory under totally different environmental conditions, generating results that do not fully represent the actual concentrations of the sample [25,41]. A second consideration is that if the analysis objective is to be more precise, both the calibration curve and the sample analysis must be done in the laboratory (under the same conditions). However, one more step is added to the analysis since the sample needs to be collected and limits its in-field application on the paper device.

An alternative to using external calibration is colorimetric analysis employing the standard addition method. The advantages of using this method are: minimizing the matrix effect (the sample is in the same medium as the standards). The complete analysis can be performed on a single device, and the results are not affected by environmental conditions, such as luminosity and reproducibility of the fabrication method [42].

The last step of the analysis of colorimetric results is the data processing. Data processing for this type of analysis can apply different color space modes, including RGB (Red, Green, and Blue); greyscale; CMYK (in this mode, the color is described as a subtraction from the complementary RGB and white, which are Cyan, Magenta, Yellow, and Black or Key); HSV (Hue, Saturation, and Value, this mode uses a conical representation of the color). The color values in each mode can be related to the intensities of color change, making it possible to construct a calibration curve [43]. The data processing's main challenge is that each type of data processing and the entire analysis play an essential role in the differences observed in each proposed analysis method, making it difficult to compare the analytical results.

11.1.3 Challenges of electrochemical paper-based devices

It is known that paper has been applied for analytical purposes, as discussed in this book for a long time, using a colorimetric method for testing acidity and basicity in solutions [44]. Since then, colorimetry has been used in the paper to fabricate low-cost and straightforward devices due to their advantages, implying the improvement of the colorimetric PADs since their first use.

Some technologies, including smartphones and cameras, have been applied to identify the pattern of the colors with the precision and accuracy that human eyes could not provide. However, this technique's sensibility and selectivity are still limited, which could compromise analyses when the analyte is present in low concentration, and a lower limit of detection is required [45].

In this scenario, the first electrochemical paper-based analytical device (ePAD) was fabricated by Dungchai and collaborators in 2009 to overcome the problematic conditions reported in colorimetric tests for biological samples analysis, which requires more sensitivity and selectivity [46]. Thus, the main challenge that this type of device revolutionized at that time was the integration of a very sensitive, low-cost, miniaturized, and rapid technique with a low-cost platform (paper). Merging the characteristics of the electrochemical techniques and the properties of the paper, devices with lower limits-of-detection, low-cost, disposability, and the possibility of miniaturization could be fabricated using straightforward tools [45,46].

Looking at the opportunity that the new detection attached to a paper brought, since 2009, the ePADs have attracted more attention from the researchers. Several different fabrication methods have been developed to overcome the challenges that this type of analytical device still offers and improve its quality. The improvements have focused on the fabrication and modification of the conductive tracks called electrodes, besides the different ways to isolate the reaction zone and detection area to facilitate mass production [47]. Then, devices with more selectivity, sensibility, robustness, and mechanical resistance could be developed [24]. Additionally, a few studies have been advanced in the detection instruments field for the portable measurements [48] and the interpretation of the data, which is one of the challenges faced in the area.

Thus, after the first reported ePAD, a paper-based microfluidic device was developed and described to be integrated with commercial electrochemical readers [49]. In this work, Nie and coworkers presented for the first time an ePAD that could be coupled with a glucometer. This equipment is widely commercially used for rapid electrochemical analysis of glucose in blood samples [49]. Using this idea, an ePAD was fabricated to be appropriately used in this equipment since it is already a well-known, portable, easy-to-use, and commercially available technology. Furthermore, the author highlighted the possibility of adapting the method to a range of analytes [49]. Recently, other commercialized and portable electrochemical readers have been developed to be coupled with ePAD sensors and enable the use of the devices without the necessity of very qualified professionals [50]. This shows critical progress aiming to achieve commercial applications of the ePADs in different areas in the next few years.

In general, electrochemical methods lack a high integration degree between the analysis steps (connection of the electrodes to electrical circuit/potentiostat, absence of sampler, automatic renovation/polishment of the electrodic surface, among others), and the interpretation of the result usually requires a specialized person to obtain the quantitative result. These drawbacks are common barriers to the expansion of electrochemistry in industrial applications, and on-site analysis makes it a challenge for everyone's usage. Therefore, the combination of the ePADs with commercial equipment like glucometers is attractive and can popularize these devices.

Following the challenges of providing high sensitivity and selective disposable device, in 2012, an electrochemical paper-based sensor modified with a DNA structure was described for the first time [51]. However, the fabrication of the device using paper as a simple material platform, a very sensitive, low-cost, and simple DNA sensor, reaching limits-of-detection of 0.2 fM for easy application in point-of-care analyses, was obtained. It was possible due to the electrochemical technique and the electrode material made of carbon ink modified with AuNPs/graphene and DNA structure [51]. This kind of modification coupled with ePADs opened new possibilities and opportunities to fabricate more complex and sensitive systems using a disposable and low-cost material. Other recognition elements with excellent stability, such as aptamers and molecularly imprinted polymers, have also been developed and applied in the area [52,53].

Another important breakthrough using this low-cost and disposable material approach integrated with electrochemical techniques was reported to cover an enormous range of applications in the area. In 2013, microelectrodes were fabricated for the first time in paper-based devices to improve the sensitivity of the system and to open the possibility to aggregate the advantages of microelectrodes with a simple, low-cost, and disposable platform [54]. Microelectrodes could improve the sensor's sensitivity and allow analyses that demand a fast response time, using a two-electrode assemble and the miniaturization of the system due to the small sizes of the electrodes [54]. This work showed the versatility of adapting the PADs for different requested analysis types.

Furthermore, until 2013 the electrodes were fabricated using three main techniques: silk-screen technology, sputtering and screen-printing. Looking at this scenario and thinking about carbon as the widely used material in electrode fabrication, Santhiago and collaborators developed a new, simple, and low-cost method to fabricate them [55]. In this method, a graphite pencil was used as a working carbon electrode, and silver ink was used as a reference and auxiliary electrode. Although, in this case, it is necessary to pay attention to the different commercial graphite types, easy ways to obtain the electrodes using graphite pencil were described for the first time [55]. The method applied used a carbon material, which is easy to get without the need for more complicated or advanced techniques to produce the electrodes system, although the drawback of not choosing other materials than carbon in this case.

Talking about low-cost methods, in 2014, the substrate of the ePADs, the paper, was the object of attention. Araujo and coworker developed an electrochemical paper-based device decreasing the cost of fabrication by 96.7% when compared with others at that time [56]. In that work, the authors reported the use of the white paper-office, for the first time, as an alternative substrate due to its abundance and inexpensiveness compared with a filter or chromatographic paper. Silver ink was used to fabricate the three-electrode system; however, the device can support other electrode materials, including graphite pencil, depending on the application [56]. It opened the possibility to obtain devices thinking even more about cost/effectiveness relationships depending on the application. It is primarily important in underdeveloped countries with no conditions for specialized machines or professionals to point-of-care analysis.

The search for better electrode fabrication methods induced other advances in the area, such as using microwires coupled to the paper platform. The first device proposed microwires as working electrodes were reported by Fosdick and collaborators in 2014 [57]. In work, the focus was on Au microwires and carbon fiber, but it could be extended to other materials. The advantage of this approach is the simplicity of obtaining electrodes made of conductive materials. Besides, there is the possibility to reuse the microwires by cleaning them or modifying them before use [57].

Other electrode fabrication methods emerged using different platforms. In 2017, for example, another groundbreaking work was reported when cardboard was used as a platform to fabricate carbon conductive tracks using the localized carbonization of the material by a CO_2 laser machine [47]. The idea is to carbonize the cardboard to obtain a carbon three-electrode system directly on device platform. This method allows for a simple, reagentless, single-step, low-cost, and disposable electrochemical sensor fabrication.

The laser-scribing technique comprises an excellent alternative approach to fabricate Ing the electrochemical tracks on paper-like platforms without reagents or binders, as the printing methods that make the use of conductive inks containing these species, which can remain after the curing steps and provide sluggish electron transfer kinetics. However, laser-scribing is limited to few paper-like substrates (high grammage) due to the degradation of the platform during the photo-thermal process induced by laser radiation. A possible route is to fabricate the electrodes using polymeric materials and transfer the conductive tracks to the paper substrate in a subsequent step.

One of the significant challenges in the development of ePADs is related to the reproducible pattern of the electroactive area of the electrodes since it depends not only on the resolution of technique as the homogeneous composition of the conductive material and the type of paper used. The area control is crucial to obtain a reproducible analytical signal in all electrochemical techniques, except potentiometry. The paper's porosity degree directly affects the smoothness of the electrodic surface because of the need to fulfill these regions with the conductive material during the printing or transfer processes. Thus, the effective electrodic area needs to be carefully calculated, taking into account the roughness and porosity of the tracks created. The batch-to-batch variation in the fabrication of the ePADs can be minimized by adjusting the composition of the inks (viscosity and superficial tension), the fabrication time, and choosing appropriate paper material. High reproducible fabrication is essential for developing and implementing disposable (single use) ePADs enabling the data comparison and obtention of accurate results.

As shown above, many advances have been made in these types of devices, where the main challenge overcome was the limited sensitivity of the colorimetric PADs. The ePADs brought the possibility to ensure lower detection limits due to the improvement of the methods, and even simultaneous detection, using both colorimetric and electrochemical measurements [58]. Nevertheless, some points are still circumvented, which opens space for development and enhancement in the area. One example is the multiplex analyses, which have been studied to guarantee more than one analysis in a unique device to monitor different markers simultaneously [59]. Although this kind of analysis is advantageous, it is still a big challenge for ePADs. More sophisticated apparatus and the need for more development in the area are needed, including in the engineering involved.

Other examples are wearable sensors, which have been gaining power in the last years due to the importance of real-time metabolites and markers monitoring to verify human biological functions or other purposes. In this context, the ePADs have also been used in this field due to the possibility to obtain analyses with low limits of detection attached to the low-cost and the intrinsic properties of the platform [24,60–62]. However, a challenge is that the paper still presents some limitations, such as mechanical resistance and liquid absorbance, which could compromise this kind of continuous analysis [63]. This problem could be circumvented using paper modifications when relevant and developing new designs, as origami-like and others, besides using multidisciplinary knowledge to move forward in the area.

11.2 Conclusions

Nowadays, PADs received many efforts to turn into a promising technological, comercializable, and accessible product for point-of-need and point-of-care applications. Hence, researchers devoted a significant part of their efforts to enhance the analytical devices' performance to reach low limits of detection using low-cost devices. Based on the pump-free properties of the paper, multiplexing detection has also been developed to reduce the analysis time and increase the number of analytes detected, as reported previously. However, the efforts now need to be redirected for advanced in the manufacturing process to obtain commercial success. The paper type used is crucial for the type of applications and to bring different flow properties. It is foreseen that more advanced work on PADs will cover a way for their commercial success in different areas that could radically increase the point-of-need and point-of-care detections in the future.

Given all the challenges related to colorimetric analysis in PADs, it is worth pointing out the prospects for future developments in this field. One of the critical things that should be explored is creating equipment-free devices, distance-based paper devices [39], representing more simplicity and versatility, especially concerning decentralized analysis and in places with limited resources. Another relevant and challenging aspect is to improve the accuracy of the analysis, understand better how to intensify the color change on paper to enhance the analytical properties [64], notably when we think of devices that can be used for point-of-care analysis.

Despite all the advances presented above for the ePADs, and the relevant applications demonstrated in the period, little evolved in the commercialization direction of these tests but seems with the more innovative detection couple to paper compared with colorimetric one. Hence, in our opinion, the ePAD could be the closer device for commercialization use since it can be coupled to handheld and user-friendly equipment, like the glucometers, facilitating the acceptance and use by the final users. However, the relationship between academic discoveries and the industries needs to be improved.

Acknowledgments

Financial support for this project has been provided by Brazilian agencies FAPESP (Grant Numbers: 2019/15065−7, 2018/08782−1, 2018/14462−0 and 2018/16250−0), CAPES, CNPq (grant numbers: 302839/2020−8, and 438828/2018−6), and INCTBio (573672/2008−3).

References

[1] A.W. Martinez, et al., Patterned paper as a platform for inexpensive, low-volume, portable bioassays, Angew. Chem. Int. (Ed.) Engl. 46 (8) (2007) 1318−1320.

[2] D.M. Cate, et al., Recent developments in paper-based microfluidic devices, Anal. Chem. 87 (1) (2015) 19−41.

[3] H. Lim, A.T. Jafry, J. Lee, Fabrication, flow control, and applications of microfluidic paper-based analytical devices, Molecules 24 (16) (2019).

[4] S. Kar, et al., Microfluidics on porous substrates mediated by capillarity-driven transport, Ind. Eng. Chem. Res. 59 (9) (2020) 3644−3654.

[5] J.A. Adkins, E. Noviana, C.S. Henry, Development of a quasi-steady flow electrochemical paper-based analytical device, Anal. Chem. 88 (21) (2016) 10639−10647.

[6] E. Carrilho, A.W. Martinez, G.M. Whitesides, Understanding wax printing: a simple micropatterning process for paper-based microfluidics, Anal. Chem. 81 (16) (2009) 7091−7095.

[7] P. de Tarso Garcia, et al., A handheld stamping process to fabricate microfluidic paper-based analytical devices with chemically modified surface for clinical assays, RSC Adv. 4 (71) (2014) 37637−37644.

[8] W. Dungchai, O. Chailapakul, C.S. Henry, A low-cost, simple, and rapid fabrication method for paper-based microfluidics using wax screen-printing, Analyst 136 (1) (2011) 77−82.

[9] P. Spicar-Mihalic, et al., CO_2 laser cutting and ablative etching for the fabrication of paper-based devices, J. Micromech. Microeng. 23 (6) (2013).

[10] D.L. Giokas, G.Z. Tsogas, A.G. Vlessidis, Programming fluid transport in paper-based microfluidic devices using razor-crafted open channels, Anal. Chem. 86 (13) (2014) 6202–6207.

[11] I. Jang, et al., Pump-free microfluidic rapid mixer combined with a paper-based channel, ACS Sens. 5 (7) (2020) 2230–2238.

[12] K.M. Schilling, et al., Fully enclosed microfluidic paper-based analytical devices, Anal. Chem. 84 (3) (2012) 1579–1585.

[13] C. Renault, et al., Hollow-channel paper analytical devices, Anal. Chem. 85 (16) (2013) 7976–7979.

[14] T.H. Kim, Y.K. Hahn, M.S. Kim, Recent advances of fluid manipulation technologies in microfluidic paper-based analytical devices (muPADs) toward multistep assays, Micromachines (Basel) 11 (3) (2020).

[15] E. Fu, et al., Controlled reagent transport in disposable 2D paper networks, Lab. Chip 10 (7) (2010) 918–920.

[16] S. Jahanshahi-Anbuhi, et al., Creating fast flow channels in paper fluidic devices to control timing of sequential reactions, Lab. Chip 12 (23) (2012) 5079–5085.

[17] S. Mendez, et al., Imbibition in porous membranes of complex shape: quasi-stationary flow in thin rectangular segments, Langmuir 26 (2) (2010) 1380–1385.

[18] B. Lutz, et al., Dissolvable fluidic time delays for programming multistep assays in instrument-free paper diagnostics, Lab. Chip 13 (14) (2013) 2840–2847.

[19] B.J. Toley, et al., A versatile valving toolkit for automating fluidic operations in paper microfluidic devices, Lab. Chip 15 (6) (2015) 1432–1444.

[20] X. Li, P. Zwanenburg, X. Liu, Magnetic timing valves for fluid control in paper-based microfluidics, Lab. Chip 13 (13) (2013) 2609–2614.

[21] T.H. Kim, et al., Solenoid driven pressure valve system: toward versatile fluidic control in paper microfluidics, Anal. Chem. 90 (4) (2018) 2534–2541.

[22] T. Kong, et al., A fast, reconfigurable flow switch for paper microfluidics based on selective wetting of folded paper actuator strips, Lab. Chip 17 (21) (2017) 3621–3633.

[23] G.M. Whitesides, The origins and the future of microfluidics, Nature 442 (7101) (2006) 368–373.

[24] V.N. Ataide, et al., Electrochemical paper-based analytical devices: ten years of development, Anal. Methods 12 (8) (2020) 1030–1054.

[25] G.G. Morbioli, et al., Technical aspects and challenges of colorimetric detection with microfluidic paper-based analytical devices (muPADs) - a review, Anal. Chim. Acta 970 (2017) 1–22.

[26] E. Carrilho, et al., Paper microzone plates, Anal. Chem. 81 (15) (2009) 5990–5998.

[27] G.I. Salentijn, M. Grajewski, E. Verpoorte, Reinventing (bio)chemical analysis with paper, Anal. Chem. 90 (23) (2018) 13815–13825.

[28] E.W. Nery, L.T. Kubota, Sensing approaches on paper-based devices: a review, Anal. Bioanal. Chem. 405 (24) (2013) 7573–7595.

[29] M.O. Salles, et al., Explosive colorimetric discrimination using a smartphone, paper device and chemometrical approach, Anal. Methods 6 (7) (2014) 2047–2052.

[30] A.W. Martinez, et al., Simple telemedicine for developing regions: camera phones and paper-based microfluidic devices for real-time, off-site diagnosis, Anal. Chem. 80 (10) (2008) 3699–3707.

[31] H. Wang, et al., Semiquantitative visual detection of lead ions with a smartphone via a colorimetric paper-based analytical device, Anal. Chem. 91 (14) (2019) 9292–9299.

[32] T. Kong, et al., Accessory-free quantitative smartphone imaging of colorimetric paper-based assays, Lab. Chip 19 (11) (2019) 1991–1999.

[33] G.O. da Silva, W.R. de Araujo, T. Paixao, Portable and low-cost colorimetric office paper-based device for phenacetin detection in seized cocaine samples, Talanta 176 (2018) 674–678.

[34] L.F. Capitan-Vallvey, et al., Recent developments in computer vision-based analytical chemistry: A tutorial review, Anal. Chim. Acta 899 (2015) 23–56.

[35] F.C. Böck, et al., PhotoMetrix and colorimetric image analysis using smartphones, J. Chemometrics 34 (2020) 12.

[36] A.K. Yetisen, M.S. Akram, C.R. Lowe, Paper-based microfluidic point-of-care diagnostic devices, Lab. Chip 13 (12) (2013) 2210–2251.

[37] N.A. Meredith, et al., Paper-based analytical devices for environmental analysis, Analyst 141 (6) (2016) 1874–1887.

[38] G.G. Lewis, M.J. DiTucci, S.T. Phillips, Quantifying analytes in paper-based microfluidic devices without using external electronic readers, Angew. Chem. Int. (Ed.) Engl. 51 (51) (2012) 12707–12710.

[39] D.M. Cate, et al., Simple, distance-based measurement for paper analytical devices, Lab. Chip 13 (12) (2013) 2397–2404.

[40] A.K. Ellerbee, et al., Quantifying colorimetric assays in paper-based microfluidic devices by measuring the transmission of light through paper, Anal. Chem. 81 (20) (2009) 8447–8452.

[41] A.W. Martinez, et al., Diagnostics for the developing world: microfluidic paper-based analytical devices, Anal. Chem. 82 (1) (2010) 3–10.

[42] C.A. Chaplan, H.T. Mitchell, A.W. Martinez, Paper-based standard addition assays, Anal. Methods 6 (5) (2014) 1296–1300.

[43] K. Grudpan, et al., Applications of everyday IT and communications devices in modern analytical chemistry: a review, Talanta 136 (2015) 84–94.

[44] R.P. Carlisle, Scientific American: Inventions and Discoveries: all the milestones in ingenuity—from the discovery of fire to the invention of the microwave oven, Wiley, 2004.

[45] M. Baharfar, et al., Engineering strategies for enhancing the performance of electrochemical paper-based analytical devices, Biosens. Bioelectron. 167 (2020) 112506.

[46] W. Dungchai, O. Chailapakul, C.S. Henry, Electrochemical detection for paper-based microfluidics, Anal. Chem. 81 (14) (2009) 5821–5826.

[47] W.R. de Araujo, et al., Single-Step Reagentless Laser Scribing Fabrication of Electrochemical Paper-Based Analytical Devices, Angew. Chem. Int. (Ed.) Engl. 56 (47) (2017) 15113–15117.

[48] P. Teengam, et al., NFC-enabling smartphone-based portable amperometric immunosensor for hepatitis B virus detection, Sens. Actuators B: Chem. (2021) 326.

[49] Z. Nie, et al., Integration of paper-based microfluidic devices with commercial electrochemical readers, Lab. Chip 10 (22) (2010) 3163–3169.

[50] K. Pungjunun, et al., Enhanced sensitivity and separation for simultaneous determination of tin and lead using paper-based sensors combined with a portable potentiostat, Sens. Actuators B: Chem. (2020) 318.

[51] J. Lu, et al., Electrochemical DNA sensor based on three-dimensional folding paper device for specific and sensitive point-of-care testing, Electrochim. Acta 80 (2012) 334–341.

[52] M. Amatatongchai, et al., Highly sensitive and selective electrochemical paper-based device using a graphite screen-printed electrode modified with molecularly imprinted polymers coated Fe_3O_4@Au@SiO_2 for serotonin determination, Anal. Chim. Acta 1077 (2019) 255–265.

[53] H. Yu, et al., Fabrication of aptamer-modified paper electrochemical devices for on-site biosensing, Angew. Chem. Int. (Ed.) Engl. 60 (6) (2021) 2993–3000.
[54] M. Santhiago, et al., Construction and electrochemical characterization of microelectrodes for improved sensitivity in paper-based analytical devices, Anal. Chem. 85 (10) (2013) 5233–5239.
[55] M. Santhiago, L.T. Kubota, A new approach for paper-based analytical devices with electrochemical detection based on graphite pencil electrodes, Sens. Actuators B: Chem. 177 (2013) 224–230.
[56] W.R. de Araujo, T.R. Paixao, Fabrication of disposable electrochemical devices using silver ink and office paper, Analyst 139 (11) (2014) 2742–2747.
[57] S.E. Fosdick, et al., Wire, mesh, and fiber electrodes for paper-based electroanalytical devices, Anal. Chem. 86 (7) (2014) 3659–3666.
[58] A. Apilux, et al., Lab-on-paper with dual electrochemical/colorimetric detection for simultaneous determination of gold and iron, Anal. Chem. 82 (5) (2010) 1727–1732.
[59] E. Noviana, C.S. Henry, Simultaneous electrochemical detection in paper-based analytical devices, Current Opinion in Electrochemistry 23 (2020) 1–6.
[60] S. Anastasova, et al., A wearable multisensing patch for continuous sweat monitoring, Biosens. Bioelectron. 93 (2017) 139–145.
[61] Q. Cao, et al., Three-dimensional paper-based microfluidic electrochemical integrated devices (3D-PMED) for wearable electrochemical glucose detection, RSC Adv. 9 (10) (2019) 5674–5681.
[62] N. Colozza, et al., A wearable origami-like paper-based electrochemical biosensor for sulfur mustard detection, Biosens. Bioelectron. 129 (2019) 15–23.
[63] S. Wang, et al., Flexible substrate-based devices for point-of-care diagnostics, Trends Biotechnol. 34 (11) (2016) 909–921.
[64] S.V. de Freitas, et al., Uncovering the formation of color gradients for glucose colorimetric assays on microfluidic paper-based analytical devices by mass spectrometry imaging, Anal. Chem. 90 (20) (2018) 11949–11954.

Index

Note: Page numbers followed by "*f*" refer to figures.

D

E

N

O

P

Printed in the United States
by Baker & Taylor Publisher Services